Artificial Intelligence and Human Enhancement

Wiener Reihe

Themen der Philosophie

Herausgegeben von
Cornelia Klinger, Herta Nagl-Docekal,
Ludwig Nagl und Alexander Somek

Band 21

Artificial Intelligence and Human Enhancement

Affirmative and Critical Approaches in the Humanities

Edited by
Herta Nagl-Docekal and Waldemar Zacharasiewicz

DE GRUYTER

Gedruckt mit freundlicher Unterstützung der Österreichischen Akademie der Wissenschaften und ihrer Kommission „The North Atlantic Triangle".

ISBN 978-3-11-135787-4
e-ISBN (PDF) 978-3-11-077021-6
e-ISBN (EPUB) 978-3-11-077029-2
ISSN 2363-9237

Library of Congress Control Number: 2021950693

Bibliographic information published by the Deutsche Nationalbibliothek
The Deutsche Nationalbibliothek lists this publication in the Deutsche Nationalbibliografie; detailed bibliographic data are available on the Internet at http://dnb.dnb.de.

© 2023 Walter de Gruyter GmbH, Berlin/Boston
This volume is text- and page-identical with the hardback published in 2022.
Cover image: Grafik © Bernd Ganser-Lion ganserlion.com
Printing and binding: CPI books GmbH, Leck

www.degruyter.com

Table of Contents

Acknowledgements —— VII

Herta Nagl-Docekal and Waldemar Zacharasiewicz
Introduction: Affirmative and Critical Approaches to Artificial Intelligence and Human Enhancement —— 1

Part 1: Challenging "Strong AI" from the Perspective of Human Agency

Sybille Krämer
The Artificiality of the Human Mind: A Reflection on Natural and Artificial Intelligence —— 17

Ludwig Nagl
Merits and Limits of AI: Philosophical Reflections on the Difference between Instrumental Rationality and Praxis-Related Hermeneutical Reason —— 33

Hille Haker
Experience, Identity and Moral Agency in the Age of Artificial Intelligence —— 51

Sabine Sielke
Outsourcing the Brain, Optimizing the Body: Retrotopian Projections of the Human Subject —— 79

Cornelia Klinger
Life Care/Lebenssorge and the Fourth Industrial Revolution —— 101

Part 2: Examining Merits and Limits of Applied AI

Darren Abramson
AI's Winograd Moment; or: How Should We Teach Machines Common Sense? Guidance from Cognitive Science —— 127

Regina Schober
Passing the Turing Test? AI Generated Poetry and Posthuman Creativity —— 151

Reinhart Kögerler and Klaus Viertbauer
Why Neuroenhancement is a Philosophical Issue —— 167

Julia M. Puaschunder
The Future of Artificial Intelligence in International Healthcare: An Index —— 181

Part 3: Encounters with Artificial Beings in Film, Literature, and Theater

Jörg Türschmann
Dark Ecology and Digital Images of Entropy: A Brief Survey of the History of Cinematic Morphing and the Computer Graphics of Artificial Intelligence —— 209

Ulfried Reichardt
Sentience, Artificial Intelligence, and Human Enhancement in US-American Fiction and Film: Thinking With and Without Consciousness —— 225

Carmen Birkle
"I, Robot": Artificial Intelligence and Fears of the Posthuman —— 237

Piet Defraeye
AI on Stage: A Cross-Cultural Check-Up and the Case of Canada and John Mighton —— 261

Johanna Pitetti-Heil
Artificial Intelligence from Science Fiction to Soul Machines: (Re-)Configuring Empathy between Bodies, Knowledge, and Power —— 287

List of contributors —— 309

Index of Authors —— 315

Index of Subjects —— 319

Acknowledgements

The present volume is based on talks that were presented (partly online) at the International Conference "Artificial Intelligence and Human Enhancement. Affirmative and Critical Approaches in the Humanities from Both Sides of the Atlantic," held in Vienna on October 29–31, 2020, that was convened by the members of the commission of the Austrian Academy of Sciences "The North Atlantic Triangle: Social and Cultural Exchange between Europe, the USA and Canada." In a long series of previous conferences and workshops, this commission has invited scholars from both sides of the Atlantic, representing a variety of disciplines of the humanities, to join efforts in exploring two major fields of interest: Firstly, the reciprocal adoption and transformation of key ideas from overseas that typically was initiated and facilitated by personal interrelations. The volumes *Transatlantic Elective Affinities: Traveling Ideas and Their Mediators* (Zacharasiewicz/Nagl-Docekal 2021) and *Narratives of Encounters in the North Atlantic Triangle* (Zacharasiewcz/Staines 2015) document this research endeavor, among a number of other books. Secondly, the focus is on shared cultural phenomena ranging from challenges created by the rapidly changing living conditions in the contemporary world to the aesthetics of the cinematic experience. An international symposium in the latter field, convened in Vienna by "The North Atlantic Triangle," is presented in *Ein Filmphilosophie-Symposium mit Robert B. Pippin: Western, Film Noir und das Kino der Brüder Dardenne* (Nagl/Zacharasiewicz 2016), while a most pressing current issue will be discussed in the forthcoming collection of papers presented at the international conference held in Vienna on November 18–20, 2021, *Polarization in the North Atlantic Triangle* (Zacharasiewicz/Birkle/Prisching).

The Austrian Academy of Sciences provided the necessary funds which permitted the invitation of scholars from abroad and overseas and covered the costs of the technical team in charge of the online presentation of the conference and of the preparation of this publication. Additional support was granted by the City of Vienna and the Society for the Promotion of North America Studies at the University of Vienna. The task of formatting and proofreading the manuscripts was undertaken by Anna Zalto, which is here gratefully acknowledged.

Herta Nagl-Docekal and Waldemar Zacharasiewicz
Introduction: Affirmative and Critical Approaches to Artificial Intelligence and Human Enhancement

The manifold technological innovations made possible by research in what is called "Artificial Intelligence", and the prospect of their enormous future increase, have been widely acclaimed – one may expect, for instance, that robots will eventually carry out many forms of work traditionally performed by humans, such as delivering parcels, advising clients of banks, supporting the medical teams in hospitals, etc., and that they will execute tasks which are very dangerous for humans (such as clearing minefields). The prospects for such future achievements have, however, also prompted serious concerns in various regards. Worries have been generated not only by the perspective of far-reaching social changes that might result from the digitalization in the spheres of work, but also with regard to the anthropological foundations of society. The most pressing concerns have been provoked by theories which suppose that humans might eventually lose control over their lives, as the calculating abilities of so-called "learning" computers, that prove, in some regards, to be clearly superior to human intelligence, become more and more sophisticated.[1] Such sweeping claims have been expressed in two different, yet complementary, ways that have both found considerable media resonance. We find, on the one hand, assertions – made without providing sound empirical corroboration – that eventually software systems will have consciousness and be capable of "autonomous" decision making, including the heeding of ethical principles, and of artistic production, etc. On the other hand, we find the idea that such bots might turn their assumed super-human competences against humans – an intuition that sets off emotions like fear and horror, as they have inspired science fiction in literature and film.

In recent discourse, strong claims for AI have found support in several trends of thought, most notably "post-humanism". The broad range of theories covered by the umbrella term "post-humanism,"[2] including "new materialisms,"[3] share the concern to challenge "anthropocentric ideals of the human as uniquely

---

[1] An assessment of this kind is formulated in: Habermas, *Auch eine Geschichte der Philosophie*, vol. 2, 593 (note 2).
[2] For an overview see: Bolter, "Posthumanism".
[3] Cf. Coole and Frost, eds., *New Materialisms*.

sovereign over the world, and binary thinking that separates nature from culture, human from animal, the animate from the inanimate, subject from object, self from environment, the living from nonliving."[4] What is also significant is the way in which ideas assembled under the heading "trans-humanism"[5] view technologies that seek to enhance the physiological capabilities of humans as providing evidence for the porous character of the boundaries on the "continuum human-machine-animal" – a view that is paradigmatically laid out in the widely received concept of "cyborg"[6]. It is obvious that projects of "neuroenhancement" that seek to improve the human brain by means of digital chips so as to achieve, for instance, reliable moral behavior or aggressiveness in combat, correspond to trans-humanist views.

As public opinion today is, to a large extent, captured by the journalistically sharpened antagonism between Utopian hopes – which may even suggest that, by linking up the human brain with an artificial body, immortal existence might be made possible[7] – and apocalyptic apprehensions, it does seem evident that a carefully argued examination of AI that rejects any excessive speculation is called for. Significant efforts to this effect have been made recently. One important approach is expounded in the concept of "digital humanism", first elaborated by Julian Nida-Rümelin and Nathalie Weidenfeld. Seeking to challenge what is sometimes called "the Silicon-Valley-Ideology", the two authors identify one key danger which many assessments of AI have in common: that simulation becomes mixed up with realization. As they point out, this misunderstanding finds expression in unwarranted language that amounts to a "mystification", and that needs to be confronted with the sober insight that "software systems do not have intentions, feelings and thoughts, and are not capable of decision making."[8] The main thrust of "digital humanism" is formulated in the thesis that digitalization does not automatically initiate a process of humanization, but rather needs to be examined in the light of the question "in which way economic, social and cultural benefits might be generated, and where dangers are looming"[9] – a question

---

4 Ringrose, Warfield, Zarabadi, eds. *Feminist Posthumanism, New Materialism and Education.*
5 See, for instance: Ferrando, "Posthumanism, transhumanism, antihumanism, metahumanism, and new materialisms: Differences and relations."
6 This abbreviation for the notion "cybernetic organism" was coined in: Haraway, *Simians, Cyborgs, and Women. The Reinvention of Nature.* For a carefully investigated survey of the current strands of thought see: Loh, *Trans-und Posthumanismus.*
7 See, for instance: Moravec, *Robot: Mere Machine to Transcend Mind.*
8 Nida-Rümelin and Weidenfeld, *Digitaler Humanismus. Eine Ethik für das Zeitalter der Künstlichen Intelligenz,* 200. The absurd consequences of this kind of mistaking simulation for reality are evident in current suggestions for granting human rights to bots.
9 Ibid., 205.

that implies issues of morality as well as justice. These concerns are further underscored in the "Vienna Manifesto on Digital Humanism" (2019)[10]. Today, moral and legal issues are being discussed, with a sense of urgency, regarding a variety of specific forms of employing AI. One case in point is the use of "big data" as an instrument for decision-making in the public sphere: given that machine learning models include common biases, as a result of which AI-based systems may re-enforce social stereotyping, accountability mechanisms are called for. The term "digitalocracy" has been coined with the intention of addressing the moral issues that result from the deployment of algorithms on the state level – *inter alia* from algorithms for the job market that provide ideal profiles. Critics blame such programs for tending to result in discrimination against significant segments of unemployed people. Others call into question the way in which globally operating internet companies employ the data of their users, without asking their permission, so as to optimize their advertizing.[11]

The essays collected in the present volume seek to contribute to a sober and careful investigation of the merits and limits of AI, as employed in a variety of different spheres of life and perceived in aesthetic productions. For this purpose, the authors – from Canada, the USA, Germany, and Austria – apply methodological tools of various fields of the humanities as well as philosophy, theology and theoretical physics. The overall focus of the essays is on three distinct spheres: Part 1 of the book explains the need to distinguish between "strong" and "weak" readings of AI and to consider in which way "strong" readings miss crucial features of the human existence; Part 2 examines some concrete applications of AI as well as calls for legal regulation; Part 3 discusses the imaginative presentation of AI in popular fiction and film, exploring the relationship between science fiction and conceptions of post-humanism.

Part 1: Challenging "Strong AI" from the Perspective of Human Agency begins with an essay by *Sybille Krämer* – "The Artificiality of the Human Mind: A Reflection on Natural and Artificial Intelligence" – that contends that the term "artificial intelligence" has all too easily obscured the genuinely social and symbolic constitution of human intelligence. Two assumptions form the basis of this claim: the first is that human intelligence is not "natural" but rather acquired through social interaction and the use of signs; the second is that, in contrast

---

10 See: "The Vienna Manifesto on Digital Humanism." This Manifesto was laid down at the "First International Workshop 'Digital Humanism'", organized by the Faculty of Computer Science at the University of Technology, Vienna, in May 2019, as explained in: Werthner, "The Vienna Manifesto on Digital Humanism", 336–55.
11 Current endeavors toward fair alternative forms of AI usage are discussed in: Heuser, "Digital aber gut. Eine neue Bewegung entwickelt die Vision einer besseren Datenwelt."

to the widespread assumption that AI technology aims to replace human activities, this technology is based on a joint performance by humans and apparatus. Elaborating her argument, Krämer points out that machine learning, in which algorithms inductively "abstract"' regularities from large data sets, must not be viewed as automated "self-learning", but remains dependent on feedback in the form of interaction between humans and machines. The specificity of the learning procedure, she explains, is that the system is trained inductively with large quantities of examples: by means of error and feedback. Discussing the phenomenon of "black boxing" – that the output does not make transparent what the internal model looks like – Krämer notes that trained software adopts the biases that are implicit in our practices and may thus support common forms of unjustifiable discrimination. Regarding the question "Can machines understand meaning?" she calls for a clear distinction: while AI models of text analysis are capable of providing overviews of relations in text corpora that human eyes cannot perceive by reading, the understanding of meaning can only be reconstructed praxeologically, i.e., connected with situated embedding, emotional responsiveness, etc., no element of which can be implemented by machines.

The essay by *Ludwig Nagl* – "Merits and Limits of AI: Philosophical Reflections on the Difference between Instrumental Rationality and Praxis-Related Hermeneutical Reason" – addresses the importance of distinguishing clearly between "weak" and "strong" readings of AI. In part one of his paper Nagl argues that the manifold useful innovations made possible by machine-instantiated algorithms rest on "weak AI": they are instances of an "instrumental rationality" which can be explained without the excessively Utopian claim that digital machines lead us in the direction of a technology-produced post-humanistic "super-intelligence". Algorithms (even so-called "learning" algorithms) are, as Nagl argues, at their core "tools": i.e., programs invented by humans, which – with regard to their social impact – are in permanent need of public control. Part two of Nagl's essay analyzes the Utopian fantasy of "strong AI" – of (as John Searle pointed out early on) the "false" claim "that appropriately programmed computers literally have cognitive states." After a brief reference to Leibniz's rejection of the idea "that human perception is explicable in terms of mechanical reason", Nagl focuses on the ways in which Hubert Dreyfus, Hilary Putnam and Charles Taylor take issue with today's fashionable tendency to overrate "instrumental rationality", and, as a result, insists on the embeddedness of algorithms in non-algorithmic, praxis-related, hermeneutical "reason". Nagl's key argument is that a rich, non-reductive conception of praxis implies the ability of humans to act in accordance with their moral judgment: an ability which – as may be convincingly pointed out along the lines of Kant's thought – cannot be fully simulated by AI's "heteronomous" mechanical execution of programmed "norms".

That AI is unable to simulate fully the moral competence of human beings also takes center stage in the essay by *Hille Haker*, "Experience, Identity and Moral Agency in the Age of Artificial Intelligence". The first part of this essay suggests reconsidering the early twentieth-century critique of technologies, for instance, Walter Benjamin's reflections on the way in which modern forms of instrumental rationality impact on our experiences so that "a completely new poverty has descended on mankind". Haker also highlights the thesis that "barbarism is the other side of modernity", as elaborated in the Critical Theory of the Frankfurt School. She employs the ideas provided by these critical analyses as she explicates the need for the relationship between AI technologies and human identity and agency to be viewed from a moral perspective. Taking issue with the fact that, currently, ethical deliberations about the "risks and chances" of AI technologies, as well as about "prohibitions and permissions", are mostly left to scientific expert commissions, parliaments and governments, Haker proposes an alternative approach: it is the vulnerability of humans, she contends, that distinguishes them categorically, and not only gradually, from AI devices. The difference between a vulnerable human and a destroyable robot is evident in the difference between damage and harm: things may be neglected, broken and destroyed, but they cannot be harmed in a moral sense. The essay then discusses the way in which AI systems tend to enforce a new kind of social disciplining, as AI comes with the risk of being programmed in a manner that directs human behavior toward a new level of conformity. In order to protect the vulnerable agency that enables moral interactions and mutual recognition, Haker argues, we need to give up attributing human competences to humanoid machines.

The ways in which the social disciplining implications of digital technologies have a detrimental impact on human agency are further investigated in the essay by *Sabine Sielke*, "Outsourcing the Brain, Optimizing the Body: Retropian Projections of the Human Subject". Sielke emphasizes that the discourses – of science as well as of visual culture – which drive current visions of the future of human subjectivity, typically do not fulfil their Utopian claim but rather reproduce traditional clichés. Under the heading "Outsourcing the Brain", the first part of the essay addresses the way in which popularized versions of cognitive science and AI revitalize mechanistic notions of humans as "brain machines". Sielke argues that any project that conceptualizes the human brain as a supercomputer tends to override more holistic approaches to human subjectivity. In general, she claims that trans-humanism has eliminated the difference between brain and mind altogether. The second part of the essay points out that this mechanistic notion in turn correlates with a phantasm of physical self-perfection that sells extremely well. Because current digital technologies and new knowledge in the fields of the bio-sciences and medicine seem to enable us to track and optimize our bodies' physiology and shape, many people are adopting rigid egocentric and time-consuming regimes of "self-perfection", Sielke observes. She contrasts these efforts with the fact that, all over the world, bodies have become increasingly overweight, so that obesity has turned into a global "epidemic". The final part of the essay draws attention to research that highlights that the synchronic dynamics of outsourcing brain power and enhancing bodies boosts the neo-liberal framework that deprives subjects of sociality and solidarity. In more general terms, Sielke concludes that these dynamics tend to interfere with human agency.

The genesis and broader context of the issues discussed here are examined by *Cornelia Klinger* under the heading "Life Care/Lebenssorge and the Fourth Industrial Revolution". Klinger argues that the breathtaking progress in the field of digitization that is deemed to amount to a new, i.e., fourth, industrial revolution needs to be seen in relation to the "Care Crisis" that is much debated today. As she maintains, "the proliferation of caring-for-life-services and facilities" represents the cutting edge of the current phase of industrialization. Explaining the basis of her reflections, she introduces a comprehensive conception of "life" that highlights the intertwining of σῆμα&σῶμα, i.e., of *soma* (body) and *sema* (sign): "the somatic and the semantic are what life is (about)." While emphasizing that "Lebenssorge" deals with all aspects of life, Klinger reconstructs the way in which, since about 1800, this complex form of relationship has increasingly been eroded. Whereas, "in the first wave of industrial revolutions", both nature and culture were relegated to the sphere of so-called "re-production", the further process of modernization has expanded into those areas, with the result that the entire sphere of reproduction is integrated into the process of science-and-technology driven industrialization. In combination with the globalized capitalist economy, high-tech achievements have brought about the implosion of traditional dualisms, as "production and reproduction are dressed in monetary garb". The utmost process of merging is obvious, Klinger notes, when the somatic is perceived as legible (by means of the genetic 'code'), while the semantic starts materializing (e.g., by means of a 3-D printer). Her essay closes by noting that "there seems to be no more inner space", as "reality has become a flat screen in the relentless process to quantify all qualities".

Part 2: Exploring Merits and Limits of Applied AI first deals with the issue of automatic text translation. *Darren Abramson* – in his essay "AI's Winograd Moment, or: How Should We Teach Machines Common Sense? Guidance from Cognitive Science" – discusses the recent debate, focusing on the question "What would it mean for machine learning to achieve human level performance at natural language tasks?" He elaborates his approach with reference to Yehoshua

Bar-Hillel, who was born in Vienna and studied philosophical logic with Rudolf Carnap, citing his provocatively titled "A Demonstration of the Nonfeasibility of Fully Automatic High Quality Translation" (1960), which argues that disambiguation of individual words in simple sentences seems to require unlimited knowledge of the human world. Abramson confronts this thesis with a number of state-of-the-art results that indicate that Winograd-related problems can, indeed, be resolved,[12] emphasizing that there have apparently been sudden, giant leaps in the ability of computers to process texts, most notably in natural language processing (NLP). In the final segment of his essay, Abramson reflects on possible future developments, suggesting that additional innovations in the "what to learn" and "how to measure learning for language models" might produce greater benefits than focusing on "how to learn". He pleads for a form of interdisciplinary cooperation that takes into account the knowledge of the hermeneutic character of the meaning of language that is provided by the humanities and social sciences. The scientific goals of machine learning based on the humanities and social sciences, he notes, may prove decisively different than those of a competitive, for-profit research paradigm.

A different, post-humanist approach to programs of NLP is expounded in the essay by *Regina Schober* – "Passing the Turing Test? AI Generated Poetry and Posthuman Creativity" – that focuses on the issue of "computational creativity". One of Schober's key intentions is to challenge the narrow idea of rational intelligence on which AI technology is typically based. Schober addresses the issue of whether artificially generated poetry might remind us of the nonsensical, the non-functional, and the non-economical that both machine and human creativity share? Rather than looking at AI generated poetry to identify the shortcomings of the machine, "may we not also be prompted to recognize and even appreciate the imperfection of human intelligence?" Schober asks. Based on these considerations, she argues for an examination of the possibilities for productive new ways of combining human creativity with modular selections of text samples generated by machine learning algorithms. In this manner, "a poet working with AI rather than being replaced by AI might be the future of generative poetry." In the concluding segment of the essay Schober raises, inspired by post-humanist thoughts, the question whether there is good reason to assume that aesthetic experience is exclusively reserved for the human species.

Neuroenhancement, which has provoked heated debates across the globe, is addressed in the essay by *Reinhart Kögerler and Klaus Viertbauer*, "Why Neuroenhancement is a Philosophical Issue". In an introductory section, the authors

---

[12] A Winograd schema presents a problem of ambiguous reference.

specify that "NE encompasses any procedure to improve (amplify) certain mental capacities of a healthy person by means of bio-technical interventions applied to the living organism", and emphasize that their focus is on NE that is not applied with therapeutic objectives but for the improvement of healthy persons. The main aim of this essay is to provide an introduction to the most contentious issues that have been discussed with regard to the diverse modes of NE interventions (that are categorized in terms of pharmacological/chemical, physical/technical, surgical, and genetic methods). Surveying the most common objections to these programs, Kögerler and Viertbauer distinguish dilemmas regarding genus ethics, autonomy, and authenticity. One issue they stress, with reference to Peter Singer, concerns the received prioritizing of the human species. Does the potential of NE techniques suggest that we need to abandon the concept of a boundary between humans, animals and machines? Furthermore, citing the dispute on "hard determinism", Kögerler and Viertbauer raise the question whether it is conceivable that, through the targeted generation of certain conditions of the brain, autonomy gradually dissolves. The final segment of the essay discusses the concept of "moral enhancement" (ME), referring to the work of Thomas Douglas, Ingmar Persson and Julian Savulascu, who argue that the moral capabilities of humans can and need to be improved by pharmacological interventions with substances such as Serotonin and Oxytocin.

A further field of application is explored by *Julia Puaschunder*, whose essay – "The Future of Artificial Intelligence in International Healthcare: An Index" – emphasizes that the call for global solutions in the monitoring of pandemic outbreaks and in crisis risk management has gained unprecedented momentum. Puaschunder's point of departure is the fact that, every day, healthcare professionals, biomedical researchers and patients produce vast amounts of data that are processed by means of electronic health records, genome sequencing machines, and high-resolution medical imaging. While emphasizing the benefits of AI-led modes of gathering actionable insights – such as tailored personal medicine, strategic data-driven interventions on medical prevention and health crisis management – Puaschunder draws attention to the way in which such benefits threaten to be undermined by political and economic corruption. She maintains that, in this age of IT governance, a quantification of the interrelation of corruption and AI-driven innovation in global healthcare is highly relevant, and seeks to contribute to this task by presenting three indices that highlight the impact which this interrelation is likely to have on future attempts to find solutions to health problems. Particular attention is paid to a "novel anti-corruption artificial healthcare index" that "outlines countries in the world that have vital AI growth in a non-corrupt environment". Such empirical research on the differences between countries across the globe is highly relevant today, Puaschunder argues, given the need for the global community to collaborate on public healthcare solutions. Furthermore, the essay addresses the ethical boundaries of digitalization in healthcare, citing, for instance, the risk of stigmatization that comes with diagnoses that set patients up on a path of discriminatory disadvantages.

Part 3: Encounters with Artificial Beings in Film and Literature first addresses the fact that, since its beginning, the cinema has been telling stories of ingenious persons inventing artificial beings with human skills. In his essay "Dark Ecology and Digital Images of Entropy: A Brief Survey of the History of Cinematic Morphing and the Computer Graphics of Artificial Intelligence", which provides a historical sketch of the cinematic inclusion of manifestations of AI that does not seem to have one (single) obvious representational model, *Jörg Türschmann* traces the emergence and function of recurrent motifs which anticipate their role in digital cinema, primarily in dystopian and science fiction horror movies. In his detailed historical survey, which also draws on several concepts of the French philosopher Michel Serres, Türschmann refers in detail to the long tradition of cultural motifs in classical myths (e. g. Pygmalion, Penelope's web) and their use in literature and in early films, which have thus prepared the way for diverse episodes or shots, especially in Cyberpunk movies. From the wide field of pertinent films, which includes the scenarios of a brilliant inventor creating an android, an individual who discovers in himself an artificial being, and a person who merges with his environment or his creature, Türschmann focuses on examples of the third scenario. It implies the dissolution of the boundary between the human and the non-human, a process labeled "dark ecology" by T. B. Morton, because romantic notions of nature are abandoned. After describing the concept of digital morphing used in cinema production for transforming anthropomorphic phenomena, Türschmann speculates on the ramifications of the filmic images of entropy, when humans dissolve into their surroundings, and its opposite, negentropy, when new artificial beings emerge, and presents instances of disconcerting morphing in the cinematic medium. The traditional motifs mentioned above have given rise to renditions of autophagic processes as well as haunting presentations of metamorphoses, which in the cinema include the transformation of amorphous bodies and the construction of sexually appealing (mostly female) robots in dozens of films. The fluidity of forms to be observed in the full range of manifestations of AI reflects, as Türschmann finally argues, the human preoccupation with the frightening search for the source of life.

In his essay "Sentience, Artificial Intelligence, and Human Enhancement in US-American Fiction and Film: Thinking With and Without Consciousness", *Ulfried Reichardt* analyzes two science fiction novels and a film against the background of technological innovations which have transformed our lives and provoked debates about the uniqueness of human intelligence and consciousness in the face of the impact of AI and the role of learning machines in the generation of knowledge. Reichardt refers to Thomas Metzinger's radical contestation of the concept of the self, and presents Katherine Hayles's post-humanist arguments, including the presence of non-conscious cognition and sentience in the entire biological realm, and the stress she places on the potential of technical systems with regard to cognitive capabilities. These recurrent claims for the emergence of a "post-humanist" phase have not been fully corroborated in narrative forms, which continue to stage interactions between humans and non-human cognition and digitally-based technological systems without eliminating the difference between them. In his analysis of the film *Her* (2013) Reichardt shows that while an advanced female AI can successfully simulate emotions and erotic feelings for a lonely man for a while, eventually the semblance of a true relationship ends, as the learning machine cannot be limited to real-world contexts and human needs. In the novel *Galatea 2.2* (1995) by Richard Powers the emotional bond the protagonist and trainer of a computer system develops with "Helen," a learning neural network, remains similarly one-sided. The knowledge acquired through literary texts and thus simply based on language is not anchored in lived experience, and the AI without an organic body cannot serve as a substitute for human interaction. In the science-fiction novel *Blindsight* (2006) by Canadian writer Peter Watts, however, Watts offers a radical interrogation of the distinctly human features by projecting an encounter between aliens and extremely modified humans with an impaired consciousness and a reduced faculty of empathy in their spaceship. The aliens, who are not conscious of themselves and completely lack empathy, are seemingly less aggressive than humans, thus implicitly questioning human exceptionalism.[13]

The extensive use of AI with the integration of robots into the fabric of human lives and the fears of the post-human, as forecast by radical advocates of AI, are the subject of the essay by Carmen Birkle "'I, Robot': Artificial Intelligence and Fears of the Posthuman". She carefully examines the relevant terms of AI, including the allegedly perfect ASI, the Artificial Super Intelligence, the post-human, and the ostensible breakdown of the boundaries to the non-human. She traces the introduction of robotics, which as early as the 1920s had led to anxiety that machines would replace the human workforce[14], and which in fiction is associated with the name of Isaac Asimov. His pattern-setting robot stories are il-

---

[13] The novel is thus in line with Thomas Metzinger's radical postulates. Some science-fiction texts go so far as to claim legal status for robots as advocated by economically powerful corporations involved in promoting AI.
[14] Cf. Elmer Rice, *The Adding Machine*, and Karel Čapek's *U.I.R.*

lustrated with the story "Satisfaction Guaranteed", which depicts the evolving emotional ties between Clare and a deep-learning humanoid, which are controlled by the three laws of robotics Asimov had formulated. Birkle also shows how writers have imagined dystopian societies in which androids appear as simulacra, as is the case in the film *AI*, in which the android "David" is fashioned as the perfect substitute for a son parents have lost because of an illness. The android's perfectly programmed proper behavior is expected to elicit love from his "parents.". The hostility humans show toward such "Mechas" lasts until an apocalypse destroys humankind, and the simulacra alone inhabit the world, in which they furnish a technologically created, cloned "mummy" for "David" and even engender emotional ties in him. The antagonism between robots and humans, despite Asimov's three laws of robotics, is also projected in the film *I, Robot* (2004), in which the original supporters of human beings have turned against their masters and caused deaths. It is only through an advanced human simulacrum, Sonny, that the menace is brought under control, though the robot also inaugurates a parallel robot society, another threat to humanity. In the film *Ex Machina* (2014) the resemblance between humans and robots is taken even further, as the simulacrum named Ava is given "logical and emotional intelligence" by her creator Nathan, and develops self-awareness, empathy and a fascinating sexuality. Eventually she achieves her freedom after killing Nathan, whose hubris in creating god-like a perfect human being is thus punished. The warnings of prominent scientists such as Stephen Hawking that we might lose control of SuperAI, which could emancipate itself and supersede human beings, are thus dramatized in fiction and films, which appear to present alarming visions of the "post-human" era.

Piet Defraeye provides in his wide-ranging essay "AI on Stage: A Cross-Cultural Check-up and the Case of Canada and John Mighton" a historical overview of the impact of AI on live theater and considers examples from Capek's play *U.I.R.* to advanced anthropomorphic robots. He deals with the development of chat-bots, whose construction echoes the training of figures such as Eliza Doolittle in G. B. Shaw's *Pygmalion*, and relates the learning aspect of robots to the use of scripts in the theater. He illustrates the application of computer science elements by performance artists and avant-garde theater producers[15] presenting robotic doubles of characters on the stage. These practices offer a bridge to Defraeye's investigation of the way in which in Canada experiments in the the-

---

15 Cf. performances of the Cypriot-Australian Stelarc and productions such as Annie Dorsen's introduction of communicating chat-bots in the avant-garde Steirischer Herbst autumn festival in Graz.

ater with new technologies are funded to promote the advancement of artists' skills in art installations. The main focus of the essay, however, is on the exploration of interpersonal relationships against the background of scientific insights in(to) human and AI in plays by John Mighton, especially in his most successful play *Possible Worlds*. This was also filmed by the prominent director Robert Lepage, whose own play *Needles and Opium* is also briefly considered, which uses computerized projection technology though no AI proper, which cannot easily manifest the motif of human desire that is a source of major interest for the theater.[16] Mighton's earlier play *Scientific Americans* presents the dysfunctional relationship between a physicist and a computer specialist studying AI in order to understand "how humans think"[17]. His later play continues the exploration of problematical relationships while taking its point of departure from (controversial) ideas linked to brain physiology experiments. It raises questions of personal identity by presenting incarnations of two characters, whose identities are unstable and shifting. Their lot is linked to two detectives who are eager to solve the riddle of body thefts in homicides in which the brains of very intelligent people have been removed, some of which may have ended up in a neurological research collection focused on AI. The turbulent events in the play have prompted intricate speculations by critics[18] as indicating a dystopian dissolution of human identities into endless possibilities. It has also induced elaborate reflections on the collapse of several worlds in this play, and emblematic scenes like "the brain in the vat" suggest a depressing reading of the world and the brain presented as an enigma.

In her essay "Artificial Intelligence from Science Fiction to Soul Machines: (Re-) Configuring Empathy between Bodies, Knowledge and Power" *Johanna Pitetti-Heil* presents the main issues in the debate about human exceptionalism linked to the assumption of the absence of empathy in AI, and the challenge to this alleged demarcation between the human and the non-human. When she revisits Richard Powers' *Galatea 2.2.*, with a close reading of the two creators and teachers of the neural network Helen, she draws on arguments of disability studies suggested by the inclusion of physically and/or mentally challenged human beings in the novel. She stresses the irony of the results of the application of the Turing Test based on a passage from Shakespeare's *The Tempest*, to which Helen responds more satisfactorily than a human student, which reveals the lim-

---

16 See, however, the arguments in other essays in this section of the collection.
17 The major focus in the play is, however, on ethical questions resulting from work for the arms industry.
18 The various critics cited draw on ideas by Baudrillard, Lacan and Zizek on human existence in a virtual reality and computational world.

itations of human "empathy", the ostensibly chief evidence for human exceptionalism. The resignation of Helen in the face of the lack of a living body, which prevents "her" from sharing in human emotions, is for New Materialists, such as Jane Bennett, whom Pitetti-Heil cites, no evidence of human exceptionalism. In her analysis of Philip K. Dick's *Do Androids Dream of Electric Sheep* Pitetti-Heil similarly questions the affective response of empathy as the accepted proof for a solely human capacity, with the Voigt-Kampff test being administered by the bounty hunters to distinguish between humans and dangerous humanoid androids. Pitetti-Heil provides circumstantial evidence for the questionable dividing line, as humans rely often on computer-assisted empathy boxes and trust some dubious spiritual leaders, while, she argues, androids seemingly care for each other. The dilemma of the "chickenhead" John Isidore, with a low IQ and severely limited options, prompts her reading of the novel as a satire on American eugenics[19], with the practice of segregation of "disabled" individuals also challenging to some extent human exceptionalism. A challenge to the dividing line between humans and non-human entities is also mounted in connection with the development of fully synthesized chat-bots with corporeal experience by the company Soul Machines, which claims that the neural system of Baby X, for example, gains experience and will exhibit not only artificial empathy, which neuroscientists aligned with radical New Materialism locate in distinct segments of the neural system, but also creativity and virtual imagination. The post-humanist thinkers thus argue that empathy is not necessarily connected to human morality but a performative act also feasible for AI.

Thus, fictional and cinematic visions of the role of advanced AI run counter to many of the ideas affirmed by philosophers advocating a sober analysis of the merits and limits of AI, often tending to the presentation of a blurring of the borderline between humans and sophisticated products of human ingenuity.

Works cited

Bolter, David Jay. "Posthumanism." *The International Encyclopedia of Communication Theory and Philosophy*. Ed. Klaus Bruhn Jensen et al. Hoboken, NJ: Wiley, 2016.
Čapek, Karel. *R.U.R.: Rossum's Universal Robots*. 1920. Trans. David Wyllie. Rockville, MD: Wildside Press, 2010.
Coole, Diana, and Samantha Frost, eds. *New Materialisms: Ontology, Agency, and Politics*. Durham: Duke University Press, 2010.

---

19 Her reading echoes Adam Pottle's interpretation in the *Disability Studies Quarterly*, 2013.

Ferrando, Francesca "Posthumanism, Transhumanism, Antihumanism, Metahumanism, and New Materialisms: Differences and Relations." *Existenz* 8.2 (2013): 26–32.

Habermas, Jürgen. *Auch eine Geschichte der Philosophie*. 2 vols. Berlin: Suhrkamp, 2019.

Haraway, Donna J. *Simians, Cyborgs, and Women: The Reinvention of Nature*. London: Free Association Books, 1991.

Heuser, Uwe Jean. "Digital aber gut: Eine neue Bewegung entwickelt die Vision einer besseren Datenwelt." *Die Zeit* 31 March 2021: 21–22.

Loh, Janina. *Trans- und Posthumanismus*. Hamburg: Junius, 2018.

Moravec, Hans. *Robot: Mere Machine to Transcend Mind*. Oxford: Oxford University Press, 2000.

Nida-Rümelin, Julian, and Nathalie Weidenfeld. *Digitaler Humanismus: Eine Ethik für das Zeitalter der Künstlichen Intelligenz*. Munich: Pieper, 2018.

Pottle, Adam. "Segregating the Chickenheads: Philip K. Dick's *Do Androids Dream of Electric Sheep?* and the Post/humanism of the American Eugenics Movement." *Disability Studies Quarterly* 33.3 (2013): n.pag.

Rice, Elmer. *The Adding Machine. Three Plays*. New York: Hill and Wang, 1965.

Ringrose, Jessica, Katie Warfield, and Shiva Zarabadi, eds. *Feminist Posthumanism, New Materialism and Education*. London: Routledge, 2018.

"The Vienna Manifesto on Digital Humanism." *Digital Transformation and Ethics*. Ed. Markus Hengstschläger and the Austrian Council for Research and Technology Development. Salzburg: Ecowin, 2020. 356–60.

Werthner, Hannes. "The Vienna Manifesto on Digital Humanism." *Digital Transformation and Ethics*. Ed. Markus Hengstschläger and the Austrian Council for Research and Technology Development. Salzburg: Ecowin, 2020. 336–55.

Part 1: **Challenging "Strong AI" from the Perspective of Human Agency**

Sybille Krämer
The Artificiality of the Human Mind: A Reflection on Natural and Artificial Intelligence

Abstract: The paper aims to correct the problematic narrative that the human mind has its 'natural' place in the mental inner life of individuals. Two assumptions are the starting point: (1) Our 'human intelligence' is not 'natural', but is acquired and exercised through social interaction and the use of signs (language, writing, images etc.). (2) Our fundamental relationship with technology is less substitution than assistance. What does it mean to start from the concept of a collective intelligence, which is a collaborative interrelationship of humans, symbols and technology? The argument proceeds in five steps: 1) *Operativity* is an indispensable and productive dimension of the human mind. 2) *'Digitality'* can be understood independently of computers. 3) As *'forensic devices'* computers make manifest as patterns what often remains hidden in our cognitive and aesthetic activities. 4) Machine learning, in which algorithms inductively 'abstract' regularities from large data sets, is not automated self-learning. 5) We do not have to fear the progress of artificial intelligence, but the stagnation of human intelligence. The fear of an almighty super-intelligence ('strong AI') distracts us from the contemporary 'weak AI' and its currently possible abuse of data.

1 Artifical intelligence – beyond the myth

If humanities scholars are interested in artificial intelligence (AI), it is usually as a myth, a vision of super intelligence or as a spectacular transhumanist ideology. They focus on something that, as a technology, is not real but fictional and circulates in a yo-yo game of illusion and disillusion. The reference point of the 'visionary AI', which is either euphorically welcomed (Kurzweil 2006) or apocalyptically exaggerated (Bostrom 2014), is the idea of a *universal* and *autonomously* acting machine intelligence. But the effectiveness and reality of AI does not consist in its visions, myths and ideologies, but in its contemporary, incidental and rather unremarkable applications. The real existing 'prosaic AI' is the result of the ongoing synthesis of sensor technology, network architectures, big data and deep learning processes and is omnipresent in the form of search engines, speech and facial recognition, navigational devices, curated shopping, transla-

https://doi.org/10.1515/9783110770216-003

tion software, etc. The gravitational centre of the socially effective forms of AI are mostly invisible data-processing technologies. Are the humanities too fascinated by 'visionary AI,' and do they fail to reflect on the potential and the threats of the currently applied, inconspicuous but highly effective 'prosaic AI'?

The well-known bipolarity between a visionary and a prosaic artificial intelligence was not discovered by John R. Searle (1980) when he split a 'strong AI' from a 'weak AI'. Yet it can be found – avant la lettre – already in the work of philosopher Gottfried Wilhelm Leibniz (1646–1716).[1] On the one hand, Leibniz was looking for a universal thinking machine ('calculus ratiocinator') to generate and examine all possible true statements and, on the other hand, he invented particular domain calculi, which develop feasible procedures for specific logical and mathematical problems. While his *universal* thinking machine, as Gödel's (1931) incompleteness theorem has shown, is logically not realizable, his *partial, narrowed-down* calculi are functional. Philosophy tends to discuss the epistemic questions of AI on the basis of the spectacular search for the automatization of *universal* thinking abilities. We want to follow Leibniz's second path. Our suggestion is to think of the achievements of AI in terms of Leibniz's domain specific calculi. Beyond myths and visions, AI has arrived in everyday life and in professional uses alike, and it is *this* phenomenon we need to describe, understand and analyse critically.

2 Natural versus artificial thinking?

The locus of thought is not only in the mental inner life in the heads of individuals, but also in forms of interaction between people, sign systems and technical instruments. Human intelligence is not 'natural' or 'unprocessed': it is a collective and social commodity, bound to linguistic and non-linguistic signs, to cultural techniques, instruments and tacit routines.

Yet sign systems can be designed as cognitive devices in such a way that the rules of sign manipulation can be applied *independently* of interpretation. The pioneering example is the written calculation within the decimal system, where rules take on the character of algorithms. It was Leibniz who used the proper name al-Hwarizmi (al-Ḫwārizmī 1967), the Islamic scholar who introduced decimal arithmetic to Europe, as a general term for calculation rules in the sense of rules for syntactic sign manipulation. Algorithms emerge from the operative fusion of language and technology, of symbolism and operativity. In

---

[1] See Krämer, *Berechenbare Vernunft* 220–79.

written arithmetic, the *language* of decimal number representation is at the same time the *tool* for number calculation. Algorithms are not bound exclusively to computers. Long before the invention of the digital computer, we developed the computer within ourselves: human hands, eyes and brains working with pencil and paper to solve number problems. The symptom of a highly developed culture of knowledge consists in the 'paradoxical' fact that parts of intellectual work can be realized in a meaningless, so to speak 'mindless' way. Artificial intelligence not only participates in this principle: it radicalizes it. The mathematician and philosopher Alfred North Whitehead (1911, 35) notes: "Civilization advances by extending the number of important operations that can be performed without thinking about them."

One of the most effective cognitive devices is the "cultural technique of flattening" (Krämer, *Figuration* 65–7), which constantly accompanies the evolution of the human mind. From ornaments, pictures and writing to diagrams, graphs and maps up to the computer screens and smartphones (Wöpking 2016): the Ariadne's thread of artificial flatness runs through all these inventions throughout the entire human history. Sciences, arts, complex technology and architecture would be unthinkable without the use of inscribed and illustrated surfaces. Everything that is, that is not yet, and even that which can never be – such as logically impossible objects, can be projected into two dimensions. Be that as it may, empirically there are no surfaces, but we treat inscribed surfaces *as if* they had no depth.

What is the reason for the huge cognitive (and aesthetic) potential of artificial flatness? Our body with its three perpendicular axes creates a basal orientation for humans in their environment: right/left, top/bottom, front/back. But everything that lies 'back' is – without back mirror! – removed from visibility and control. With the artificial surface, a medium has been created that uses the two-dimensional formatting left/right, top/bottom, while subtracting the third dimension, its depth. Thus a bird's eye view on painted, inscribed, printed, electronic surfaces is opened. Flatness, as a device for displaying unobservable, complex facts and figures, enables us to observe, process and make collectively accessible that which would otherwise remain hidden. A kind of cognitive transparency is created. Written arithmetic or letter algebra is paradigmatic for this: mathematical ingenuity and computational intuition, previously reserved for mathematical talents, become algorithmically regulated procedures, which can be executed by anyone who is willing to learn the appropriate symbolic languages and rules of transformation.

The epistemic potential of artificial flatness lies in the visualization and operationalization of the invisible: what is conceptually abstract, withdrawn and imperceptible can be made visible, controllable and socially dividable. A workplace of thought, a laboratory of technical design and an experimental space of artistic creation and composition have been created. To sum up: we have to interpret the cultural technique of flattening not as a deformative, but a formative, creative potential.

3 From 'readability of the world' to the 'machine-redability of the data universe'

Electronic networking creates a constant stream of data that parallels the world with a digitalized shadow world. The 'readability of the world,' to use a phrase coined by Hans Blumenberg (1986), is transformed into the 'machine-readability of the data universe' (Krämer, "Kulturtechnik Digitalität"). Computers are set-ups for recognizing and processing patterns in extensive data corpora (Nassehi 2019). To use an optical analogy: networked computers function like microscopes and telescopes in the data universe. Something implicit in the inscribed and illustrated surfaces is made explicit and accessible by the machine.

Electronic networking, digitization of the cultural heritage, the ubiquity of sensor technology, the use of social platforms, and the general use of smartphones produce an uninterrupted stream of data, which duplicates the world – sometimes in real time – in the form of a digital universe. Computers reveal traces in this data universe that remain invisible to the human eye. Their functional area is an *'epistemology of latency.'*

What we have said about the function of artificial flatness increases with digitization: everything that is, that has been, that will be in the future can be transformed into a data structure and searched, analysed and processed by algorithms. It is no longer understanding, but rather searching for data that becomes the "ideal path" within the data universe.

The computer was already characterized as a writing machine by Jay David Bolter (2001). But with Adrian Mackenzie (2017) this aspect can be extended: computers are diagrammatic machines. It is characteristic of this kind of machine that it only becomes functional through software, an operative form of textuality. Moreover, human-machine interaction in information technology still takes place largely via the flatness of screens, despite the tendency towards speech recognition and the disappearance of interfaces in ubiquitous computing.

Here, a transition from control to its loss is already becoming apparent: on the one hand, large parts of knowledge will be searchable and accessible. But anything that is not in the form of digitized material falls outside the search space. At the same time, the criteria for personalized data presentation and page ranking remain hidden to users. And if the algorithms are trained with extensive data sets, as in Deep Learning – we will come back to this point – it remains concealed which stereotypes, templates, prejudices and discriminations, implicit in the training data, are transmitted to and 'incorporated' in the learning algorithm.

Datification is characterized by a dialectic of transparency and opacity, and a transition from transparency to intransparency accompanies almost all network activities. The ambivalence of data power and data disempowerment is the breeding ground for critically reflecting the current situation of artificial intelligence.

4 Artificial intelligence in everyday assistance

The history of the AI debate is often described in seasonal metaphors[2] (Schuchmann 2019; Hendler 2008): twice already – in the early 1970s and late 1980s – the exaggerated expectations generated by artificial intelligence were terminated by a so-called 'winter of artificial intelligence:' extensive funding, diverse project and congress activities, led to the focus of media attention migrating to other areas. In the cycle of boom and bust, an upswing is currently unmistakable. Contemporary AI demonstrates spectacular results, such as algorithms that produce expertly deceptive Rembrandt style paintings (Tietgen 2016), or high-resolution photographs of non-existent persons (Schauer-Bieche 2019). The LIBRATUS program (Brown/Sandholm 2019) has beaten the four best players in the world in the Texas Hold'em poker game; it is remarkable that poker, unlike chess or go, requires players to act in circumstances of incomplete knowledge and to deceive other players.

Yet, despite such results, the current upswing of AI is rooted in its everyday and efficient use in almost all areas of society: in the sales strategies of online commerce, in the risk assessments of lending, in the diagnoses of image-based medicine, etc. In addition to these professional applications, Artificial Intelligence acts as an 'everyday assistant': the individualized playlists of streaming services, the voice recognition of Alexa, Cortana or Siri, automatic parking aids, the image settings of smartphone cameras, navigation, individualized product placements – the list is almost endless. Andrew Ng, a chief engineer of artificial intelligence, calls this 'new electricity' (Lynch 2017). Yet it is precisely the

---

2 Critical to these metaphors: Floridi 2020.

diversity of AI applications that makes it clear that it is *special* expertise that is optimized. Something like general AI is nowhere in sight.

Two aspects are important to understand the everyday assistance function of AI.

(1) Scale effect through feedback:

One driving force behind all these advances is to be found in feedback methods. This means that information from the application is incorporated into subsequent behavior. Made prominent in Norbert Wiener's cybernetics in the 1940s, feedback data is now the most important resource of technical optimization. When translation machines improve with every accepted translation; when navigation systems diagnose and communicate traffic jams caused by driving speeds: in such cases perfection in function is achieved in the process of product application. The more often the software is used, the better it becomes at carrying out its function. This principle of 'the more popular something is, the better it gets' also causes a highly problematic side of digitization: the market power of large data groups and social platforms, which – through their mass use – offer their users ever more convenient and powerful services. "Feedback creates data monopolies" (Ramge, 49).

(2) Co-performance instead of substitution

In contrast to the widespread assumption that technology aims to replace human activities, AI technologies are based on a co-performance (Kuijer and Giaccardi 2018) by humans and equipment, on sustained human-machine interaction (Kuijer 2014): during the nights of the poker competition, the winning LIBRATUS program was further trained by its developers. The interaction with a translation program offers a prototypical example: the software provides alternatives, while the human being selects and thus reinforces and optimizes the program's future proposals. In area-specific AI technologies, it is less a matter of replacing the function of people, but rather of assisting them in co-performance. A relation of mutual assistance is established. Artificial intelligence is Artificial Agency based on collaboration between humans and machines (Blackwell 2015). It is about 'distributed artificial intelligence' in real-world situations for very specific purposes (Vlassis 2007; Ferber 1999).

5 Artificial intelligence as a toolbox

The ongoing boom in artificial intelligence is primarily associated with machine learning processes. However, despite this progress in machine learning driven by the large volumes of data available, artificial intelligence is and will remain a toolbox (Lenzen 2020, 29) containing symbol processing approaches (GOFAI), with machine learning procedures and all forms of hybrids in between. In 2011, IBM's 'Watson' system beat the two award-winning human players in the *Jeopardy!* quiz game for the first time (Markoff 2011). It is a reverse quiz, where the answers are given and the appropriate questions are to be found. Watson's potential lies in the fact that it *combines* techniques of machine learning, or more precisely Deep Learning, with classical symbolic knowledge representation and automatic inference procedures.

An electric drill in a toolbox does not make hammer and pliers superfluous, nor does machine learning replace automatic symbol processing. 'Watson' is a natural-language oriented, semantic search engine whose sophistication is based on an optimal combination of the elements of the AI toolbox. This search engine finds just the right thing; however, it does not understand what it finds.

In order to explore the scope and limits of current AI, the so-called autonomous machine learning in the Deep Learning variant is now to be analysed in more detail. However, it should not be forgotten that there are different machine learning methods, even if Deep Learning currently seems to be the most promising.

6 Deep learning

Cognition theory distinguishes between two modalities (Kahnemann 2011). On the one hand, there is slow, laborious, time-consuming thinking, practiced in reasoning and arguing, calculating and problem solving, i.e., in all forms of complex sign use. On the other hand, there is the fast, mostly automatic and sub-symbolic recognition, whose key phenomenon is perception in identifying faces, voices or objects. This happens instantaneously, almost instinctively and independently of intentional use of signs.

It is precisely in the latter case that machine learning methods are used, especially the technique of Deep Learning (Goodfellow et al. 2016). Machine learning beyond the symbol-processing approach (GOFAI: Boden 2014; Haugeland 1985) has a long tradition, starting with Warren McCulloch's and Walter Pitt's 'Neuron Model' (1943) and Frank Rosenblatt's Perceptron (1958). But it was only with the contemporary transformation of the connected world into data streams, combined with the extreme increase in the power of processors, that the conditions were created for machine learning methods to to give AI a remarkable boost.

It is almost a commonplace, especially in the Humanities, that deep learning replaces the explicit programming typical of the symbol processing approach of AI with *self-adaptation* of the machine, i.e., *autonomous learning* through experience (Domingos 2015). But this view is problematic, if not wrong. The principle of co-performance between humans and machines also applies to machine learning methods. Deep learning is based on the availability of large amounts of training and test data that people have prepared, cleaned and annotated. Be it through hosts of clickworkers, who are mediated via crowdsourcing platforms (Mühlhoff, "Human-aided artificial intelligence"); be it through involuntary contributions from users, such as the well-known ReCaptcha, where it must be shown that 'you are not a robot' (Ahn et al. 2008): without extensive extraction of user data and without human voluntary or involuntary labelling activities, the data hunger of learning algorithms cannot be satisfied. The more labelled data, the better the algorithm.

Machine learning is a statistical method for recognizing patterns as numerical relations. Its technical architecture is based on interconnected artificial neurons arranged in staggered layers (Shalev-Shwartz/Ben David 2014). There is an input layer, a varying number of hidden intermediate layers, and an output layer. This layered neuron architecture has no biological counterpart, but can be considered as a folded diagrammatic network in machine implementation. The connections between the neurons of the different layers are weighted, so they can be gradually strengthened or weakened during the learning phase. The hidden layers can be interpreted as multiplying operating surfaces in between input and output. The term 'depth' only refers to the number of intermediate layers, not to the quality of pattern recognition. Each of these layers realizes a sub-function: with respect to image recognition for example, the first hidden layer identifies brightness values, the second identifies nodes and edges, the third identifies figurative links, the fourth identifies colour contrasts, etc.

But we must not forget that algorithms do not have eyes; they have no literal vision. The function of the Artificial Neural Network in its learning phase is to generate a mathematical mapping function where a given input leads to a desired output: handwritten digits should be identified as well-defined numbers (LeCun et al. 1989), an animal picture as a cat picture. But, regardless of whether digits or cats are to be diagnosed, for the machine these are always numerical values in their statistically analyzable relationship and not real-world objects. The statistical methods mean that machine identification has a probability character: the system identifies a cat (97%) and a dog (3%) as image objects. The training of the algorithm is finished when the human instructors consider the percentage reached to be sufficient. In most cases this rate already exceeds the humanly realizable diagnostic capability.

Hence we see that the specificity of the learning procedure is that the system is not instructed by giving a rule for pattern recognition, but is trained inductively through large amounts of examples: by means of error feedback (backpropagation: Bengio et al. 2016; Dreyfus 1990; Schmidhuber 2015) a kind of reinforcement learning emerges, in which the weights successively assigned to individual neuronal connections change with each training step.

But the price for not programming is an extremely high consumption of data resources. Even the data universe is limited and finite. Not incidentally, it was the visual data sets acquired through image practices in social media as well as through digitization of image and photo archives that favored automatic image recognition through learning algorithms (Krizhevsky/Sutskever/Hinton 2017). However, a small child learns the difference between a cat and a dog image after two or three examples; a machine learning system with seven hidden layers needs one million examples. And the learning system of course is still not aware that cats are not flat! This is because it lacks the 'embodied experience' in the practical handling of pictures in books or of living animals, through which children learn that cats can scratch, but also feel pain.

The artificial intelligence that a machine system acquires by learning from training data is always a socially distributed intelligence based on hybrid human-machine interactions. Deep learning does not embody autonomous machine learning. Deep learning makes it clear that where the vision for intelligence is to be at work, the 'brute force' of computational power still remains the main actor. Moreover, the stupidity, concentration and untiring patience to play millions of games against oneself or to search millions of clinical pictures for deviations, or to examine hundreds of digitized novels for word neighbourhoods, is something that humans cannot even do. Behind the great achievements of current artificial intelligence lies the datification of the world, the enormously increased computing power combined with tireless operative dullness.

7 Blackboxing

Now, there are obvious problems concerning the mathematical-technical side as well as the interpretation of the methods. Knowledge engineers are mostly aware of these problems (Marcus 2018; Richbourg 2018). But this is just not true for the public, i.e., for society in general.

A machine learning system builds an internal model from training data in order to generate the output desired. Since the model is built in the hidden layers, the output does not make transparent what the internal model looks like. This is where the phenomenon of blackboxing comes into being: "When a machine runs efficiently, when a matter of fact is settled, one need focus only on its inputs and outputs and not on its internal complexity. Thus, paradoxically, the more science and technology succeed, the more opaque and obscure they become," Bruno Latour (1999, 304) noted.

Where self-correction by training is built into an algorithm, it becomes almost impossible to reconstruct the complexity of the trained algorithm as an explicit knowledge structure. The algorithm constructively embodies a knowing-how that can hardly be transformed into a human accessible knowing-that. What is implicit in the algorithm can no longer be made explicit by humans. However, blackboxing is not an insoluble problem: with a great deal of technical effort it is possible to explore the modeling in the hidden layers ('DARPA explainable AI effort'). Nevertheless, the black-box problem, i.e., the hiddenness and opacity of what a machine extracts from the sample data, remains a complication, if not a pitfall. It forms an interesting antithesis to that principle of cognitive transparency originally associated with artificial flatness in the tradition of alphanumeric literacy.

Only two facets of the blackboxing problem are to be mentioned:

(a) Small deviations in the data input, invisible to human eyes, can lead to completely different results in the machine output: a change of only 0.004% in the input pixels in a learning system that previously correctly diagnosed 'panda' now indicates as output 'gibbon with 99%' (Richbourgh 2020). Thus, minimal changes invisible to humans can turn a joystick into a chihuahua, a hot air balloon into a labrador (Preusse/Wick 2018) – pretty funny as long as it is not a matter of driving assistants or medical diagnostic systems.

(b) The opacity of the acquired learning model also includes the instantiation of social discrimination implicit in the training data. Trained software adopts the biases and stereotypes that are implicit in our practices. The fact that AI-based systems, making proposals for credit allocation or job applications, unjustifiably discriminate people has meanwhile been recognized as a central problem (Eubanks 2018, Noble 2018). Paradoxically, one might say that, by adopting both familiar attitudes of discrimination and unconscious biases and templates, the learning system can also help make them explicit – by putting discrimination to use – in the first place, thus displaying what is implicit in our practices, often subverted in them unnoticed.

8 Excursus: Leibniz, the fold and folded neural networks

Although this is a somewhat daring association of ideas, a remarkable analogy stands out: it refers to a similarity between the role played by the technique of folding in Leibniz's *Metaphysics of the Monad* (Leibniz 1971) and folding as a mathematical principle fundamental to Convolutional Neural Networks.

Folding as a specific practice is an interesting phenomenon: by bending and pushing together, it allows to create a larger surface in a smaller space. Now the principle of folding is also central to Leibniz's metaphysics of the monad (Krämer, "Was bedeutet 'Perspektivität'" 337–40). Monads form a system among themselves, so that each monad represents the whole universe of monads, broken by its respective location in the system, i.e. in its individual perspective (Leibniz 1966, 435–56). Monads have, according to Leibniz, no windows. But all monads (except the monad of God) have a body in which all events in the monadic world are imprinted as clear or blurred traces, graduated according to proximity or distance from the event. Traces are surface structures that are interpreted as signs of the withdrawal and passing of something. In relation to the formation of traces in the monadic world, this means: Leibniz conceptualizes the tracks, which characterize each monad individually by virtue of its corporality, in the form of folds. Those monads, which have self-consciousness, and the ability to reflect, can recognize world affairs by unfolding the traces folded into their bodies, although only in their particular perspective. What matters here is the idea of folding. The layered structure of artificial neural networks can be interpreted as a form of increasing surface volumes through folding.

Folding as a mathematical-technical principle, to produce a third function from two functions, is a numerical procedure which has nothing to do with practical folding in everyday life. But when LeCun/Begio/Hinton (2015) explain the functioning of a Convolutional Neural Network in their now classic essay by means of a diagram, it becomes clear how much the technique of folding can be interpreted as a technology to increase and multiply surfaces, each of which, with its individual interconnected neurons, brings a local partial aspect (to a picture, to a speech utterance) to representation, so that through their interconnection the objects become identifiable in the form of pictures, faces or speech utterances. There are many examples of successful application of this technique of artificial intelligence.

Leibniz is a pioneer of the digital in many respects by inventing the binary alphabet, a four-species computing machine, and also the idea of the net, understood not as a trap but as a productive network (Krämer, "Leibniz ein Vordenker"). With his folded, so to speak 'convolutional monads' he introduces a principle of knowledge representation in distant analogy to the Convolutional Neural Networks!

9 Do machines understand meaning?

The basic question in critical commentary on AI is not 'Are machines intelligent?' but: *'Can machines understand meaning?'* We have to distinguish between two forms of meaning: an intrinsic, operative meaning and an extrinsic understanding and comprehension of meaning. Intrinsic means: the meaning of the sign consists in the system-internal operations to be performed. For extrinsic understanding it is possible to go beyond the system to access a system's exterior world, to change perspectives, to consider ambiguities and paradoxes. AI aims to model meaningful understanding in the form of operative meaning, because its potential is surface technology. For example: finding topics in texts is a form of quantitative content analysis in which the meaning of words is attributed to the occurrence of word neighbourhoods. (Heyer et al. 2018, 353) Latent regularities implicit in the text surfaces are analysed. Topics can only be indications of thematic structures of texts, and that only for humans who interpret the computationally calculated topics (Horstmann 2018). Topic modeling is useful in the digital humanities, but also with search engines, where large amounts of text have to be searched through. All the results found with this method are not based on anything that resembles the human reading and understanding of texts but creates by calculation overviews of intrinsic relations in text corpora that human eyes can no longer perceive by reading.

'Understanding meaning' in contrast to 'operational meaning' does not mean being able to distinguish cats from dogs, but rather means knowing that cats are not flat, can scratch and feel pain; understanding recognizes that a situation takes on a new shape and new content from different perspectives; it relates a perceptible pattern to something that is not available in the ontology of patterns. Diagnosing patterns means ignoring ambivalences and paradoxes.

The understanding of meaning can only be reconstructed praxeologically, i.e., in a way connected with physicality, situational embedding, emotional responsiveness and empathic resonance. Nothing in current technology indicates that something like this can be implemented in machines: the ideal route to machine semantics is and remains the tracing back of the understanding of meaning to intrinsic pattern identification and manipulation. However, in view of machine computing power and innovative data technologies, there is no definite, absolute limit. Moreover, it is important to remember that even with people emotional responsiveness and emphatic resonance, not to mention cognitive talents, are all distributed quite differently!

What we have to fear is less artificial intelligence on the part of machines, but irrationality on the part of people. What we have to hope for is less the omnipotence of our machines than the enlightened reason and maturity of people in dealing with their machines.

Works Cited

Ahn, Luis von, Benjamin Maurer, Colin McMillen, et al. "reCAPTCHA: Human-Based Character Recognition via Web Security Measures." *Science* 321, 5895 (2008): 1465–68.

Bengio, Yoshua, Dong-Hyun Lee, Jorg Bornschein, Thomas Mesnard, and Zhouhan Lin. "Towards Biologically Plausible Deep Learning." Cornell University. Ithaca, New York, 2016. arxiv:1502.04156, Accessed: 01.12.2020.

Blackwell, Alan. (2015): "Interacting with an Inferred World: The Challenge of Machine Learning for Humane Computer Interaction." *Aarhus Series on Human Centered Computing* 1,1 (2015), https://doi.org/10.7146/aahcc.v1i1.21197, Accessed: 20.11.2020.

Blumenberg, Hans. *Die Lesbarkeit der Welt*. Suhrkamp: Frankfurt am Main, 1986.

Boden, Margaret A. "GOFAI." *The Cambridge Handbook of Artificial Intelligence*. Ed. Keith Frankish, William M. Ramsey. Cambridge: Cambridge UP, 2014. 89–107.

Bolter, Jay David. *Writing Space: Computers, Hypertext, and the Remediation of Print*. Routledge: London, 2001.

Bostrom, Nick. *Superintelligence. Paths, Dangers, Strategies*. Oxford: Oxford UP, 2014.

Brown, Noam, and Tuomas Sandholm. "Superhuman AI for multiplayer poker." *Science* 365, 6456 (2019): 885–90, DOI: 10.1126/science.aay2400, Accessed: 24.08.2020.

Domingos, Pedro. *The Master Algorithm. How the Quest for the Ultimate Learning Machine Will Remake our World*. New York: Basic Books, 2015.

Dreyfus, Stuart. "Artificial Neural Networks, Back Propagation and the Kelley-Bryson Gradient Procedure." *Journal of Guidance, Control and Dynamics* 13,5 (1990): 926–28.

Eubanks, Virginia. *Automating Inequality: How High-Tech Tools Profile, Police, and Punish the Poor*. New York: Melia Publishing Services, 2018.

Ferber, Jacques. *Multi-Agent System: An Introduction to Distributed Artificial Intelligence*. Harlow: Addison Wesley, 1999.

Floridi, Luciano. "AI and its New Winter: from Myths to Realities." *Philosophy & Technology* 2020 (33), 1–3, Springer Nature, Cham 2020, https://doi.org/10.1007/s13347-019-00345-y, Accessed: 01.12.2020.

Goodfellow, Ian, Yoshua Bengio, and Aaron Courville. *Deep Learning*. Cambridge: MIT Press, 2016.

Gödel, Kurt. "Über formal unentscheidbare Sätze der Principia Mathematica und verwandter Systeme I." *Monatshefte für Mathematik und Physik* 38 (1931): 173–98.

Ḫwārizmī, Muḥammad Ibn-Mūsā al-. *Die älteste lateinische Schrift über das indische Rechnen nach al-Ḫwārizmī*. Ed. Menso Folkerts und Paul Kunitzsch. München: Verlag der Bayrischen Akademie der Wissenschaften, 1997.

Haugeland, John. *Artificial Intelligence: The Very Idea*. Cambridge: MIT Press, 1985.

Hendler, James. "Avoiding Another AI Winter." *IEEE Intelligent Systems* 2008, 1541–1672, see also DOI: 10.1109/mis.2008.20, Accessed: 24.08.2020.

Heyer, Gerhard, Gregor Wiedemann, and Andreas Niekler. "Topic-Modelle und Ihr Potenzial für die philologische Forschung." *Digitale Infrastrukturen für die Germanistische Forschung*. Ed. Henning Lobin, et al. Berlin: De Gruyter, 2018. 351–68.

Horstmann, Jan. "Topic Modeling." *forTEXT. Literatur digital erforschen*, 2018, https://fortext. net/routinen/methoden/topic-modeling, Accessed: 22.07.2020.
Kahnemann, Daniel. *Thinking, Fast and Slow*. New York: MacMillan Publishers, 2011.
Krämer, Sybille. *Berechenbare Vernunft. Kalkül und Rationalismus im 17. Jahrhundert*. Berlin: De Gruyter, 1991.
Krämer, Sybille. *Figuration, Anschauung, Erkenntnis. Grundlinien einer Diagrammatologie*. Berlin: Suhrkamp, 2016.
Krämer, Sybille. "Leibniz ein Vordenker der Idee des Netzes und des Netzwerkes?" *Vision als Aufgabe. Das Leibniz-Universum im 21. Jahrhundert*. Martin Grötschel, et al., ed. Berlin: Berlin-Brandenburgische Akademie der Wissenschaften, 2016. 47–60.
Krämer, Sybille."Was bedeutet 'Perspektivität'? Eine Erörterung mit Blick auf Leibniz." *Allgemeine Zeitschrift für Philosophie* 44,3 (2019): 325–43.
Krämer, Sybille."Kulturtechnik Digitalität. Über den sich auflösenden Zusammenhang von Buch und Bibliothek und die Arbeit von Bibliotheken unter den Bedingungen digitaler Vernetzung." *Künstliche Intelligenz und Bibliotheken*. Ed. Christina Köstner-Pemsel, Elisabeth Stadler, and Markus Stumpf. Graz: Grazer Universitätsverlag, 2020. 57–74. https://doi.org/10.25364/guv.2020.voebs15.7, Accessed: 19.11.2020.
Krizhevsky, Alex, Ilya Sutskever, and Geoffrey Hinton. "ImageNet Classification with Deep Convolutional Neural Networks." *Communications of the ACM*, May 2017, https://doi.org/10.1145/3065386, Accessed: 24.08.2020.
Kuijer, Lenneke, and Elisa Giaccardi. "Co-performance: Conceptualizing the Role of Artificial Agency in the Design of Everyday Life." *CHI '18: Proceedings of the 2018, CHI Conference on Human Factors in Computing Systems*, March 2018, DOI: 10.1145/3173574.3173699, Accessed: 24.08.2020.
Kuijer, Lenneke. "Automated artefacts as co-performers of social practices: washing machines, laundering and design." *Social Practices and More-than-humans: Nature, materials and technologies*. Ed. C. Maller and Y. Strengers. Palgrave Macmillan, 2018. DOI: 10.1007/978-3-319-92189-1_10, Accessed: 24.08.2020.
Kuutti, K., and L.J. Bannon. "The turn to practice in HCI: towards a research agenda." *Proceedings of the 32nd annual ACM conference on Human factors in computing systems*. ACM, Toronto: 2014. 3543–52.
Kurzweil, Ray. *The Singularity Is Near*. New York: Penguin Books, 2006.
Latour, Bruno. *Pandora's hope: essays on the reality of science studies*. Boston: Harvard UP, 1999.
LeCun, Yann, et al. "Backpropagation applied to handwritten zip code recognition." *Neural Computation* 1 (1989): 541–51.
LeCun, Yann, Yoshua Bengio, and Geoffrey Hinton. "Deep learning." *Nature* 521 (2015): 436–44.
Lenzen, Manuela. *Künstliche Intelligenz. Fakten, Chancen, Risiken*. München: C. H. Beck, 2020.
Leibniz, Gottfried Wilhelm. *Hauptschriften zur Grundlegung der Philosophie*, Band II. Ed. Ernst Cassirer. Hamburg: Meiner Verlag, 1966. 435–56.
Leibniz, Gottfried Wilhelm. *Neue Abhandlungen über den menschlichen Verstand*, Buch II. 1703/1704. Ed. Ernst Cassirer. Hamburg, Meiner Verlag: 1971. 126–128.
Lynch, Shana, and Andrew Ng. "Why AI Is the New Electricity. A computer scientist discusses artificial intelligence's promise, hype, and biggest obstacles." *Insides by Stanford*

Business, 2017. See: https://www.gsb.stanford.edu/insights/andrew-ng-why-ai-new-electricity, Accessed: 12.09.2020.

Mackenzie, Adrian. *Machine Learners. Archeology of Data Practice.* Cambridge: MIT Press, 2017.

Marcus, Gary. "Deep learning: A critical Appraisal." *Computer Science*, 2018. See: ArXiv, abs/1801.00631, Accessed: 20.11.2020.

McCulloch, Warren, and Walter Pitts. "A logical calculus of the ideas immanent in nervous activity." *Bulletin of Mathematical Biophysics* 5 (1943): 115–33.

Mühlhoff, Rainer. "Menschengestützte Künstliche Intelligenz. Über die soziotechnischen Voraussetzungen von "deep learning"." *Zeitschrift für Medienwissenschaft* 21: "Künstliche Intelligenzen" 11,2 (2019): 56–64. See https://doi.org/10.25969/mediarep/12633, Accessed: 02.12.2020.

Mühlhoff, Rainer. "Human-aided artificial intelligence: Or, how to run large computations in human brains? Toward a media sociology of machine learning." *new media & society* (2019): 1–17. See DOI: 10.1177/1461444819885334, Accessed: 24.08.2020.

Nassehi, Armin. *Muster. Theorie der digitalen Gesellschaft.* München: Beck, 2019.

Noble, Safiya. *Algorithms of Oppression: How Search Engines Reinforce Racism.* New York: Combined Academic Publishing, 2018.

Parisi, Luciana. "Das Lernen lernen oder die algorithmische Entdeckung von Informationen." Ed. Christoph Engemann and Andreas Sudmann. *Machine Learning. Medien, Infrastrukturen und Technologien der Künstlichen Intelligenz.* Bielfeld: Transcript, 2018. 93–116.

Preusse, Thomas, and Hanna Wick. "Blick in die Blackbox." *Republik*, 26.06.2018, see https://www.republik.ch/2018/06/26/blick-in-die-blackbox, Accessed: 06.09.2020.

Ramge, Thomas. *Mensch und Maschine. Wie Künstliche Intelligenz und Roboter unser Leben verändern.* Stuttgart: Reclam, 2018.

Richbourg, Robert (2018): "'It's either a Panda or a Gibbon': AI Winters and the Limits of Deep Learning." https://warontherocks.com/2018/05/its-either-a-panda-or-a-gibbon-ai-winters-and-the-limits-of-deep-learning/, Accessed: 24.08.2020.

Rosenblatt, Frank. "The Perceptron: A Probabilistic model for information storage and organization in the brain." *Psychological Review* 65,6 (1958): 386–410.

Schauer-Bieche, Florian. "Deepfakes – was darf man heute noch glauben? " *t3n-newsletter*, 2019. See https://t3n.de/news/deepfakes-darf-heute-noch-1185369/, Accessed: 09.08.2020.

Schmidhuber, Jürgen. "Deep Learning." *Scholarpedia* 10,11 (2015), 328–32.

Schuchmann, Sebastian. "Analyzing the Prospect of an Approaching AI Winter." 2019. https://www.researchgate.net/publication/333039347_Analyzing_the_Prospect_of_an_Approaching_AI_Winter; DOI: 10.13140/RG.2.2.10932.91524, Accessed: 24.08.2020.

Searle, John R. "Minds, Brains, and Programs." *The Behavioral and Brain Sciences* 3 (1980): 337–56.

Shalev-Shwartz, Shai, and Shai Ben-David. *Understanding Machine Learning. From Theory to Algorithms.* New York: Cambridge UP, New York, 2014. DOI: https://doi.org/10.1017/CBO9781107298019, Accessed: 24.08.2020.

Sudmann, Andreas. "Szenarien des Postdigitalen. Deep Learning als MedienRevolution." *Machine Learning. Medien, Infrastrukturen und Technologien Künstlicher Intelligenz.* Ed. Christoph Engemann and Andreas Sudmann. Bielefeld: Transcript, 2018. 55–73.

Tietgen, Madita. "3-D-Drucker erschafft ein neues Rembrandt-Gemälde." 2016. See https://www.welt.de/kultur/kunst-und-architektur/article154421365/3-D-Drucker-erschafft-ein-neues-Rembrandt-Gemaelde.html", Accessed: 05.09.2020.

Vlassis, Nikos. *A Concise Introduction to Multiagent Systems and Distributed Artificial Intelligence.* Williston, VT: Morgan and Claypool Publishers, 2007.

Whitehead, Alfred North. *An Introduction to Mathematics.* London: Williams and Norgate, 1911.

Wöpking, Jan. *Raum und Wissen. Elemente eines epistemischen Diagrammgebrauches.* Berlin: de Gruyter, 2016.

Ludwig Nagl
Merits and Limits of AI: Philosophical Reflections on the Difference between Instrumental Rationality and Praxis-Related Hermeneutical Reason

Abstract: *Part one* looks at some innovations made possible by machine-executed algorithms that undoubtedly prove useful, such as the digitization of libraries and the gains in precision in medical diagnosis. These forms of AI may be called "weak": they are instances of an "instrumental rationality" which can be explained without the Utopian claim that digital machines are leading us towards a post-humanistic "singularity". They all rest on algorithms that, in their core, are "tools" which – with regard to their social consequences – need permanent public control. *Part two* analyzes the Utopian ideal of "strong AI" – of (as John Searle pointed out) the "false" claim "that appropriately programmed computers literally have cognitive states.". After a brief reference to a classical critique of the idea "that human perception is explicable in terms of mechanical reason" (Leibniz, *Monadology*), the essay examines core arguments developed by Hubert Dreyfus, Hilary Putnam and Charles Taylor – philosophers who all challenge the reductionist overrating of "instrumental rationality". A rich, non-reductive conception of praxis addresses the ability of humans to act according to their moral judgment: an ability which – to speak with Kant – cannot be fully simulated by AI's "heteronomous" mechanical execution of programmed "norms".

1 Introduction: Setting the stage with Jürgen Habermas and Sybille Krämer

In a long footnote in his book *Auch eine Geschichte der Philosophie* Jürgen Habermas addresses the "agitated discussions surrounding Artificial Intelligence" – discussions, as he notes, that are instigated by "post-humanists" who dramatize the Utopian idea that human intelligence will soon be superseded by the superior calculation potentials of probabilistically leaning computers.[1] This agitated discourse, Habermas critically remarks, is, however, a hasty overreaction which

---

[1] Habermas 2019, vol. 2, 593, footnote 2 (English translation L.N.).

https://doi.org/10.1515/9783110770216-004

results from the incapacity to deal adequately with the perplexities caused by digital innovations. It re-affirms an abstract scientism that shuns away from all careful, philosophical analysis of the complex, intersubjectively structured life of human society: from an investigation, that is, of the non-deterministic action horizons in which AI innovations are embedded.[2]

In a similar vein Sybille Krämer – in an article published 2005 in Sandbothe/Nagl, *Systematische Medienphilosophie*[3] – distanced herself from the excited claims raised in connection with today's "media turn". She addressed the assertion that this turn will be able to deconstruct and re-read core problems connected with human symbol use and action. The fashionable talk of a "media apriori", Krämer wrote, lacks scrutiny. What a reflecting public really needs is a careful investigation of the role that today's media are actually able to play in modern societies. "We want to stimulate a sensitive philosophical discourse in regard to our dealings with the media", she writes, "a sensitivity, that is, which does not upgrade the media to substitutes for a program which is well established within the respectable post-Kantian, Wittgensteinian, Austinian and Peircean tradition."[4] Thus investigations are needed – as Krämer argues – that, again, bring

---

[2] Philosophical analyses of AI, its benefits and problems, are thus well advised to clearly distinguish between (mathematics-based) instrumental "rationality" (*Verstand*) and full human reason (*Vernunft*). Emphasizing this important (but often overlooked) distinction, the German philosopher Dieter Mersch, in his essay "Ideen zu einer Kritik 'algorithmischer' Rationalität" ["Ideas on a critique of 'algorithmic' rationality"], argues that it is of the utmost importance to speak of "rationality", and not of "reason", in connection with AI, in order to make visible the operative character of algorithmic processes and to distinguish them from human logos. (English translation L.N.) [In its German original Mersch's text runs as follows: [Es ist von größter Wichtigkeit] "im Falle der Algorithmik [...] von 'Rationalität' und nicht von 'Vernunft' [zu sprechen] um den operativen Charakter dieser Praxis gegenüber dem humanen Logos zu unterstreichen." (Ibid., 853)] While algorithmic "rationality" is chained to its "instrumental" character, human "reason" ("Vernunft im eigentlichen Sinn") is manifest in human judgment, understanding, and justification ("Urteilsbildung, Verstehen und Begründung"). (Ibid.)

Similarly, the American cognitive scientist Melanie Mitchell, in her book *Artificial Intelligence. A Guide for Thinking Humans*, points out that AI rationality faces a "barrier of meaning. This phrase [used by the mathematician and philosopher Gian-Carlo Rota]" – Mitchell writes – "perfectly captures an idea that has permeated [my] book: humans, in some deep and essential way, understand the situations they encounter, whereas no AI system yet possesses such understanding." (Ibid., 235)

[3] Sybille Krämer, "Medienphilosophie der Stimme".

[4] English translation L.N. The German original of Krämer's text runs as follows: "Wir möchten [...] eine Sensibilität des philosophischen Diskurses für Fragen und Probleme, die sich im Umgang mit Medien stellen, anregen": eine Sensibilität, "welche Medien *nicht* zu Platzhaltern eines Programms macht, das innerhalb der – transzendental orientierten [nach-kantischen, Wittgen-

to the fore two core insights of this post-Kantian discourse: *firstly*, that media are not at all self-sufficient, and, *secondly*, that they are nowhere fully "self-organizing", since the very talk of a "medium" – as Peirce showed in his triadic sign concept – always implies that there is *something* which is *"mediated"* to *someone*. "The characteristic feature of the media is their heteronomy," Krämer writes, " – the fact, that is, that in the production of those outcomes that they are able to produce they follow not their own organization, but a 'foreign law', an external direction."[5] It is thus obvious that today's exaggerated AI-discourse cannot plausibly answer, or deconstructively "transform", the basic questions of our human existence: "The media are unable," Krämer emphasizes, "to occupy convincingly the role of an 'ultimate foundation'".[6] This has to be kept in mind at the very moment when the digital net – as some theologians, ironically, have written – more and more seems to become the *"sacrament of permanent presence"*: the post-transcendent, self-celebratory instantiation of a quasi-divine "omnipresence".[7] On sober reflection, however, even top-end AI-structures like auto-cor-

---

stein-Austinschen, Peirceschen] – Philosophie wohl etabliert ist und über eine respektable Tradition verfügt." (Ibid., 221)

5 "Die Besonderheit von Medien liegt in ihrer Heteronomie," Krämer writes," – in dem Sachverhalt, dass Medien in dem, was sie leisten, nicht selbst organisiert sind, sondern einem 'fremden Gesetz', einer äußerlichen Vorgabe und Auflage folgen." (Ibid., 222)

6 "Medien sind nicht geeignet", Krämer argues, "die Stelle eines 'Unhintergehbaren', 'Vorgängigen' und 'Letztbegründenden' zu besetzen." (Ibid.)

7 See Gianluca De Candia, "Von digitaler Gegenwart. Über Medialität, Repräsentation und Sakrament" [On digital presence. Mediality, representation, and sacrament]. De Candia, Professor of Systematic Theology at the University of Siegen, Germany, points out that in the digital world "communicative mediation is, in itself, message, dynamics and modality. The net appears to be a sacrament of permanent presence." [English translation L.N. The German original runs as follows: "Die kommunikative Vermittlung [ist] an und für sich Botschaft, Dynamik, Modalität. Das Netz erscheint als Sakrament ständiger Präsenz, das indes keine anderen Sakramente mehr neben sich zu dulden scheint." (De Candia, https://www.academia.edu/42175822, Accessed: December 7, 2021)]

In our "secular age", even the religious ideas of immortality and resurrection are algorithmically (quasi-)inherited. Moritz Riesewieck and Hans Block, in their book *Die digitale Seele. Unsterblich werden im Zeitalter Künstlicher Intelligenz* (2020), describe (so-called) "digital resurrections", that is to say AI-steered avatars that simulate the behavior of deceased people. Is such a simulation significantly different, however, from the use of items of remembrance like photos and films in analogue times? Silicon Valley ideologues tout that this is indeed the case: according to them, the new, AI-mediated "digital immortality" will, in a secular age, help to close the space that is left open after the collapse of a religious belief in "resurrection". Critics of such claims point out, quite convincingly, however, that AI-based quasi-"interactions" with a "bot" – far from digitally "resurrecting" the deceased – destroy, in precarious ways, not only the acknowledgment of death, but also the acknowledgment of the basic fact that the (so called) "re-

recting (so called "learning") machines – i.e.: software systems with probabilistic functions that (with regard to reach and speed) have, indeed, amazing capacities – *are nothing but tools created by humans*. The benefits which they make possible are accompanied by a Foucauldian shadow: by the, quite realistic, fear that a full implementation of today's advanced technological possibilities will organize a surveillance system driven by economic and political considerations which affects and controls our every utterance and move.

2 "Weak" versus "strong" AI

A brief glance at some of the innovations made possible by machine-executed algorithms shows that they doubtlessly prove useful in many respects. AI-related benefits entail, in the field of education, for instance, the swift and universal availability of books made possible by the digitalization of libraries; digital teaching innovations and research data management; or, in a different field, the execution by robots of tasks that are dangerous for humans (like minesweeping); the possibility of speech and facial identification;[8] the gains in the precision of medical diagnoses such as software-based "polyp detectives" in coloscopy, for instance; AI-mediated support of juridical procedures via the automated collection and comparison of laws and edicts; the use of Virtual Reality in planning infrastructure. All these AI instantiation – innovative as they are – *may be called in one important regard "weak"*: they all are instances of an "instrumental rationality" whose means-ends-logic can be analyzed without any recourse to the excessively Utopian claim that digital machines are on the verge of implementing a "strong" version of AI that leads us in the direction of a post-humanistic "singularity".[9]

---

surrected person" is nothing but a computer (without any self-awareness or feeling). (See in this context also Moritz Riesewieck's and Hans Block's essay in *Die Zeit*.)

8 "Facial recognition has many upsides," the cognitive scientist Melanie Mitchell writes, but "it is just as easy to imagine applications that many people find offensive or threatening." (Mitchell 2019, 122)

9 For basic information on the recent debate on "singularity" see Mitchell 2019, "Singularity Skeptics and Adherents", 58–9.

Calculating rationality rests on algorithms (or, in more complex modes, on – metaphorically so called – "learning" algorithms: i.e. algorithms that are probabilistically dimensioned). All such "tools" – even if they are able to outdo humans by far in regard to speed, efficiency and extent of their performance – remain finite and fallible, and are, as programs invented by humans, in

The term "weak AI" is thus used in this paper as a descriptive category for these "*limited task* algorithms".[10] It is not used, as it is often the case, as a misleading label for the problematic idea – which Nida-Rümelin and Weidenfeld rightly criticize[11] – that (in spite of the fact that human and artificial intelligence differ in principle) in the end "all human thought, all modes of experiencing and decision making can be simulated by software systems."[12] The "difference between our algorithmic tools and human authorship" is, indeed, substantial.[13]

This point is often overlooked in one important, innovative field of AI – in "neuroenhancement", which – in principle – clearly belongs to the realm of "weak AI". Recent debates in this field often fail to focus on the *actual improvements* that AI-mediated alterations of brain performance are able to produce for the human patients, and rather support the "strong" claim that the modulations of the brain made possible by pharmacological and computer-mediated interventions (metaphorically called modulations of the "self") are on the verge of critically dissolving the (substantial) conceptual differences between man, animal, and machine. Critics of these inflated claims (like Michael Pauen[14] and Dieter Sturma[15] in the Viertbauer/Kögerler volume on *Neuroenhancement*[16]) argue that neuromodulation should not be called "moral enhancement", since moral action is tied to free decision, and is thus not a result of causal alteration. "Autonomy-supporting, authentic 'enhancement'", Sturma writes, "takes place through labor or education" and is thus "an expression of the fact that the 'enhancing' person is *a human subject*. In contrast to this, those forms of enhancement which produce improvements of physical states via external causation or passive endurance are *manifestations of an instrumentalization syndrome* which treats humans like objects." [17]

---

need of permanent public control with regard to their positive as well as negative social consequences.
10 By contrast, the term "strong AI" refers to the "false" claim (as John Searle argued) "that appropriately programed computers literally have cognitive states." (John Searle, "Minds, Brains, and Programs", 18)
11 Nida-Rümelin / Weidenfeld 2018.
12 Ibid., 59–60.
13 Ibid., 60.
14 Pauen 2019.
15 Sturma 2019.
16 See Viertbauer / Kögerler 2019.
17 English translation L.N. The German original runs as follows: "Autonomiefördernde Verbesserung der menschlichen Eigenschaften = authentisches 'Enhancement' stellt sich durch Arbeit oder Bildung ein und ist Ausdruck des Sachverhalts ein Subjekt zu sein. Demgegenüber sind For-

Similar concerns were articulated in 2019 by Martha Nussbaum in her short text "We are no machines. We are humans"[18], where she hails – in opposition to today's rampant AI reductionisms – the genesis of a new, *anti*-reductionist philosophical debate that critically subverts digital "instrumentalisms". In the United States, Nussbaum writes, a complex philosophical discourse is about to re-emerge which – while affirming the importance and usefulness of science – rejects, at the same time, dogmatic scientism.[19] In our age of digitally mediated interactions, a philosophical analysis of the non-instrumental modes of *reason* that allow us to regulate digitalization processes socially and legally is of increasing importance. In *The Offensive Internet*, a book that Nussbaum co-edited in 2010 with Saul Levmore,[20] a discourse of this kind is started with regard to the question how "privacy" can be protected in our digital age in which the invasion of privacy tends to become the rule.[21]

---

men von Enhancement, die Zustandsverbesserungen auf dem Wege von äußerer Einwirkung und passivem Erleiden erzeugen, Ausdruck eines Instrumentalisierungssyndroms, das Personen wie Sachen behandelt." (Sturma 2019, 143)
In a similar vein, Lutz Wingert, in his essay "Grenzen der naturalistischen Selbstobjektivierung", points out that in neurobiological discourses the fact that causally dimensioned "self-objectivations" are problematic (and, if they are generalized and routinized, can turn out to be dangerously deficient) is marginalized. AI-based "enhancements" do not improve "morality", they alter "behavior". They are dangerous, if they overhastily leave aside the complex question, whether, in a particular case, an intervention not only seems to be (medically or socially) "justified", but is also acceptable to its addressee.
18   English translation L.N. (Nussbaum 2019, 40)
19   This discourse – which, as Nussbaum hopes, will be able to challenge the instrumentalism-related "hegemony of analytic philosophy" – was nourished, in recent decades, by the philosophical work of thinkers like John Rawls, Jürgen Habermas and Charles Taylor. All of them defend the idea that "the interpretations of human action as well as of societal phenomena are not less 'scientific' than those of natural phenomena, the explanation of which follows a positivistic interpretation of science. Humanistic themes can thus be analyzed on the basis of a comprehensive conception of human reason which is not narrowed down to 'calculating rationality'". (English translation L.N. The German text runs as follows: [Martha Nussbaum schreibt] "dass die Interpretation menschlichen Handelns und gesellschaftlicher Phänomene nicht weniger 'wissenschaftlich' ist als die von Naturphänomen, deren Erklärung einem positivistischen Wissenschaftsverständnis folgt [...] All diese Themen könnten mit einem umfassenden Verständnis menschlicher Rationalität untersucht werden." [Nussbaum 2019, 40]
With regard to Horkheimer's and Adorno's critique, in *Dialectics of Enlightenment*, of the career of "instrumental reason", as well as to Habermas's concept of "communicative action", Nussbaum writes: "The European Enlightenment has not only set free instrumental rationality but also communicative reason." ["Die europäische Aufklärung hat eben nicht nur die instrumentelle, sie hat auch die kommunikative Vernunft freigesetzt."] (Ibid.)
20   Levmore / Nussbaum 2010.
21   See in this context also Becker / Seubert 2020.

Of course nobody of sound mind would deny that contemporary AI programs are efficient, fast, accurate and extremely useful, if they (correctly) execute *limited tasks:* for instance – to pick a random example presented by Thomas Ramge in his 2019 book *Who's Afraid of AI?* – the limited task to arrange, via a telephone conversation with a "bot", a haircut appointment. If the conversation between human and machine does not transgress the set communication limits, algorithm-steered programs are able to arrange an appointment "with such perfection", Ramge points out, "that the person on the other end of the line has no idea that [he or she is] talking to a data-rich IT system."[22] Weak AI swiftly performs *limited* tasks.

However, doesn't AI (in other fields of innovation – in "algorithm-mediated art", for instance) *creatively transgress* the limits of "weak AI"? Pioneers of computer art (Max Bense, Abraham Moles) have shown that, by random variation, unexpected objects can be produced that the recipients of art – if they find them appealing – are willing to call "art". This acceptance of AI-produced objects into the artistic realm depends ultimately, though, *always* on human judgments: on a background understanding, that is, of where this "surprisingly new" object is able to be placed in the long history of innovations that characterize human artistic expressions. Dieter Mersch, in a recent analysis of *stochastic aesthetics*, rightly pointed out that the "new" which is produced by (so-called) "creative" AI-art programs does nowhere "evaluate" itself: without human recogni-

---

[22] Ramge 2019, 2.
This form of digital communication cannot be meaningfully expanded, however, even if today's feuilleton-based AI propaganda dreams of such an expansion (as, for instance, Stefan Huck's article "Wir brauchen mehr künstliche Freunde. Je besser lernende Maschinen uns kennen, desto mehr profitieren wir von ihrer Intelligenz" shows).
We return, however, at that point to safer ground – that is to say to "weak AI" – and look at another, indeed promising, digitalized task. In medicine, AI-systems are of invaluable help in identifying certain cancer cells. They significantly outperform, with respect to accuracy, medical analyses that are carried out without algorithmic help. This efficiency gain should not nourish the exaggerated wish, however, that medical diagnosis *in toto* will soon be automated. In order to be safe, machine results need to be (and, it is to be hoped, will in the future still be) ultimately controlled through a holistic human judgment: a judgment which does not lose sight of the fact that any illness is embedded in the complex life situation of the patient.
Or, to pick another example: the risky business of financial investment allows, in some regards, its transformation into ultra-fast digitized calculation processes. In spite of the AI-supported modes of "stock trading", the danger of a financial crisis remains: the transformation of "risk" into an algorithm is *unable* to eliminate the possibility of a devastating systemic failure in economic planning.

tion, Artificial Creativity is nothing but a series of "meaningless leaps"[23]. Or, as Melanie Mitchell puts it: while computer programs generate "things that its programmer never thought of", "*the human provides judgment of the resulting work*, which comes from the human's understanding of abstract artistic concepts. The computer has no understanding whatsoever, so it alone is not creative."[24] Hence Artificial Intelligence is, even in AI-art, unable to leave behind its "weakness", its status, that is, of being a "tool."

3 The Promethean Utopianism of "strong AI," critically re-assessed

Self-improvement through the use of tools is a general human practice. Since humans are beings without an "invariant essence", their actions entail, at all times, the possibility of "enhancement". The invention, for instance, of a basic "analog" object, the eyeglass, was, indeed, able to "enhance" human vision. Enhancement is *thus not at all a new category related to digitalization*. Due to the fact that humans are capable of re-defining themselves via individual action as well as via socio-institutional reform, humanity is, and has always been, *an open project*. Only if this openness remains out of sight the *false* picture of a "rigidly fixed" humanness can hold us captive – a deficient picture that invites an excessive, *equally defective counter-picture*: the idea of a (post-) humanness which dreams (in a mode of hubris) of the self-production of an "*Übermenschentum*". Such an idea of negating finiteness is quite old: it is the idea of Prometheus.[25] In a new technological guise it is revived in Ray Kurzweil's Utopian scenario[26] of an AI-enabled "superintelligence"; or (in negative inversion) in dystopias in the manner of Nick Bostrom.[27] The claim that a "singularity"-orient-

---

[23] Mersch 2019, 866–7. Mersch sums up his considerations in the following thesis: Creativity, as a "poetics of invention", cannot be imitated by algorithms: "Die Kreativität als eine 'Poetik der Findung' widersteht jeder Kalkülisierung durch einen Schematismus." (Ibid., 869)
[24] Mitchell 2019, "Could a Computer be Creative?", 272–73.
[25] In its modern form, this idea seeks to get its intellectual backing in recourse to "immanence"-focused, post-Hegelian (Feuerbachian and Nietzschean) deconstructions of religion.
[26] Kurzweil 2012.
[27] Bostrom 2014. "If some day we build machine brains that surpass human brains in general intelligence", Bostrom writes, "then this new superintelligence could become very powerful. And, as the fate of the gorillas now depends more on us humans than on the gorillas themselves, so the fate of our species would depend on the actions of the machine superintelligence. […] Once unfriendly superintelligence exists, it would prevent us from replacing it or changing

ed "post-humanism" is about to arrive, is a favorite idea of today's scientism-nourished feuilleton scene. Its inversion, the dystopian idea that our tools will ultimately turn against us, their inventors, (a core idea of science fiction), is the negative *complement* to the affirmative, Prometheus-inspired utopianism. There are good reasons to argue, that both of these images fail to be convincing[28].

Thus the *philosophical question* has to be raised: "What are the limits of computerized modes of 'instrumental rationality'?"

In the following it will be argued that important clues for a non-reductive analysis of this question can be found in *four* different philosophical discourses: in thoughts by Leibniz (3.1), by Hubert Dreyfus (3.2), by Hilary Putnam (3.3), and by Charles Taylor (3.4)

3.1. The first of these discourses took place in the early modern period. In paragraph 17 of his *Monadology*, Gottfried Wilhelm Leibniz formulated his famous critique of the idea that human perception is explicable in terms of mechanical reason. Leibniz's so-called "mill argument" (*Mühlengleichnis*) runs as follows: "If one imagines a machine which is able to feel and perceive, one could think of it as being enlarged to an extent that one could walk into it as into a mill. If one thus visits its interior one would find separate parts that push each other, but nothing at all from which perception could be explained."[29]

When we describe the neurophysiological events in a brain the (re-applied) Leibnizian argument would run today: we are *not* exploring consciousness itself, but only its material basis. Algorithms can, at best, simulate these *material* pro-

---

its preferences. Our fate would be sealed". In my book, Bostrom continues, "I try to understand the challenge presented by the prospect of superintelligence, and how we might best respond." Thus, in Bostrom's dystopia, the difference between machine superintelligence and human thinking and planning is preserved (albeit in an unexplained, matter-of-fact way): AI-intelligence – Bostrom knows very well – is nowhere its own product: it is ultimately produced by us, the humans: "We do have one advantage," Bostrom rightly points out: "We get to build the stuff", and we thus, "in principle, could build a kind of superintelligence that would protect human values." (Ibid., V)

28 See Nida-Rümelin / Weidenfeld 2018, "Die transhumanistische Versuchung", 188–207.
29 The German original of Leibniz's *Monadologie*, paragraph 17, runs as follows: "Denkt man sich etwa eine Maschine, die so beschaffen wäre, dass sie denken, empfinden und perzipieren könnte, so kann man sie sich derart proportional vergrößert denken, dass man in sie wie in eine Mühle eintreten könnte. Dies vorausgesetzt, wird man bei der Besichtigung ihres Inneren nichts weiter als einzelne Teile finden, die einander stoßen, niemals aber etwas, woraus eine Perzeption zu erklären wäre."

cesses, but they are, in no way, able to become *full* substitutes for the (free, and intersubjectivity-related) thought processes of self-aware human beings.[30]

In *contemporary* philosophy, elements of this Leibnizian anti-reductionistic thought were taken up by Hubert Dreyfus, by Hilary Putnam (with reference to Wittgenstein's late philosophy), and by Charles Taylor.

3.2 In his essay "Coping with Change: Why People Can and Computers Can't", published in 1986 in Volume 1 of *Wiener Reihe. Themen der Philosophie: Wo steht die Analytische Philosophie heute?*[31], Hubert Dreyfus and his brother, the Professor of Industrial Engineering at Berkeley, Stuart Dreyfus, argue that the following problem severely limits AI: "Human-like intelligence requires understanding of the everyday physical and social world which for cognitivists requires representing a huge number of facts, while for human beings such background understanding seems to take the form of skills rather than of explicit knowledge." Thus, for human beings, "situations show up with relevance or salience built in, along with anticipations, based on past experience of similar meaningful situations, of how what is relevant will change as the situation develops." AI faces unsurmountable problems when it tries to simulate *fully* this complex human *"background understanding."*[32]

3.3 A similar philosophical critique of exaggerated AI-claims was voiced by the Harvard philosopher *Hilary Putnam*, who is well known in AI circles through his thought experiment "Brains in a Vat"[33]. In his essay "The Project of Artificial Intelligence" Putnam takes issue with a core, "positivism-informed" assumption of AI, that is to say with the reductionist idea that the mind is "a sort of 'reckoning machine'". [34] This idea, Putnam points out, goes back to the "birth of the scientific world view in the seventeenth and eighteenth centuries" – back to Hobbes, who claimed that "thinking is a manipulation of signs according to rules (analogous to calculating rules)", as well as to La Mettrie's materialist conception of man in *L'Homme Machine.*[35] "Today" – "hardly well thought out", Putnam crit-

---

30 Leibniz – the inventor of the first calculating machine – saw clearly the limits of mathematical calculation: algorithms are helpful, but merely tools: "[E]s ist ausgezeichneter Männer unwürdig, ihre Zeit mit sklavischer Rechenarbeit zu verbringen", Leibniz pointed out. (See Reichl 2020, 37)
31 Dreyfus / Dreyfus 1986.
32 Ibid., 169–70.
33 Putnam 1981, 1–21.
34 Putnam 1992, 1–18.
35 Ibid., 3.

ically remarks – "the notion of a Turing machine", in AI discourse, is seen "as a way of making this materialist idea precise."[36] Artificial intelligence is, however, a misnomer, Putnam writes: AI does not "really try to simulate intelligence at all [...], its real activity is just writing clever programs for a variety of tasks. This is an important and useful activity, although it does not sound as exciting as 'simulating human intelligence' or 'producing artificial intelligence'."[37]

In his book *Renewing Philosophy*, Putnam shows that Ludwig Wittgenstein convincingly argued that "understanding people" (*Menschenkenntnis*) is a very complicated skill.[38] *Menschenkenntnis*, Wittgenstein writes in *Philosophical Investigations*, "is not a technique; one learns correct judgments. There are also rules, but they do not form a system, and only experienced people can apply them right. *Unlike calculating rules*"[39], Wittgenstein concludes.

This Wittgensteinian point was re-emphasized, in 2019, by the computer scientist Melanie Mitchell in her book *Artificial Intelligence:* "While state-of-the-art AI systems have nearly equaled (and in some cases surpassed) humans on certain narrowly defined tasks", she writes, "these systems all lack a grasp of the rich meanings humans bring to bear in perception, language, and reasoning.

---

36 Ibid., 4.
37 Ibid., 13. "There is no reason," Putnam continues, "why the study of human cognition requires that we try to reduce cognition either to computations or to brain processes." (Ibid, 18) While AI, as Putnam writes, undoubtedly has its merits, the "strong" claim that a reductionist instrumental approach to human action can explain how we understand the world, and how we creatively interact with each other, rests on a blatantly dogmatic form of scientism.
In his essay "Plädoyer für die Verabschiedung des Begriffs 'Idolatrie'", Putnam argues that the reductionist pictures which inform modernity [*inter alia:* the excessive idea of strong AI: of a superintelligence which is a "machine" substitute for God, created by us] are based on a secularism-focused, post-religious thought figure that circles around the excessive idea of a "deification of humanity": around the claim, i. e., that humanity, via science and technology, will, *in toto*, be able to overcome its finiteness. "Die Vergötterung der Menschheit (welche stets mit der Umwandlung der Humanität in eine bloße Abstraktion einhergeht)", Putnam writes, "stellt eine beängstigende Tendenz dar.[...] Mein Freund Benjamin Schwarz [...] bemerkte mir gegenüber einmal, 'Der Mensch ist der schlechteste Gott, den es gibt' – und diese Bemerkung hat einen tiefen Eindruck in mir hinterlassen." (Ibid., 50)
38 Putnam 1992, "Materialism and Relativism", 211. "Human understanding", in its full, intersubjectively dimensioned sense, can thus not be completely reenacted by algorithms (even by those which – governed by rules – are, metaphorically speaking, "learning", and can thus execute – like the AI-steered "assistant" Alexa – limited [quasi]"communications"). See in this context also Otto Neumaier's essay "A Wittgensteinian View of Artificial Intelligence", in which he argues that AI, for Wittgenstein, would "have been one element of what he called the 'sickness of a time'" : – of "*our* time", that is, "to whose therapy he devoted a great deal of his reflections." (Ibid., 167)
39 Wittgenstein 1968, 227 (Emphasis L.N.)

This lack of understanding is clearly revealed by the un-humanlike errors these systems can make; by their difficulties with abstracting and transferring what they have learned; by their lack of commonsense knowledge; and by their vulnerability to adversarial attacks. The barrier of meaning between AI and human-level intelligence still stands today."[40]

3.4 A well-argued philosophical critique of strong AI is presented also in Charles Taylor's and Hubert Dreyfus's book *Retrieving Realism*[41], published in 2015. Similar to Hilary Putnam, Dreyfus and Taylor criticize the public career of scientistic reductionisms. However, they base their critique not, like Putnam, on post-analytical considerations, but on thoughts that are inspired by phenomenology and hermeneutics. Dreyfus and Taylor defend a *"pluralist robust realism"*[42] that questions "the vogue in recent decades for accounts of thinking based on the idea that the brain operates in some respects like a computer."[43] In such accounts – Dreyfus and Taylor argue – "the kind of not totally transparent intuitions that humans have as embodied, social, and cultural agents" are routinely excluded – like "knowing whether I can jump this ditch", or "whether you are mad at me", or realizing "that the atmosphere of the party has suddenly become tense", etc.[44] All these examples show that human understanding is not limited to instrumental reason. The "full shape of human linguistic capacities" – which Taylor, 2016, investigated carefully in his book *The Language Animal*[45] – encompasses *many modes of non-calculating reason:* genuine artistic creativity, for instance, as well as attempts to provide a religious interpretation of

---

40 Mitchell 2019, 235–36.
41 Dreyfus / Taylor 2015.
42 This "pluralist robust realism" avoids "a *reductive realism*, which holds that science explains all modes of being", as well as "a *scientific realism*, which holds that there is only one way the universe is carved up into kinds so that every user of such terms must be referring to what our natural-kind terms refer to." (Ibid., 160)
43 Ibid., 15. They describe the core elements of this deficient view of human thought and practice as follows: "The computer model [...] (*first*) speaks of the mind as receiving 'inputs' from the environment and producing 'outputs'". (*Secondly*), computations proceed on the basis of bits of clearly defined information, which get processed. The brain computes explicit pieces of information. (*Thirdly*), the brain is a purely 'syntactic engine'; its computations get their 'reference' to the world through these 'inputs'. And (*fourthly*) the account proceeds on the materialist basis that these mental operations are to be explained by the physical operations of the underlying engine, the brain." (Ibid.)
44 Ibid., 15–16. Unlike algorithm-executing bots, human beings are able to read and re-interpret complex situations creatively: "Our everyday experience of our direct embodied contact with an independent reality", Dreyfus and Taylor write, "opens a space for a whole range of accounts of our essential nature and the nature of the universe." (Ibid., 160)
45 Taylor 2016.

"being", and experiments in philosophy to systematically explicate human *praxis* and thought.⁴⁶

All four of these critical analyses of AI call for its efficient socio-practical regulation. The necessary political and legal controls cannot be "executed" by the "learning loops" of AI-programs "themselves". Efficient controls depend on *system-external* ethical considerations – on a *human discourse* about the "use and abuse of digitalization". This discourse cannot be delegated to the computer-savvy "Big Data" companies (and their vested interests), but has to take place in the public, socio-political sphere.

This sphere is itself, however, today significantly (and in some regards quite dangerously) re-structured by AI-driven "social media". What started as the project to democratically "disrupt" the established "gate keeping" systems in journalism and publishing, has in recent decades more and more shown the tendency to morph into a culture of permanent (economics-driven) distraction, and into new political modes of communicative simplification. One-sided, affectively tuned-up communication bubbles start to destroy – by insistently re-affirming a narrowly closed world view – the open, consensus-oriented discourse which is essential for any flourishing democracy.⁴⁷ Charles Taylor, Patrizia Nanz and Madeleine Beaubien Taylor, in their essay "Reconstructing Democracy", thus point out that while, on the one hand, "digitalization provides citizens with easy and broad access to information", the "largely anonymous social networks" distance, on the other, "citizens from the political sphere. With its focus on finding sympathetic others within 'echo chambers' that reject dissenting opinion, this form of media consumption acts as a barrier to collective learning and meaningful deliberation. Instead, it provides fertile ground for electronic populism."⁴⁸

---

46 See in this context Nagl 2019.
47 See Marantz 2019. For John Dewey's classical argument that any functioning democracy rests on open and unrestricted discourses, see the chapter "John Dewey: Demokratie und Erziehung" in Nagl 1998, 128–37.
48 Taylor / Nanz / Beaubien Taylor 2020.
See in this context also Becker and Seubert, who point out in their essay "The Self-endangerment of Autonomy" that today's political situation is further aggravated by the fact that in our digital age even the (possible) *loci* of resistance – "autonomy", "individuality" and "authenticity" – are increasingly subverted by a culture industry that, via digital "influencers" and AI-supported political *acteurs*, mass-distributes *simulacra* of "authentic, individualized freedom" which destroy what they pretend to offer: critical reflection and self-determination. Thus, today – Becker and Seubert write – a much neglected early classic of Critical Theory, Horkheimer's and Adorno's *Dialectic of Enlightenment* (in particular its chapter "The Culture Industry: Enlightenment as Mass Deception") "gains unexpected actuality", since the sharp-sighted reflec-

4 Conclusion: Two modes of AI-related ethical discourse

For the regulation and control of AI *two* modes of ethical discourse are required:

4.1 First, a close examination of the "inbuilt rules" that direct machines/bots. Analytic shortcuts will not do their job here. The following example taken from the debate about "self-driving cars" will illustrate this. In view of the unexpected traffic situations that automatic cars could face, the so called "trolley problem" (a "staple of undergraduate ethics courses in Analytic Philosophy Departments ", as Melanie Mitchell writes) has become "a central talking point in discussing AI ethics."[49] Reductionist "ethical" discourses tend to terminate in debates that center, for instance, on the question whether automatic cars are to be allowed, in the situation of an unavoidable collision, to run over senior citizens in order to protect young lives?[50] Argumentation of this type calculates, in a utilitarian manner, the "overall advantage" that a particular algorithmic "performance" will produce. Any well-argued philosophical view of ethics has, however, to reject such quasi-ethical shortcuts. Computer programs that calculate a trans-individualistic "general use" operate along the lines of a utilitarian "consequentialism", and are, thus, incompatible with *the basic right of every human individual to live*, a right which is encoded in the constitution of all modern states. Nida-Rümelin and Weidenfeld formulate this maxim in their study *Digitaler Humanismus* correctly as follows: "The violation of basic rights cannot

---

tions on the economy-driven production of "(quasi)-subjective needs" which Horkheimer and Adorno provide substantially contribute to an analysis of the social pathologies that accompany today's ever-increasing "commodification and reification of privacy". (English translation L.N.) The German original runs as follows: [Becker and Seubert analysieren die] "Formen einer freiheitseinschränkenden Selbstbedrohung des Privaten, die mit der Digitalisierung des Alltags einhergehen". [Sie zeigen] "dass es gerade die ältere Kritische Theorie um Horkheimer und Adorno ist, die im Kontext der Digitalisierung eine neue, unvermutete Erklärungskraft gewinnt. Aus ihren Werken, vor allem der ‚Dialektik der Aufklärung' lassen sich Grundfiguren einer Kritik modernen Erkenntnis- und Kulturproduktion herausarbeiten, die sich zur Grundlage einer sozialphilosophisch informierten Zeitdiagnose der Digitalisierung und ihrer sozialpathologischen Effekte erweitern lässt." [Ibid., 230–31])
49 See Mitchell 2019, 127–28.
50 Post-humanists even suggest that "bots" [= (quasi-)"intelligent" machines] are entitled to "human rights". See in this context Nida-Rümelin's and Weidenfeld's critique ("Roboter als neue [digitale] Sklaven", 2018, 23–31).

be offset by the advantages gained by third parties, however great they may be. No human being ought to be treated merely as a means. Humans do not 'optimize'. In emergency situations we act in accordance with moral intuition and not an optimizing calculation."[51] Does this imply that self-driving cars have – in principle – to remain under the supervision of human drivers? In regard to the five levels of autonomy defined by the US Traffic Safety Administration[52], only "partial autonomous driving" (a mode of cruising in which the car does the driving in "certain circumstances", but the human driver has to take over if needed) seems to be (relatively) unproblematic. "Full autonomy" driving would depend, in order to be sufficiently safe, *inter alia*, on the organization of "geofenced areas": that is to say of mapped safe zones, which (as AI researchers suggest)[53] require, in addition, that pedestrians are educated to behave in a more predictable way in the vicinity of self-driving vehicles. Large-scale autonomous driving thus seems to require something which many will hold to be quite undesirable: that our entire urban and extra-urban habitats are rigidly restructured along new, AI-induced safety criteria.[54]

4.2 What, *secondly*, is called for, is a discourse about the external – juridical and socio-political – criteria governing the public regulations of AI: a philosophical, in-depth debate, that is, on the ethical principles of human action.

Most importantly, any non-reductive conception of praxis that critically reflects on the limits of computerized calculation has to focus on *the human ability to act in accordance with moral judgment:* on an ability, that is, which – to speak with Kant – cannot be fully simulated by "heteronomous" mechanical execu-

---

[51] Ibid. 96–7 (English translation L.N.). The German original runs as follows: "Die Verletzung von Grundrechten lässt sich durch Vorteile für Dritte, wie groß diese auch sein mögen, nicht aufwiegen. In kantischen Begriffen ausgedrückt: Ein Mensch darf niemals als bloßes Mittel behandelt werden. [...] Menschen optimieren nicht. In Notsituation handeln wir nach moralischer Intuition und nicht nach einem Optimierungskalkül."
[52] See Mitchell 2019, 267–71. The five levels of autonomous driving start with level 0, where "the human driver does all the driving". On level 1, the vehicle sometimes assists the driver (with steering, or vehicle speed); on level 2, "the vehicle controls both steering and vehicle speed '*under some circumstances*' (usually highway driving)"; on level 3, the vehicle "can perform all aspects of driving under certain circumstances, but *the human driver must pay attention at all times.*" On level 4, "the vehicle can do all the driving *under certain circumstances* and *the human does not need to pay attention.*" On Level 5, "the vehicle can do all the driving in all circumstances" and "the human occupants are just passengers and never need to be involved in driving."
[53] The AI researcher Andrew Ng, for instance.
[54] See Mitchell 2019, "Regulating AI", 124; as well as Nida-Rümelin / Weidenfeld 2018, 99.

tions of programmed "norms" (even if these norms, as in advanced AI, are able probabilistically to "re-adapt").

Spelled out in fuller philosophical terms, human ethics has *two* constitutive elements. It rests, *first (formally* described), on the basic maxim to respect the ability of every human being to act in an autonomous and self-determined manner. Hence every human subject deserves "respect", as Kant convincingly argued: it is to be treated as an *end in itself,* not merely as a *means.* All human beings have "dignity", not just an (economically calculable) "value".

This *formal* principle – as Kant was well aware – must *secondly,* in order to be implementable, be situationally specified. All human *praxis* depends on the agent's *hermeneutical sensitivity for the context* in which his action is embedded. It thus depends on *the agent's capacity for a judgment adequate to the situation.* This complex judgment cannot be attained – in an algorithmic mode – by means of the *mere "subsumption"* of a particular instance *under a general rule.* Situation-adequateness rather presupposes the human ability to judge *"reflectively".* Only in this manner can the formal maxim be instantiated. (If this second, hermeneutical element of Kant's elucidation of *praxis* is overlooked, his ethics can be *misread* – as is still done by many analytic as well as post-analytic philosophers – *as an abstract "formal" procedure* [55].)

In conclusion: since human moral judgment is *not* the "subsumption" *of a case under an algorithm-defined rule,* it *cannot be fully simulated* by computers (not even by those which, due to their probabilistically engendered ability to make alterations, are, *metaphorically,* called "deep learning" machines). This is the thorn in the flesh of the Utopian fantasies of those prophets of "strong AI" who propagate that the arrival of "singularity" is near. Useful *weak* modes of Artificial Intelligence *are to be fully welcomed, however,* as important new tools, which – *if properly controlled by society* – have the potential to speed up substantially the ongoing process of human self-improvement.

Works cited

Becker, Carlos, and Sandra Seubert. "Die Selbstgefährdung der Autonomie. Eckpunkte einer Kritischen Theorie der Privatheit im digitalen Zeitalter". *Digitale Transformation der Öffentlichkeit.* Ed. Jan-Philipp Kurse and Sabine Müller-Mall. Weilerswist: Velbrück Wissenschaft, 2020. 229–61.
Bostrom, Nick. *Superintelligence: Paths, Dangers, Strategies.* Oxford: Oxford UP, 2014.

---

[55] See Nagl 1983, "Der kategorische Imperativ, eine leere Tautologie?", 41–2.

De Candia, Gianluca. "Von digitaler Gegenwart. Über Medialität, Repräsentation und Sakrament". Paper presented at Katholische Akademie Berlin, 18 March, 2020. https://www.academia.edu/42175822/ Accessed: December 7, 2021.
Dreyfus, Hubert, and Stuart Dreyfus. "Coping with Change: Why People Can and Computers Can't". *Wo steht die Analytische Philosophie heute?* Ed. Ludwig Nagl and Richard Heinrich. Vienna: Oldenbourg, 1986. 150–70.
Dreyfus, Hubert, and Charles Taylor. *Retrieving Realism*. Cambridge, Massachusetts: Harvard UP, 2015.
Habermas, Jürgen. *Auch eine Geschichte der Philosophie*. Berlin: Suhrkamp, 2019.
Huck, Stefan. "Wir brauchen mehr künstliche Freunde. Je besser lernende Maschinen uns kennen, desto mehr profitieren wir von ihrer Intelligenz". *Die Zeit*, 15 October 2020. 41.
Krämer, Sybille. "Medienphilosophie der Stimme". *Systematische Medienphilosophie*. Ed. Mike Sandbothe and Ludwig Nagl. Berlin: Akademie, 2005. 221–37.
Kurzweil, Ray. *How to Create a Mind: The Secret of Human Thought Revealed*. New York: Viking, 2012.
Leibniz, Gottfried Wilhelm. *Monadologie*. Hamburg: Felix Meiner, 1956.
Levmore, Saul, and Martha Nussbaum, eds. *The Offensive Internet. Speech, Privacy and Reputation*. Cambridge, Massachusetts: Harvard UP, 2010.
Marantz, Andrew. *Antisocial. Online Extremists, Techno-Utopian, and the Hijacking of the American Conversation*. New York: Viking, 2019.
Mersch, Dieter. "Ideen zu einer Kritik 'algorithmischer' Rationalität". *Deutsche Zeitschrift für Philosophie,* 2019, No. 5. 851–73.
Mitchell, Melanie. *Artificial Intelligence. A Guide for Thinking Humans*. New York: Farrar, Straus and Giroux, 2019.
Nagl, Ludwig. *Gesellschaft und Autonomie. Historisch-systematische Studien zur Entwicklungsgeschichte der Sozialtheorie von Hegel bis Habermas*. Vienna: Verlag der Österreichischen Akademie der Wissenschaften, 1983.
Nagl, Ludwig. *Pragmatismus*. Frankfurt a.M.: Campus, 1998.
Nagl, Ludwig. "What is it to be a Human Being? Charles Taylor on 'the Full Shape of the Human Linguistic Capacity'". *Re-Learning to be Human in Global Times: Challenges and Opportunities from the Perspective of Contemporary Philosophy of Religion*. Ed. Brigitte Buchhammer. The Council for Research in Values and Philosophy: Washington, DC., 2019. 117–36.
Neumaier, Otto. "A Wittgensteinian View of Artificial Intelligence". *Artificial Intelligence. The Case Against*. Ed. Rainer Born, London and Sydney: Croom Helm, 1987. 132–73.
Nida-Rümelin, Julian, and Weidenfeld, Nathalie. *Digitaler Humanismus. Eine Ethik für das Zeitalter der Künstlichen Intelligenz*. München: Piper, 2018.
Nussbaum, Martha. "Wir sind keine Maschinen. Wir sind Menschen". *Die Zeit*, 13 June 2019, 40.
Pauen, Michael. "Autonomie und Enhancement". *Neuroenhancement. Die philosophische Debatte*. Ed. Klaus Viertbauer and Reinhart Kögerler. Berlin: Suhrkamp, 2019. 89–114.
Putnam, Hilary. *Reason, Truth, and History*. Cambridge: Cambridge UP, 1981.
Putnam, Hilary. *Renewing Philosophy*. Cambridge, Massachusetts: Harvard UP, 1992.
Putnam, Hilary. "Plädoyer für die Verabschiedung des Begriffs 'Idolatrie'". *Religion nach der Religionskritik*. Ed. Ludwig Nagl. Vienna-Berlin: Oldenbourg-Akademie Verlag, 2003. 49–59.

Ramge, Thomas. *Who's Afraid of AI? Fear and Promise in the Age of Thinking Machines*. New York: The Experiment, 2019.

Reichl, Peter. "Gottfried Wilhelm Leibniz – mehr als ein Keks", *E-Media*, February 2020. 37.

Riesewieck, Moritz, and Hans Block. *Die digitale Seele. Unsterblich werden im Zeitalter Künstlicher Intelligenz*. München: Wilhelm Goldmann, 2020.

Riesewieck, Moritz, and Hans Block. "Wollen wir digital wiederauferstehen? Den Zugang zum Toten sollen Angehörige wie ein Netflix-Abo buchen". *Die Zeit*, 8 October 2020. 59–60.

Searle, John. "Minds, Brains, and Programs". *Artificial Intelligence. The Case Against*. Ed. Rainer Born, London and Sydney: Croom Helm, 1987. 18–40.

Sturma, Dieter. "Subjekt sein: Über Selbstbewußtsein, Selbstbestimmung und Enhancement". *Neuroenhancement. Die philosophische Debatte*. Ed. Klaus Viertbauer and Reinhart Kögerler. Berlin: Suhrkamp, 2019. 115–47.

Taylor, Charles. *The Language Animal. The Full Shape of Human Linguistic Capacity*. Cambridge, Massachusetts: Harvard UP, 2016.

Taylor, Charles, Patrizia Nanz, and Madeleine Beaubien Taylor. *Reconstructing Democracy. How Citizens are Building from the Ground Up*. Cambridge, Massachusetts: Harvard UP, 2020.

Viertbauer, Klaus, and Reinhart Kögerler, eds. *Neuroenhancement. Die philosophische Debatte*. Berlin: Suhrkamp, 2019.

Wingert, Lutz. "Grenzen der naturalistischen Selbstobjektivierung". *Philosophie und Neurowissenschaft*. Ed. Dieter Sturma, Frankfurt a. M.: Suhrkamp, 2005. 240–60.

Wittgenstein, Ludwig. *Philosophical Investigations*. New York: MacMillan Publishing Co, 1968.

Hille Haker
Experience, Identity and Moral Agency in the Age of Artificial Intelligence

Abstract: AI systems transform all sectors of society, and the ramifications of this revolution in data processing, cognition and learning, communication and social interactions are unforeseeable. For some, AI systems increase human freedom; for others, they threaten the status of human beings: machines have better memory, more efficient strategies to pursue ends, with better means, and they are as predictable as they are capable of learning from mistakes. They may not show the same susceptibility to violence, and thus may even solve the problem of evil that has haunted the history of human morality. I examine why human vulnerability in the form of openness to others, together with the unpredictability of interactions, are necessary elements of moral identity and agency. They can only be overcome at the price of human freedom. Without human freedom, however, interactions are reduced to the exchange of information, needs and desires, and the pursuit of ends that undermine the self-fulfillment associated with moral identity as well as the responsibility that arises from the claims agents make on one another. AI systems have no way to mirror relationships of recognition and responsibility, because they cannot reflect the reciprocal normative claims entailed in moral interactions.

It is hard to foresee the consequences of the cultural transformation resulting from the current technological changes associated with Artificial Intelligence, because there are so many areas that are in flux. My narrow perspective in this essay concerns some ethical reflections, especially the ramifications of artificial intelligence on *our* moral identity and moral agency. Already this is, of course, a very large question that has often been neglected in debates that rather focused on the agency and consciousness of AI devices, on data privacy, and security. My goal is therefore to shift the emphasis and to explicate the moral point of view regarding the broader question of the relationship of AI technologies and identity and agency.

Before I engage with the current developments, however, I want to take a brief look into a previous transformation at the end of the 19th century, with its radical changes in human experience and the conceptualization of identity. My hope is to learn from the early 20th century critique of technologies that emerged at a historical moment when their effects were as unknown as the current inventions are for us today. I only aim at offering a flashlight that will illuminate my own questions regarding artificial intelligence.

1 Culture and technologies – a century ago

At the beginning of the industrial revolution, Marx described the transformation of human practices and relationships, anticipating a radical cultural transformation associated with the economic, profit-driven commodification of human life:

> Finally, there came a time when everything that men had considered inalienable, became an object of exchange, of traffic, and could be alienated. This is the time when the very things which till then had been communicated, but never exchanged; given, but never sold; acquired but never bought – virtue, love, conviction, knowledge and conscience, etc. – when everything, in short, passed into commerce. It is the time of general corruption, of universal venality.[1]

In a similar vein, the early Frankfurt School turned their attention to this theory of alienation, and today, it is renewed as a critique of reification (Honneth) or alienation (Jaeggi) that is ultimately rooted in the very economic exploitation that Marx foresaw.[2] Recently, Hartmut Rosa has presented a sociological critique of alienation that is as much indebted to Heidegger's critique of technology and Charles Taylor's critique of modern rationality as to Marx and the Frankfurt School critical theory.[3] Yet, rarely is the examination of technology that Walter Benjamin pursued in his own writings featured in the reflections on the new technologies, let alone in the discourse on Artificial Intelligence.[4]

---

[1] Marx, Karl and Frederick Engels. *Marx & Engels Collected Works Vol 06: 1845–1848*. London: Lawrence & Wishart (1976), 105.(1976).
[2] Rahel Jaeggi, *Alienation* (2016); Axel Honneth, with comments by: Judith Butler, Raymond Geuss, Jonathan Lear, Martin Jay, *Reification: A New Look at an Old Idea* (2007).
[3] Cf. Hartmut Rosa, *Resonanz Eine Soziologie der Weltbeziehung* (2017). Hille Haker, *Resonanz. Eine Analyse aus Ethischer Perspektive* (2018).
[4] Andrew Feenberg, one of the most prominent readers of the Frankfurt School philosophy of technology, does not examine Benjamin's works closely. Andrew Feenberg, *Between Reason and Experience: Essays in Technology and Modernity* (2010).

1.1 Walter Benjamin's theory of experience

Walter Benjamin is well-known as one of the sharpest observers of the cultural transformation around the end of the 19th century, especially regarding changes with respect to human experience. In a text from 1933, Benjamin turned to the "poverty of experience."[5] Here, he reminded his readers that the First World War had had devastating effects on those who survived the battlefields. In a recent review essay about the development of chemical warfare, Elisabeth Kolbert quotes a soldier who testified about his experiences:

> When we got to the French lines, the trenches were empty. But in a half mile, the bodies of the French soldiers were everywhere.... You could see where men had clawed at their faces, and throats, trying to get their breath. Some had shot themselves. The horses, still in the stables, cows, chickens, everything, all were dead.[6]

Yet, the war left many soldiers, returning from the war, speechless, lacking the ability to share their stories with those who had not lived through the carnage themselves. It was not only the devastation of war as such that was so silencing, Benjamin holds, but the overwhelming force of the new arms and warfare, developed in and for the First World War of the 20th century: machine guns, tanks, airplanes, and submarines were all used for the first time; chlorine and mustard gas killed more soldiers than in any other war before. One century – and another world war as well as multiple regional wars – later, wars are now fought with unmanned drones and cyberweapons. The numbness and silence of veterans is described in medical terms rather than seeing in them the signs of the barbarism within civilization, and we rarely take the time to even think of the meaning of the acronym, PTSD (Post-Traumatic Stress Disorder), in terms of moral harms. When it comes to the most dramatic technological innovation of the 21st century so far, however, we have yet to understand how, and how much, AI systems are going to change our experiences of the world and our treatment of others, in both civil and military applications.

Benjamin's point, it seems to me, culminates in a deeper observation which he expressed in a stylistic juxtaposition: here, we have the wealth of ideas, with countless inventions at the turn of the century, due to the ingenuity of scientists and engineers who combine the modern forms of instrumental rationality with

---

[5] Walter Benjamin, *Experience and Poverty* (1999). This section builds in parts upon a lecture I gave at the International Congress of the European Society of Catholic Theology in 2019. Cf. Hille Haker, *Information or Communication – the Loss of the Language of the Human* (2021).
[6] Elisabeth Kolbert, *Chemical Warfare's Home Front* (11 February 2021).

the monetary incentive of the patent system. There we sense the impact on our experiences which Benjamin sees as "poverty." Barbarism is the other side of modernity.[7]

> With this tremendous development of technology, a completely new poverty has descended on mankind. And the reverse side of this poverty is the oppressive wealth of ideas that has been spread among people, or rather has swamped them [...] But here we can see quite clearly that our poverty of experience is just a part of that larger poverty that has once again acquired a face – a face of the same sharpness and precision as that of a beggar in the Middle Ages. For what is the value of all our culture if it is divorced from experience? Where it all leads when that experience is simulated or obtained by underhanded means is something that has become clear to us from the horrific mishmash of styles and ideologies produced during the last century – too clear for us not to think it a matter of honesty to declare our bankruptcy. Indeed (let's admit it), our poverty of experience is not merely poverty on the personal level, but poverty of human experience in general. Hence, a new kind of barbarism.[8]

Often returning to the themes of his life, namely language, identity, and the possibility of collective action under the conditions of modernity, Benjamin was interested in the ways collective experience is constituted and mediated in social practices. Experience, in German, is not only related to a person's course of life but also to traveling, and so, in traditional oral storytelling, experience spans time (in the stories of local history) and space (in the stories of travelers). Following the works of Georg Lukács, Benjamin observes how this practical and collective tradition is transformed into something else with the access to the written works, reading, and the rise of the novel.[9] The latter allows for the lonesome exploration of the inner self, but this turn to the individual renders one's subjective experiences incommensurable: "the birthplace of the novel is the individual in his isolation, the individual who … lacks counsel and can give none."[10] In the early 20th century, however, yet another medium transforms experience, centered on information rather than storytelling or inner reflection: newspapers focus on short-lived information, selected for readers who more react to the stimulus of

---

[7] This, of course, is the theme that brought together the intellectuals of the Frankfurt Institute of Social Research. It would become the defining argumentation in Horkheimer and Adorno's major work, written at the end of World War II: Theodor Adorno, Horkheimer, Max, *Dialectic of Enlightenment* (2016 (orig. 1947)).
[8] Benjamin, *Experience and Poverty*, 732.
[9] Cf. Georg Lukács, *The Theory of the Novel: A Historico-Philosophical Essay on the Forms of Great Epic Literature* (1971).
[10] Cf. Walter Benjamin, *The Storyteller. Observations on the Works of Nikolai Leskov (Orig. 1936)* (1999), 146.

suspense than to politically relevant – but often boring and complicated – news. Benjamin quotes the founder of the French newspaper *Le Figaro*, Jean Hippolyte Cartier de Villemessant: "To my readers, an attic fire in the Latin Quarter is more important than a revolution in Madrid."[11] The problem with information is, however, Benjamin notes, that the many stories one receives leave little room to imagination: "Every morning brings us news from across the globe, yet we are poor in noteworthy stories."[12]

Individual experiences are entangled with histories and stories, which is the title of Wilhelm Schapp's phenomenological study, itself already a response to Husserl and Heidegger: *In Geschichten verstrickt*.[13] Disentanglement from history may well be the effect of modern technologies – but it comes at the price of the poverty of experience, the loss of storytelling, and it introduces, as critics from Nietzsche to Heidegger had pointed out around the turn of the century, the construction of the homogeneity of the "average person."[14] Experiential knowledge, in contrast, is tied to a person's unique history and the story of their lives. And insofar as self-consciousness is corporeal and temporal, it is tied to the unique memory of an individual. Memory that entails the entanglements of the self with stories and histories is not merely the storage of experiences in the form of information. On the contrary, memory is a "cat," as Uwe Johnson, following Benjamin closely, calls it, alluding to the uncontrollability of past experiences that nevertheless are part of one's history.[15] This understanding of memory alludes to the alienness within oneself, with memories often appearing involuntarily, inaccessible in their subconscious and unconscious workings, as Sigmund Freud illuminated in his works on dreams and the unconscious. As part of the complex self-consciousness, memory is the subjective self-relation over time, complementing the narrative of oneself that can be shared, with the alienness of the inner self that is inaccessible and incommsurable.

Benjamin was far from merely regarding the new cultural inventions that transformed the everyday life of societies having access to them – from the telephone to the radio, gramophone to the film – as signs of cultural decline. Rather, he was interested in their potential to break the spell of alienation by creating

---

11 Ibid.
12 Ibid., 147.
13 Wilhelm Schapp, *In Geschichten Verstrickt. Vom Sein von Mensch und Ding* (1953).
14 Cf. Martin Heidegger, *Sein und Zeit* (1977).
15 Uwe Johnson, *Jahrestage: Aus dem Leben von Gesine Cresspahl* (2014 (orig. 1970 –1983)); *Anniversaries: From a Year in the Life of Gesine Cresspahl* (2018). Cf. Hille Haker, *Moralische Identität. Literarische Lebensgeschichten als Medium ethischer Reflexion. Mit einer Interpretation der "Jahrestage" von Uwe Johnson* (1999).

new mediations of self-consciousness, by increased access to art and culture, by creating new objects of perception, or by overcoming the public/private divide that characterized the 19th century bourgeois society. He compared the different forms of experience in traditional societies to their respective forms in modernity and aims at inventing a writing style that captures the art of storytelling within the modern context of an aura-less art. *Berlin Childhood Around 1900* in particular contains multiple folded stories, as if Benjamin needs to hide and save them inside quotes, allusions, or references that the reader needs to unfold. He alludes to works of literature and authors of the tradition, often with no regard for the canon, leaving traces and codewords here and there for the reader to decipher. In much of his writings, Benjamin excels in the short form and in fragments, in narrative vignettes that blur the line between storytelling and reflection.[16] His stories are, in fact, *Denkbilder*, thought images, stemmed against the tide of the loss of storytelling as a culture of practical wisdom. It is almost stunning how Benjamin's artful depiction of his childhood memories, often in correspondence with the experience of a new technological device such as the telephone,[17] demonstrates his reflection on subjectivity, domination and submission, violence, and self-forgetting. Here, the new devices that shape the child's experience as much as the space of the city or the bourgeois interieur of his parents' house have a mythical power over the child, and they may well serve as a source for comparison to our own – adult – encounters with new digital devices a century later.

Benjamin was especially interested in *mimesis* as a dimension of experience (and language), because the imitation games children engage in are not *perceived* by them as merely imitations: in the child's experience, mimesis blurs the lines between the agent as the subject and the tool as an object that is of instrumental value only. On the contrary, in children's experiences, the material world is often perceived and/or imagined as alive, and the agents perceived as patients and recipients rather than as agents who act upon the objects. Adults, Benjamin argues, embody traces of this mimetic relationship to the world and to "things" that transcend the neat distinction between subject and object. Children (still) have the capability to play roles in which they humanize things and animals, developing their self-consciousness in experiences that elude the modern understanding of rationality as abstraction and reification.

---

16 Walter Benjamin, *The Storyteller: Tales out of Loneliness* (2016).
17 Cf, among others, the story called "The Telephone." *Berlin Childhood around 1900* (2002), 349–50. I have interpreted this story more closely in *Haker, Information or Communication – the Loss of the Language of the Human*.

From Benjamin's thought images of his childhood we learn, on the one hand, that human encounters with others, whether these are "things," non-human living beings or human beings, are more complex and far more fluid than we often assume when we speak of subjects and objects. Especially the liberal understanding of a self-centered individual who acts autonomously within the limits of one's own and others' liberties, engaged in quasi-contractual relationships between two equal partners, turns out to be reductionist and in many cases misleading. On the other hand, we may understand better that the calculating rationality that is defined by causal relations, functional reasoning, and the representation of actions by mathematical models has become so ubiquitous that it is easy to forget that moral agency "sublates" – i.e. transforms, yet maintains the learning experiences of one's development history – the stages of one's moral development in a life-long history of encountering the "world" through unique and subjective processes. Moral agency rests upon a dialectic of experience and interpretation, in German: *Widerfahrnisse* and meaning-making.

1.2 Transcending the human or transcending the subject of modernity?

Transhumanism is perhaps the most consequentialist heir of Marx's theory of fetishism – the ultimate dialectic transformation of human beings into things. The transhumanist movement of radical life extension, for instance, sees itself in continuity with the scientific and technological paradigm that defines modern life – and redefines concepts of the *good life*.[18] The goal of a longer life entails the desire for a life with less suffering, either from declining health, premature death, or the premature loss of loved ones.[19] Ultimately, death, mortality, and the ageing body are regarded as constraints of human freedom, and as such they are not goods that a rational person can either strive for or accept. No other than Walter Benjamin saw this coming – and it is worthwhile to quote him once again, because his observation, confirmed exponentially over the course of the 20th century, situates the utopian visions of transhumanism *within* the modern tradition of self-denial:

> It has been evident for a number of centuries how, in the general consciousness, the thought of death has become less omnipresent and less vivid. In its last stages this process

---

18 Aveek Bhattacharya, Simpson, Robert Mark, *Life in Overabundance: Agar on Life-Extension and the Fear of Death* (2014).
19 For a further discussion cf. Ludwig Nagl's essay in this volume.

is accelerated. And in the course of the 19th century, bourgeois society – by means of medical and social, private and public institutions – realized a secondary effect, which may have been its unconscious main purpose: to enable people to avoid the sight of the dying. Dying was once a public process in the life of the individual, and most exemplary one; think of the medieval pictures in which the deathbed has turned into a throne that people come toward through the wide-open doors of the dying person's house. In the course of modern times, dying has been pushed further and further out of the perceptual world of the living. It used to be that there was not a single house, hardly a single room, in which someone had not died. [...] Today people live in rooms that have never been touched by death – dry dwellers of eternity; and when their end approaches, they are stowed away in sanatoria or hospital by their heirs. Yet, characteristically, it is not only a man's knowledge or wisdom, but above all of his real life – and this is the stuff that stories are made of – which first assumes transmissible form at the moment of his death.[20]

Transhumanists presuppose that moral agents are atomistic pursuers of goods that they choose and which they consider instrumental to achieving a good life.[21] Yet, the justification for life extension is tautological, immunizing itself from critique: it is good because "better is good." Proponents always refer to benefits for society: "As far as ethics is concerned, engineers in general and roboticists in particular, have always been concerned with serving the public good and delivering work that may potentially benefit humanity."[22] However, the *motivation* of engineers must not be confused with the *reasons* that may or may not justify the pursuit of a particular goal; the vague term of being "concerned with serving the public good" can be filled with any content, and certainly, the motivation cannot be questioned. But individuals as different as a billionaire who is pursuing his hobby of going to Mars and a Secretary of Defense who is pushing for the development of military drones, among others, may all claim that they are "concerned with the public good." Ethics can be seen as the endeavor to reflect upon such claims in view of criteria that determine what ought to count as "public good" or which technologies as well as which of their uses "benefit humanity."

In contrast to the enthusiastic embrace of new technologies to overcome the vulnerabilities of the human condition, other scholars critique the modern *conception* of the subject (and object respectively), arguing for a posthumanist approach. They may well echo Benjamin's critique in their turn to a new understanding of the materiality of things that are regarded neither as objects nor

---

20 Benjamin, *The Storyteller. Observations on the Works of Nikolai Leskov (Orig. 1936)*, 151.
21 Cf. among many other papers that represent this view: *Bhattacharya, Life in Overabundance: Agar on Life-Extension and the Fear of Death.*
22 Karolina Zawieska, *Disengagement with Ethics in Robotics as a Tacit Form of Dehumanisation* (2020).

as subjects but in-between, blurring the categories and rendering the modern divisions more fluid and porose than they were in that tradition.²³ The difference to Benjamin, however, must be marked: while he reflects upon the mimetic relationship as a *childhood* experience that can only be *remembered*, creating a critical distance to the aesthetically mediated experiences of the adult who reflects upon them, this is not always the case in the posthumanist approaches.²⁴

With these reminders of the history and context of the discourse on the relationship of technologies, human experiences, identity, and moral agency, I want to now turn to the digital transformation and how it is dealt with in ethics reports.

2 The digital transformation

The digital transformation radicalizes the management and rationalization of everyday life. It entails the triad of the "internet of things," "artificial intelligence," and "virtual reality,"²⁵ all of which contribute to a social transformation that is as dramatic as the one Benjamin referred to: an uprooting that may either enrich or impoverish human experience and social practices. The question is not whether a technology as such is demonized – there is no reason for doing so – but whether the intertwining of contemporary capitalism and technologies create the necessity of ever-new innovation that are disconnected from any personal or social need and which are pursued merely because it is possible. The digital transformation comes with the promise of *enhancing* human freedom and well-being almost beyond any limitation – but this promise is barely more than the hollow promises of advertisements, which must not be confused with reality. The exciting possibilities of AI have led several billionaires and myriads of scientists and engineers to explore ideas that some of them claim could solve all – or almost all

---

[23] Jane Bennett, *Vibrant Matter: A Political Ecology of Things* (2010). Cf. also Rosi Braidotti, *Posthuman Knowledge* (2019). Cf. also Rosi Braidotti and Simone Bignall, *Posthuman Ecologies: Complexity and Process after Deleuze* (2018).
[24] Posthumanists, of course, must not be confused with transhumanists who aim at or anticipate the replacement of the human by machines. Cf. Michael Hauskeller, *Mythologies of Transhumanism* (2016).
[25] Thomas M. Siebel, *Digital Transformation: Survive and Thrive in an Era of Mass Extinction* (2019).

– problems of mankind.[26] Since I am only interested in the ethical perspective, I will now focus on the overall lens through which AI is considered.

2.1 Artificial intelligence – ethical principles and ethical guidelines

Over the last decades, ethics has become just as much an instrument of rationalization and institutional management as its scientific and technological counterparts. Instead of providing perspectives that address the intertwining of science, technology, economics, politics, and civil societies' interests, the ethics reports often embrace the rhetoric that new technologies will benefit humankind when handled responsibly.[27] A large machinery of scientific, social, legal, ethical, and political expertise has been set in motion to respond to the "chances and risks" of artificial intelligence. These groups who are issuing their statements for the European Commission, the United Nations, the OECD, among others, have generated a framework that is meant to orient policy-makers, legislators, scientists, and developers. One group is often mentioned last: citizens, and if so, they are largely seen as consumers who wish to gain access to the innovations. Safety, security, privacy, non-discrimination, etc. are the posts that normatively secure the responsible use of the applications and protect against their misuse. Yet, this approach is in many ways technocratic, and retro-active rather than pro-active.

The EU High-Level Expert Group on AI (HLEG AI) has provided this definition of Artificial Intelligence in June 2020:[28]

> Artificial intelligence (AI) refers to systems that display intelligent behaviour by analysing their environment and taking actions – with some degree of autonomy – to achieve specific goals.
>
> AI-based systems can be purely software-based, acting in the virtual world (e.g. voice assistants, image analysis software, search engines, speech and face recognition systems)

---

[26] As Ruha Benjamin has recently shown, AI technologies may in fact increase the digital gap, social and racial injustices. Ruha Benjamin, *Race after Technology : Abolitionist Tools for the New Jim Code* (2019).

[27] René Von Schomberg and Jonathan Hankins, *International Handbook on Responsible Innovation: A Global Resource* (2019). The relevance for AI is explored in Alexander Buhmann and Christian Fieseler, *Towards a Deliberative Framework for Responsible Innovation in Artificial Intelligence* (2021).

[28] High Level Expert Group on Artificial Intelligence, *A Definition of AI: Main Capabilities and Disciplines. Definition Developed for the Purpose of the Ai Hleg's Deliverables* (April 2020).

or AI can be embedded in hardware devices (e.g. advanced robots, autonomous cars, drones or Internet of Things applications).

AI systems are first and foremost rational systems.

> But how does an AI system achieve rationality? [...] by: perceiving the environment in which the system is immersed through some sensors, thus collecting and interpreting data, reasoning on what is perceived or processing the information derived from this data, deciding what the best action is, and then acting accordingly, through some actuators, thus possibly modifying the environment.
>
> AI systems can either use symbolic rules or learn a numeric model, and they can also adapt their behaviour by analysing how the environment is affected by their previous actions.
>
> [...] Rational AI systems are a very basic version of AI systems. They modify the environment but they do not adapt their behaviour over time to better achieve their goal.
>
> A learning rational system is a rational system that, after taking an action, evaluates the new state of the environment (through perception) to determine how successful its action was, and then adapts its reasoning rules and decision making methods.

Having worked on ethics committees for many years myself, among them as a member of the *European Group on Ethics in Science and New Technologies* to the European Commission, I am far from claiming that ethical guidelines are not warranted. As Jon Truby says: "Big Tech has proven time and again that it cannot be trusted to behave in an ethical or responsible manner, so certainly cannot be entrusted to operate freely in a matter so important to society as AI."[29] Yet, the question is whether guidelines and reports *matter* at all when they are regarded mostly in relation to *legal* regulations. In that relation, ethical guidelines are identified as "soft law" because of their non-binding status. Unsurprisingly, academic ethics does not necessarily share this understanding of ethics. It has a much broader understanding that entails, for instance, the multiple forms of moral formation, the critical assessment of social values and social norms, obligations and human rights that offer criteria for social practices and actions, and structures and/or institutions of governance, establishing (or lacking) the frameworks that allow for equality and justice in a given space of legislation, whether it is national, transnational, or international. Ethics, this means, does not only reflect on justifications of normative claims and a normative framework but also, in a non-directive manner, on ideologies and utopian visions, offering orientations, sources, and points to consider in the existential

---

[29] Jon Truby, *Governing Artificial Intelligence to Benefit the Un Sustainable Development Goals* (2020), 947.

and social deliberation about goods and ends of societal practices as participant and "immanent critique" of social practices.[30] Only a small part of "ethics" concerns the experts' contribution to policy decisions, and these must be seen in the context of the much broader reflections – just as the scientific explanations in reports cannot replace the engagement with the scientific literature.

Ethics reports, this means, should be read with this caveat, but they do fulfil the function of inserting the language of responsibility. Almost all countries have by now adopted regulatory frameworks for the use of new technologies, either as national, transnational, or international laws, as ethical frameworks, as professional, labor, and consumer norms of safety and security, as Codes of Conduct, or educational material from kindergarten to college. Like other new technologies, AI ethics is either constructed as "top-down" ethics, elaborated by expert committees, again on the national, transnational, or international level, or as "bottom-up" ethics, brought forward by stakeholders and lobbyists, citizens, or vulnerable groups.[31] Both approaches are accompanied by scholarly works. AI ethics mostly takes a technology-centered normative approach, i.e. one that either re-acts to technological developments or pro-actively projects risks and chances, mostly envisioned in rather narrow terms. Alternative and more complex concepts of ethics are rarely considered, for instance: social ethics approaches that *begin* with societal needs and priorities – ends, goals, and goods – before entering the discussion about the means, among them but never exclusively new technologies.[32]

2.2 Principles for artificial intelligence

According to the predominant framework of political liberalism, the question of the "good" (and goods worth striving for) is mostly left to civil society – but in reality, especially under conditions of capitalist societies, the "good" is not defined ethically but economically: the good is intertwined with the economic question of what is marketable, in other words: what can be transformed into goods. According to liberal political philosophy, the so-called comprehensive doctrines must be translated into questions of the "right" to be applicable in public political deliberations – overlooking the fact that not all public debates

---

30 Titus Stahl, *What Is Immanent Critique?* (2013). Cf. also Hille Haker, *Towards a Critical Political Ethics. Catholic Ethics and Social Challenges* (2020).
31 Good first overview gives: Mark Coeckelbergh, *AI Ethics* (2020).
32 Hille Haker, "Nanomedicine and European Ethics. Part One" (2013); "Synthetic Biology – an Emerging Debate in Europe – Part Two" (2013).

are political: this means that citizens as well as corporations may pursue whatever ends and goods they want, as long as their activities do not harm others or collide with the rights of others.[33] Public ethics is then conceived as "political liberalism" (Rawls) or, in Habermas' version of a theory of communicative action, as a "discourse ethics," which either narrows ethical questions to those of negative freedom or to just procedures. It is certainly right that in democracies – represented in its institutions of governance, legislation, and jurisdiction – the state must attend to the moral pluralism of the citizenry and not favor one lifeform over another.[34] But this cannot mean that discussions are centered mostly on permissions and/or prohibitions, without further deliberation on the ends of technologies within given societies. Big Tech companies (google, facebook, amazon, apple, Microsoft as the most powerful companies) who are eager to win over people as consumers will, not surprisingly, advertise any new invention as "beneficiary." As Marx foresaw with respect to the production side of capitalism, benefits are mostly defined in economic terms: by growth of the GDP, accumulation of wealth, and private interests. In liberal societies, affluent (or ultra-rich) individuals such as Elon Musk, Jeff Bezos, Bill Gates, or Jack Ma by now determine far more the direction of technological innovation than any "deliberative" or "discourse ethics" approach would allow for. With their private investments into research and development, they have launched civil space programs, renewable energy technologies, or civil and military AI technologies. In the USA, government funding even for basic research, for instance, has fallen below the 50% threshold of the overall research investment over the last years, even though AI funding has steadily risen. State-Capitalist societies do not fare better regarding a just representation of all civil groups: here, societal goals are mostly determined by the leaders of one party or ruling class. Overall, a mix of corporate and political incentives, national interest in economic competitiveness, and the interest of scientific societies and/or universities determine the goals, research programs, and development of new technologies. Despite numerous experiments with citizen conferences and "science and society" projects, citizens rarely participate directly in the overall direction of public research agendas, at least not other than as consumers. The so-called "citizen science" that has gained

---

[33] In the United States, corporations have the legal status of persons in terms of certain rights, among them the freedom of speech and freedom or religion. For a comment on this cf. Adam Winkler, *'Corporations Are People' Is Built on an Incredible 19th-Century Lie* (5 March 2018). Cf. the critique of political liberalism raised, among others, by social philosopher Jeffrey C. Alexander, *The Civil Sphere* (2006).
[34] John Rawls held this belief up to the end of his life. Cf. John Rawls, *Justice as Fairness: A Restatement* (2001).

some prominence over the last years points to the more immediate *participation* of citizens in *scientific* projects, but not necessarily the participation in the deliberations and decision-makings of research agendas.

The ethical deliberations about the "risks and chances" on the one hand, and the "prohibitions and permissions" on the other, are then mostly left to scientific experts commissions, parliamentary debates, and governments. They are therefore mostly focused on funding of research and development or on legal restrictions. As an example, the proposed "principles for responsible stewardship of trustworthy AI" recommended by the OECD in 2019, entail five principles: growth, sustainable development, and well-being; human-centered values and fairness; transparency and explainability; robustness, security, and safety; and accountability.[35] Similarly, the HLEG AI lists the following principles: human agency and oversight, technical robustness and safety, privacy and data governance, transparency, diversity, non-discrimination, and fairness, environmental and societal well-being, and accountability.[36] Floridi and Cosh recently counted as many as 46 principles in the most relevant international reports, although they argue that many of these could be subsumed under the four traditional bioethics principles, namely autonomy, non-maleficence, beneficence, and justice. They propose to apply these four, complemented by the new principle of "explicability," which they want to understand both as intelligibility and accountability.[37] In the following, I want to propose a different approach, one that deliberates the relationship of agency and vulnerability. Both have been discussed as principles of bioethics, with vulnerability especially becoming more prominent in the last few years.[38]

3 Agency, identity, and vulnerable agency in the context of artificial intelligence

Obviously, there are conflicting theories of action and human agency – suffice it to say that any theory of agency depends on an understanding of the "striving for

---

[35] OECD, *Recommendation of the Council on Artificial Intelligence* (2019).
[36] High Level Expert Group on Artificial Intelligence (AI HLEG), *Ethics Guidelines for Trustworthy Ai* (2019).
[37] Cf. Luciano Floridi and Josh Cowls, *A Unified Framework of Five Principles for Ai in Society* (2019).
[38] Cf. Henk Ten Have, *Vulnerability: Challenging Bioethics* (2016). For a discussion of vulnerability in the context of bioethics, cf. also Hille Haker, *Verletzliche Freiheit. Zu einem neuen Prinzip der Bioethik* (2021).

a good life:" *agents aim for ends they consider to be good*. The distinction between *actors* who act according to the causality model of action and may be bots, robots, or any kind of digital actors, and *agents* is crucial for the understanding of morality – and the reason why AI learning systems such as self-driving cars, for example, need to be traced back to human agents who can be held accountable where the cars cannot. Often, the philosophical question about agency points to the "neutrality" of the technologies which agents can use in moral and immoral ways. But ethically speaking, technologies are never neutral, because they already belong to networks of actions that make it difficult to isolate "the" technologies from the constellations of collective agency – which Latour has called actor *networks*.[39]

Yet, only agents need to give meaning to their actions over time, as Charles Taylor rightly argues, and therefore must connect their self-identity with goods they strive for.[40] Only agents are responsible for their actions, and they can therefore be held accountable by others. As Nida-Rümelin emphasizes, it is exactly the understanding of identity, freedom, and agency that is currently redefined in light of an alternative rationality model, which goes back to the Cartesian understanding of cognition and the Baconian embrace of instrumental reason over against the idealism of the metaphysical tradition. This empirical and instrumental concept of reason is just as rooted in modernity as the hermeneutical model that Taylor and Nida-Rümelin, but also Walter Benjamin and the Frankfurt School defended.[41]

3.1 Agency and the moral status of AI devices

In the era of AI, we may well be able to connect with anybody through social media; we may understand texts within seconds with the help of translation tools; we may discover more about our bodily functions with the help of movement, blood pressure, or sleep trackers, and we may interact with robots in a similarly intimate and personal way as with human beings – in other words,

---

39 One does not need to follow Bruno Latour all the way into his theory to see the value of this point. Bruno Latour, *Reassembling the Social: An Introduction to Actor-Network-Theory* (2005).
40 Taylor argues that humans are not only driven by desires but have the capacity to step back and evaluate what they want in view of self-ideals. This is the reason why comparative evaluations, weak and strong, and the relation of identity and the good are inescapable. Cf. Charles Taylor, *Sources of the Self. The Making of Modern Identity* (1992).
41 Julian Nida-Rümelin and Nathalie Weidenfeld, *Digitaler Humanismus: Eine Ethik für das Zeitalter der Künstlichen Intelligenz* (2018).

the information about ourselves, insights, and even self-care options certainly increase by utilizing AI learning systems. At the same time, however, it is not so clear how our identities will be shaped by the devices, systems, and humanoid machines. Artificial intelligence relies on the translatability of intentions into causations, and causations into action chains that can be programmed, no matter how flexible their "deep learning" capacities are.[42]

Most scholars hold that robots, for example, lack some of the most important features of self-identity, among them the sense of an inner life, desires, motives, or repressed fears. They lack the sense of corporeality that marks the embodied subjectivity of human beings, and they lack the sense of uniqueness. Robots are programmed to define (and pursue) the good as an extrinsic and instrumental value rather than an intrinsic part of their identity.[43] And this is what makes them problematic. When we understand *human* practices increasingly in human-machine interface terms, with little relation to the experiential side that involves the meaning-making and life narrative of individuals, in their "entanglement with the stories," humans are perhaps merely entering a new and intensified era of alienation and self-alienation. Regarding robots' agency, scholars therefore discuss whether robots or other engineered machines merely resemble or in fact actualize human agency.[44] The philosophy of mind is not the only one interested in this question but ethics, too, attempts to determine it – in part, because agency determines the presence and absence, or the scope of accountability. As we have seen above, bioethicists often refer to the principle of autonomy, i.e. the ideal of freedom and independence, going back to the classical liberal tradition. Yet, agency cannot be reduced to the liberal interpretation of self-determination. Alternatively, autonomy in Kant's understanding concerns the self-giving of moral rules in view of universal considerations, culminating in the respect for other agents.[45] The constellation of humans and machines is often depicted in terms of an *asymmetric* relation. The problem is that too often, humans are then envisioned in the position of the "master," and the ma-

---

[42] For a discussion on the reductionism of this approach regarding the concept of the self cf. Paul Ricœur, *Oneself as Another* (1992).

[43] Taylor calls this the radical reflexivity that requires the relation of the agent to their experiences, enabling the "experience of the experience," first examined by Augustine. This, I hold, is an impossible stance for any AI device. Cf. Taylor, *Sources of the Self. The Making of Modern Identity*, 127–42.

[44] Eva Wiese, Giorgio Metta, and Agnieszka Wykowska, *Robots as Intentional Agents: Using Neuroscientific Methods to Make Robots Appear More Social* (2017).

[45] A good analysis is provided by O'Neill, applied to the context of bioethics: Onora O'Neill, *Autonomy and Trust in Bioethics* (2002).

chine in the position of the "servant."[46] But when the lines between humans and machines are blurred, as is the case, for instance, automated in telemedicine or any kind of robocalls that use voice imitation so authenticly that it is almost indistinguishable from human-to-human communications, this relational model becomes ethically problematic. The question of the moral status then turns into a question of what we owe to other humans, and how what we owe to humans is distinct from what we owe to machines.[47]

Those who believe that humanoid robots do indeed actualize agency in the sense of self-regulation are open to grant them certain rights.[48] Others compare interactive robots to animals, also in an attempt to attribute to them a certain moral status. Analogies, however, must determine similarities and differences. Robots, like pets, can support humans in their everyday life and they can become emotional companions without sharing the same features as humans or animals. The self-learning machines' difference to humans and to animals outweigh their similarities, as Deborah Johnson and Mario Verdicchio have argued, comparing features such as sentience, suffering, and consciousness in connection with legal liability and responsibility. In this, they follow the traditional lines of determining moral agency.[49]

Maybe the safest way is to ascribe *functional* agency to certain AI devices, especially those learning systems that use a humanoid voice, write messages, or otherwise resemble human actions. I believe that regulations must indeed be set up to clarify and reveal the identity of an AI device, simply because in our perception, the line between machines and humans is easily blurred. Ironically, the "master-servant" metaphor may also be misleading: it could well be that AI devices increasingly render *humans* as the servants of masters *behind* the systems who surveil us and use our data without our knowledge or understanding. We are forced to accept these practices that increasingly infringe upon our bodily integrity, because we cannot participate in multiple social prac-

---

46 Steve Petersen, *Designing People to Serve* (2011). Petersen who argued that it is indeed possible to design robots as servants, sparked a good ethical discussion, in which I agree with his critics. Cf. Bartek Chomanski, *What's Wrong with Designing People to Serve?* (2019); Maciej Musiał, *Designing (Artificial) People to Serve – the Other Side of the Coin* (2017).
47 I have discussed the question of deception in the context of medical care for patients with dementia and care robots in: Hille Haker, *Ethische Fragen des Einsatzes von Pflegerobotern* (2014).
48 Mark Coeckelbergh, *Robot Rights? Towards a Social-Relational Justification of Moral Consideration* (2010).
49 Deborah G. Johnson and Mario Verdicchio, *Why Robots Should Not Be Treated Like Animals* (2018), ibid. For an overview of AI ethics and literature cf. *Coeckelbergh, AI Ethics*.

tices without the digital technologies.[50] This heteronomy does not have its origin in the "machine" but in the agency of other persons – individuals or corporations who seek to render consumers more and more dependent on devices, who exploit the information for behavior manipulation, who control the healthcare a person may eligible to, the citizenship or asylum status, access to transportation services, etc.

Watching popular robots such as "Sophia" or "Nadia" at the current state of development, we are not (yet) tricked into confusing them with a human being. But does that also hold true when the "we" are children or elderly? Or when the learning AI systems communicate with us in writing? Or when their voices become indistinguishable from human voices, as is the case in robocalls? Ethically, it might be more relevant to ask whether AI devices make *humans* more capable, and if so, in what way, instead of asking how capable machines are. One may ask whether AI devices enable us to (better) give meaning to our lives, or whether they in fact *reduce* our freedom and *impoverish* our experience. At first sight, Alexa, Siri, or even more sophisticated systems such as social and/or monitoring robots may make everyday practices more efficient and comfortable, but they may also reduce interactions to the transportation of information, or even giving and following orders. All these questions go far beyond the current debates, and they will not warrant always a binary response.

Today, what Kant described as heteronomy is often linked to dependency and vulnerability.[51] There may be moral persons who are not humans – such as angels or, today, super-intelligent robots – but they don't matter for morality exactly because they are not vulnerable agents. I therefore want to argue that we do not need to complement the principle of agency (warranting respect) with a separate principle of vulnerability (warranting beneficence and care). Rather, we need to understand the intertwining of agency and vulnerability. I believe that it is humans' vulnerable agency that ultimately distinguishes them categorically, and not only gradually, from AI devices.

3.2 Vulnerable agency

In the *Course of Recognition*, his final book before his death that goes back to lectures given in Vienna, Paul Ricœur has offered a simple definition of "what

---

50 Shoshana Zuboff, *The Age of Surveillance Capitalism: The Fight for a Human Future at the New Frontier of Power* (2019).
51 Cf. Paula Formosa, *The Role of Vulnerability in Kantian Ethics* (2014).

makes us human:" the human being is the *homme capable*, the "capable human." An agent is capable to respond to others in a particular way, to wit, as the one who is *meant*, the one who is *addressed* by someone else with the claim or plea to be heard, to be seen, to be *recognized* and *respected*, to be *cared* for. All these capabilities are not only socially mediated but also shaped by cultural and social norms, and they are symbolically expressed in different languages that uphold the plurality and diversity of experiences. Most importantly, to understand oneself as being addressed as a moral agent presupposes the capability to understand the *moral* language of the human, i.e., one must not only know grammatical rules of morality but also, one needs to position oneself in relation to the demands made by someone else.

In the Kantian tradition, the principle of dignity refers to the complex set of moral capabilities – but in contrast to the Kantian tradition, by taking up the concept of recognition, elaborated first by Fichte in response to Kant, and then the capability to reason by Hegel, there is more to the autonomy and the dignity of agents. Not only in their self-development but over the course of their whole life, humans are – and in fact must be – addressed by others to whom they respond.[52] Mimesis is a verb rather than a noun, closely connected to an agent's capability to "attune" to others and/or one's environment: children first imitate others in their self-development, but all humans, from the early stages on, attune to each other in their inter-actions. The mimetic adaptation to an environment is mirrored in the self-learning of AI actors. Vulnerability refers as much to the social constitution of the self as to the general affectability of human beings by their environs. Building upon multiple works over the last decades,[53] I distinguish between three dimensions: ontological, moral, and structural vulnerability. This framework will help to gain a deeper understanding of what is at stake in the concept of vulnerable agency, which I consider to be a reinterpretation of human dignity.

Ontological vulnerability stresses the risk that human affectability and openness entails. The Latin verb *vulnerare* means "to wound", its passive form is "to be wounded." Paradoxically, it is also the condition for a most basic openness to the world, and this makes the concept of vulnerability more ambivalent than the negative connotation may suggest. In order to be affected by others in the pos-

---

[52] Cf. for a phenomenological account or responsivity the many works of B. Waldenfels, most prominently: Bernhard Waldenfels, *Antwortregister* (1994).
[53] For a good overview of current approaches cf. Catriona Mackenzie, *Vulnerability: New Essays in Ethics and Feminist Philosophy* (2014). The following section summarizes a more thorough examination in Haker, *Towards a Critical Political Ethics. Catholic Ethics and Social Challenges*, chapter 5.

itive sense, one must be receptive to others, taking the risk to embrace one's lack of control, which in turn holds the promise that one may be transformed by new experiences.[54]

In contrast to ontological vulnerability, *moral vulnerability* refers to the potential harm inflicted upon someone by another person or other persons – intentionally. As ontological vulnerability is indeed, in its negative sense, the susceptibility to *any* pain and suffering, moral vulnerability is, in its negative sense, the susceptibility to someone else's wrong-doing. Moral vulnerability stresses the affectability by the evaluative attitudes of others, the responsiveness to their address, and the constant struggle between conformity and non-conformity with others. A vulnerable agent, in a nutshell, is an agent who belongs to a community of others while being other to all others, and *other* to oneself.[55] "To belong" means to be recognized as being of the "same kind" as the "others," yet also as a "unique" self. Similar to the ontological dimension of vulnerability, moral vulnerability, too, is not exclusively tied to the negative interpretation. It points to the risk agents must take in their actions and interactions, which in its positive connotation is often accompanied by excitement, curiosity, and openness to adjust to the wishes and needs of the other. Moral injury, in contrast, is so harmful because it threatens the trust that is necessary for taking the risk of inter-action. With the term *moral* vulnerability, I not only refer to the possibility of harm done but also to the injured person's *moral sense* to be (morally) humiliated and not just – coincidently – injured. Without sensing and interpreting another agent's intentions, one cannot distinguish a non-intentional injury from wrongdoing. Analytical philosophy, which examined more closely the criteria for holding others accountable for actions, and included, for example, how the judgment of blame is rooted in the resentment of the other's or others' actions, demonstrates clearly how inapplicable the concept of accountability for moral harm in this sense is to artificial intelligence devices.[56]

Finally, in my understanding, the concept of vulnerability must not overlook the *structural vulnerability* that refers to particular *states* of vulnerability. Age, ill-

---

[54] Erinn C. Gilson, *The Ethics of Vulnerability: A Feminist Analysis of Social Life and Practice* (2014), 37. Cf. also Hille Haker, *Verletzlichkeit als Kategorie der Ethik* (2015). In that work, I have called the "ontological" vulnerability "anthropological" to stress its connection to the *conditio humana*; in order to create a more common language, I now take up Gilson's term.

[55] Different terms are used by different authors to point to this opacity to oneself: Ricœur calls it the "oneself as another," Butler "opacity," and Waldenfels "alienness," and Theodor Adorno calls it the "non-identity."

[56] Cf. the influential essay by Peter Strawson on Freedom and Resentment, in Peter Frederick Strawson, *Freedom and Resentment and Other Essays* (2008).

ness, disability, or similar factors that increase the risk of suffering may elevate the ontological dimension of vulnerability, and a history of discrimination may go hand in hand with ever-new forms of disenfranchisement. The question is whether AI devices may support the inclusion and well-being, for example, of persons with disabilities or remote communities and/or individuals – or whether they instead increase the vulnerability of particular groups, for instance the elderly, people with dementia, minors, minorities, or people who are excluded from using the devices.

Most interesting for me right now is moral vulnerability, i.e. the susceptibility to harm and to violence that one fears. Interaction or, to use a more precise term, inter-agency, I have said, depends on the reciprocal trust that the other will not exploit this vulnerability, and among agents, a premise of interaction seems to be that the vulnerable agency is *mutual* despite possible differences in the capabilities to act or the susceptibility to harm. In interactions with robots who are closest to human agents, however, this is not the case, and it renders the famous "Asimov's Three Laws of Robotics" inadequate.[57] Benjamin points to the quasi-mythical power that things may gain over someone, and this seems to be supported by experiences with AI devices.[58]

The difference between a vulnerable human and a destroyable robot is the moral difference between damage and harm. Things may be neglected, they may become outdated and abandoned, damaged, broken and destroyed, but with all the power things may hold over us, *they cannot be harmed in a moral sense*, whereas *we* can certainly be morally harmed by them. Ethically speaking, misrecognition – the ethical translation of the violation of dignity into intersubjective terms – means that a person is transformed into an object-thing that may be abandoned, destroyed, forgotten, disposed of or replaced. While no robot may be intentionally designed to do this, the AI systems together still enforce a new kind of *social disciplining:* in order to be predictable for the machine, a "good child" or a "good elderly" must conform with certain behaviors and be "nudged" to compliance. AI comes therefore with the risk to be programmed in a way that directs human behavior to a new level of conformity, threatening the dialectic of

---

[57] Asimov's first norm was to secure that robots must not harm human beings through action or inaction [1]; the second norm limited a robot's action to following the "orders given it by human beings" except when these would conflict with the first norm [2]; and the third norm inscribed self-preservation and self-protection as an end a robot must pursue unless in conflict with the first or the second norm [3]. Isaac Asimov, "Runaround." *I, Robot* (1950).

[58] Dix gives the example of the game Faunasphere, in which the "fauna," similar to the Tamagotchi figures, provoked strong emotions and attachments among the players. Alan Dix, *I in an Other's Eye* (2019), 68.

conformity and non-conformity that I identified as constitutive for the development of (moral) identity. Respectively, ascribing qualities and capabilities to a thing as if it entailed agency but not vulnerability – which includes the openness to the unpredictable encounter with others (experienced also, in part, in human-animal interaction) as well as the susceptibility to harm – creates a tension between the humanized "thing" and the self.[59] The practical ambivalence creates an epistemological ambiguity, too, that is followed by a moral ambiguity, because it is not clear any longer, as I said, what my responsibilities towards the "living machine" are. As long as the device is only a device, screaming at it, discarding it, or simply forgetting about it when a new gadget has caught one's (consumerist) attention, may be inherently unproblematic.[60] It becomes *inherently* problematic, however, when the human agents cannot determine whether the "other" they encounter or interact with is a device or another human being. Furthermore, indirect moral questions also arise, which may concern the relationship towards third persons: for instance, it is not clear whether one may deceive an elderly person who may be cognitively unable to see through the monitoring device that she may merely identify as a pet that reduces her emotional distress, but that caregivers or family members know has multiple other functions.[61]

Conclusion: AI ethics – the path forward?

In a recent paper, Mark Graves has suggested to develop a new kind of virtue ethics that examines the shared moral and spiritual development among human persons and artificially intelligent agents.[62] He promotes the development of a virtue ethics that "teaches" AI machines the habits of a *community* to which they belong. This view entails a communitarian virtue ethics that is uncritical and easily undermines the freedom of the individual. But morality is not a question of integration into a community. While nobody can flourish without a community, no community can flourish without the resistance of the "I" to the "we" either. The moral stance, or moral point of view, does not rest upon the habits

---

[59] This point requires a more thorough discussion in light of the concepts of posthumanism to which I referred in the first part.
[60] It is, of course, highly problematic as a capitalist mode of living, as I argued above with reference to the critique of the Frankfurt School critical theory.
[61] Cf. Alessandro Vercelli et al., *Robots in Elderly Care (2018)*; Haker, "Ethische Fragen des Einsatzes von Pflegerobotern".
[62] Mark Graves, *Shared Moral and Spiritual Development among Human Persons and Artificially Intelligent Agents* (2017).

that are meant to foster the communal values and the sense of the individuals' belonging. Morality begins with the acknowledgment of one's vulnerable agency that one shares with other vulnerable agents.

In conclusion, I take from the discussion of vulnerable agency that to acknowledge one's "selfhood" means to acknowledge one's vulnerability that is as intertwined with one's agency as, vice versa, agency is intertwined with vulnerability. It means to acknowledge that *experience* requires an openness of the self to the other and to one's own otherness.[63] One must become another to oneself in order to become oneself, and yet, one cannot purposefully aim for it. I cannot see how this complex self-other-relation can be translated into the design of learning AI systems. A robot may well speak of love and compassion, of hope, freedom, peace, happiness, solidarity, or justice, but it will still only be the semblance of the moral language that agents speak to each other. When Saudi Arabia granted the robot Sophia the right to citizenship, it was not a genuine gesture but a gesture of power by an authoritarian regime that disregards the most basic human rights of its citizens. With this gesture, it demonstrated indifference towards the most important right that agents can demand because of their vulnerability to moral harm, namely the right to have rights. A robot does not need this right, and for its designers, politics may also just be a game.[64]

When humans approach machines that resemble humans, they show the child-like mimetic capability that Walter Benjamin describes in his poetic fragments of his childhood: the capability to be open to the alien, and sometimes even the alien within oneself. This is why humans respond to robots such as "Sophia," "Siri," or "Alexa," and in the future perhaps also to an empathetic robot such as "Nadia," encountering them with an attitude of politeness and curiosity, which they have learned in their self-formation and moral formation. But robots cannot reciprocate the same openness. Humans will address them on their own terms, translating the robots' voices, gestures and expressions back into the human language and experiential knowledge, both emotional and intellectual. We are faced with the paradox that in order to use the AI systems, we need to play along – humanizing them and interacting with them *as if* they were just like us; yet, in order to protect the vulnerable agency that enables moral interactions and indeed mutual recognition, in order to *stay human*, we need to de-

---

63 *Dix, I in an Other's Eye.*
64 Cf., for example, https://www.wired.co.uk/article/sophia-robot-citizen-womens-rights-detriot-become-human-hanson-robotics and her performance at the Munich Security Conference 2018, https://securityconference.org/mediathek/asset/town-hall-meeting-msc2018-the-force-awakens-artificial-intelligence-and-modern-conflict-1720-15-02-2018/

humanize the humanoid machines. Otherwise, we may pay the price of losing what makes us human, our moral identity and vulnerable agency.

Works cited

Adorno, Theodor, and Max Horkheimer. *Dialectic of Enlightenment.* London: Verso Books, 2016 (orig. 1947).
Alexander, Jeffrey C. *The Civil Sphere.* New York: Oxford UP, 2006.
Asimov, Isaac. "Runaround." *I, Robot. The Isaac Asimov Collection.* Ed. Isaac Asimov. New York: Coubleday, 1950, 40.
Benjamin, Ruha. *Race after Technology: Abolitionist Tools for the New Jim Code.* Cambridge, UK: Polity, 2019.
Benjamin, Walter. "Berlin Childhood around 1900." Chap. 349–350 *Walter Benjamin: Selected Writings.* Ed. Michael Jennings. Cambridge, MA: Harvard UP, 2002. 349–50.
Benjamin, Walter. "Experience and Poverty." *Walter Benjamin: Selected Writings.* Ed. Michael Jennings. Cambridge, MA: Harvard UP, 1999. 731–36.
Benjamin, Walter. *The Storyteller: Tales out of Loneliness.* Verso Books, 2016.
Benjamin, Walter. "The Storyteller. Observations on the Works of Nikolai Leskov (Orig. 1936)." *Illuminations* Ed. Hannah Arendt. London: Pimlico, 1999. 83–107.
Bennett, Jane. *Vibrant Matter: A Political Ecology of Things.* Durham: Duke UP, 2010.
Bhattacharya, Aveek, Simpson, Robert Mark. "Life in Overabundance: Agar on Life-Extension and the Fear of Death." *Ethical theory and moral practice* 17, no. 2 (2014): 223–36.
Braidotti, Rosi. *Posthuman Knowledge.* Polity Press Cambridge, 2019.
Braidotti, Rosi, and Simone Bignall. *Posthuman Ecologies: Complexity and Process after Deleuze.* Rowman & Littlefield, 2018.
Buhmann, Alexander, and Christian Fieseler. "Towards a Deliberative Framework for Responsible Innovation in Artificial Intelligence." *Technology in Society* 64 (2021): 101475.
Chomanski, Bartek. "What's Wrong with Designing People to Serve?". *Ethical theory and moral practice* 22, no. 4 (2019): 993–1015.
Coeckelbergh, Mark. *Ai Ethics.* The Mit Press Essential Knowledge Series. Cambridge, MA: MIT, 2020.
Coeckelbergh, Mark. "Robot Rights? Towards a Social-Relational Justification of Moral Consideration." *Ethics and information technology* 12, no. 3 (2010): 209–21.
Dix, Alan. "I in an Other's Eye." *AI & Society* 34, no. 1 (2019): 55–73.
Feenberg, Andrew. *Between Reason and Experience: Essays in Technology and Modernity.* Inside Technology. Cambridge, MA: MIT P, 2010.
Floridi, Luciano, and Josh Cowls. "A Unified Framework of Five Principles for Ai in Society." *Issue 1.1, Summer 2019* 1, no. 1 (2019).
Formosa, Paula. "The Role of Vulnerability in Kantian Ethics." *Vulnerability: New Essays in Ethics and Feminist Philosophy.* Ed. Catriona Mackenzie, Wendy Rogers, Susan Dodds. New York: Oxford UP, 2014. 88–109.
Gilson, Erinn C. *The Ethics of Vulnerability: A Feminist Analysis of Social Life and Practice.* New York: Routledge, 2014.

Graves, Mark. "Shared Moral and Spiritual Development among Human Persons and Artificially Intelligent Agents." *Theology and Science* 15, no. 3 (2017): 333–51.

Haker, Hille. "Ethische Fragen des Einsatzes von Pflegerobotern." *Kooperation ohne Akteure? Automatismen in der Globalisierung*. Ed. Claus Leggewie. Duisburg: Käthe Hamburger Institut/Centre for Global Cooperation Research, 2014. 55–68.

Haker, Hille. "Information or Communication – the Loss of the Language of the Human." *Hope. Where Does Our Hope Lie?* (International Congress of the European Society of Catholic Theology 2019 – Bratislava, Slovakia). Ed. Miloš Lichner. Vienna-Zurich: LIT, 2020. 505–23.

Haker, Hille. *Moralische Identität. Literarische Lebensgeschichten als Medium ethischer Reflexion. Mit einer Interpretation der "Jahrestage" von Uwe Johnson*. Tübingen: Francke, 1999.

Haker, Hille. "Nanomedicine and European Ethics. Part One." *Ethics for Graduate Researchers: A Cross-Disciplinary Approach*. Ed. Cathriona Russell, Linda Hogan, Maureen Junker-Kenny. Amsterdam et al.: Elsevier, 2013. 87–100.

Haker, Hille. "Resonanz. Eine Analyse aus ethischer Perspektive." *Harmut Rosa: Resonanz*. Ed. Jean-Pierre Wils. Würzburg: nomos, 2018, 33–44.

Haker, Hille. "Synthetic Biology – an Emerging Debate in Europe – Part Two." *Ethics for Graduate Researchers: A Cross-Disciplinary Approach*. Ed. Linda Hogan, Cathriona Russell, Maureen Junker-Kenny. Elsevier, 2013. 227–40.

Haker, Hille. *Towards a Critical Political Ethics. Catholic Ethics and Social Challenges*. Studien Zur Theologischen Ethik 156. Basel: Schwabe Verlag, 2020.

Haker, Hille. "Verletzliche Freiheit. Zu einem neuen Prinzip der Bioethik." *Theologische Vulnerabilitätsforschung. Gesellschaftsrelevant und Interdisziplinär*. Ed. Hildegund Keul. Stuttgart: Kohlhammer, 2021. 99–118.

Haker, Hille. "Verletzlichkeit als Kategorie der Ethik." *Zwischen Parteilichkeit und Ethik. Schnittstellen von Klinikseelsorge und Medizinethik*. Ed. Monika Bobbert. Medizinethik in der Klinikseelsorge. Berlin/Münster: Lit, 2015. 195–225.

Hauskeller, Michael. *Mythologies of Transhumanism*. Springer, 2016.

Heidegger, Martin. *Sein und Zeit*. Historische Gesamtausgabe. Bd. 2: Abteilung 1, Frankfurt am Main: Klostermann, 1977.

High Level Expert Group on Artificial Intelligence. "A Definition of AI: Main Capabilities and Disciplines. Definition Developed for the Purpose of the AI Hleg's Deliverables." April 2020.

High Level Expert Group on Artificial Intelligence (AI HLEG). "Ethics Guidelines for Trustworthy AI." Brussels: European Commission, 2019.

Honneth, Axel, with comments by: Judith Butler, Raymond Geuss, Jonathan Lear, Martin Jay. *Reification: A New Look at an Old Idea*. New York: Oxford UP, 2007.

Jaeggi, Rahel. *Alienation*. [in English] New York: Columbia UP, 2016.

Johnson, Deborah G., and Mario Verdicchio. "Why Robots Should Not Be Treated Like Animals." *Ethics and Information Technology* 20, no. 4 (2018): 291–301.

Johnson, Uwe. *Anniversaries: From a Year in the Life of Gesine Cresspahl*. Edited by NYRB Classics. New York: NYRB Classics, 2018.

Johnson, Uwe. *Jahrestage: Aus dem Leben von Gesine Cresspahl*. Berlin: Suhrkamp, 2014 (orig. 1970–1983).

Kolbert, Elisabeth. "Chemical Warfare's Home Front." *New York Review of Books*, 11 February 2021.
Latour, Bruno. *Reassembling the Social: An Introduction to Actor-Network-Theory*. Oxford UP, 2005.
Lukács, Georg. *The Theory of the Novel: A Historico-Philosophical Essay on the Forms of Great Epic Literature*. Cambridge MA: MIT, 1971.
Mackenzie, Catriona. *Vulnerability: New Essays in Ethics and Feminist Philosophy*. Oxford, UK: Oxford UP, 2014.
Marx, Karl and Frederick Engels. *Marx & Engels Collected Works Vol 06: 1845–1848*. London: Lawrence & Wishart, 1976. 105–219.
Musiał, Maciej. "Designing (Artificial) People to Serve – the Other Side of the Coin." *Journal of Experimental & Theoretical Artificial Intelligence* 29, no. 5 (2017): 1087–97.
Nida-Rümelin, Julian, and Nathalie Weidenfeld. *Digitaler Humanismus: Eine Ethik für das Zeitalter der Künstlichen Intelligenz*. Piper ebooks, 2018.
O'Neill, Onora. *Autonomy and Trust in Bioethics*. Gifford Lectures; 2001. Cambridge, MA: Cambridge UP, 2002.
OECD. "Recommendation of the Council on Artificial Intelligence." In *OECD Legal Instruments*, edited by OECD Legal Instruments: OECD Legal Instruments, 2019.
Petersen, Steve. "Designing People to Serve." *Robot Ethics*. Ed. Patrick Lin, George Bekey and Keith Abney. Cambridge, MA: MIT Press, 2011. 283-298.
Rawls, John. *Justice as Fairness: A Restatement*. Cambridge, MA: Harvard UP, 2001.
Ricœur, Paul. *Oneself as Another*. Chicago: U of Chicago P, 1992.
Rosa, Hartmut. *Resonanz: Eine Soziologie der Weltbeziehung*. Berlin: Suhrkamp, 2017.
Schapp, Wilhelm. *In Geschichten verstrickt. Vom Sein von Mensch und Ding*. Hamburg: Meiner, 1953.
Siebel, Thomas M. *Digital Transformation: Survive and Thrive in an Era of Mass Extinction*. Edited by Condoleezza Rice. First edition. ed. New York, NY: RosettaBooks, 2019.
Stahl, Titus. "What Is Immanent Critique?" *Available at SSRN 2357957* (2013).
Strawson, Peter Frederick. *Freedom and Resentment and Other Essays*. New York: Routledge, 2008.
Taylor, Charles. *Sources of the Self. The Making of Modern Identity*. Cambridge, MA: Harvard UP, 1992.
Ten Have, Henk. *Vulnerability: Challenging Bioethics*. London, New York: Routledge, 2016.
Truby, Jon. "Governing Artificial Intelligence to Benefit the Un Sustainable Development Goals." *Sustainable Development* 28, no. 4 (2020): 946–59.
Vercelli, Alessandro, Innocenzo Rainero, Ludovico Ciferri, Marina Boido, and Fabrizio Pirri. "Robots in Elderly Care." *DigitCult-Scientific Journal on Digital Cultures* 2, no. 2 (2018): 37–50.
Von Schomberg, René, and Jonathan Hankins. *International Handbook on Responsible Innovation: A Global Resource*. Edward Elgar Publishing, 2019.
Waldenfels, Bernhard. *Antwortregister*. Frankfurt am Main: Suhrkamp, 1994.
Wiese, Eva, Giorgio Metta, and Agnieszka Wykowska. "Robots as Intentional Agents: Using Neuroscientific Methods to Make Robots Appear More Social." *Frontiers in Psychology* 8 (2017): 1–19.
Winkler, Adam. "'Corporations Are People' Is Built on an Incredible 19th-Century Lie." *The Atlantic*, 5 March 2018.

Zawieska, Karolina. "Disengagement with Ethics in Robotics as a Tacit Form of Dehumanisation." *AI & Society* 35, no. 4 (2020/12/01 2020): 869–83.

Zuboff, Shoshana. The Age of Surveillance Capitalism: The Fight for a Human Future at the New Frontier of Power. New York: PublicAffairs, 2019.

Sabine Sielke
Outsourcing the Brain, Optimizing the Body: Retrotopian Projections of the Human Subject

Abstract: This paper examines the paradoxical retrotopian dynamics, evident in research and science as much as in cultural practice, that drives current visions of the future of human subjectivity and of what it means to be human. Clearly differentiating AI from both trans- and posthumanism, I map the common ground these discourses tread: the aspiration to enhance human life and the seeming certainty that this can be achieved by way of transcending "human nature." Figuring in cognitive science as a field that has fundamentally informed this conversation, I show how AI, with its dematerialized notion of subjectivity and demise of body (and brain) markets the ongoing "optimization" of the subject and makes the perceived boundary between human agents and their technological environment increasingly porous. Current conceptions of the human subject are retrotopian, rather than utopian, not only because popularized versions of cognitive science and AI revitalize mechanistic notions of humans that go back to the Renaissance. Coinciding with current practices of physical self-optimization, the sense of humans as "brain machines," I hold, revitalizes and cements socio-economic hierarchies and inequalities that we deemed overcome some time ago.

Preface

I want to preface my essay with a quotation from the very beginning of Mark O'Connell's 2017 book *To Be a Machine: Adventures Among Cyborgs, Utopians, Hackers, and the Futurists Solving the Modest Problem of Death*. "All stories begin in our endings," O'Connell opens his exploration, "we invent them because we die. As long as we have been telling stories, we have been telling them about the desire to escape our human bodies, to become something other than the animals we are" (O'Connell 2017, 1). This desire to transcend 'human nature' – be it our existence as God's creation or as a biological species – has driven religious beliefs, philosophical thought, motivations of science and research, as well as modes of storytelling for millennia. It informs the field of artificial intelligence since its emergence in the 1950s[1] and dominates transhumanism (which some do consider a kind of religion). Both have inspired multiple utopian and dystopian narratives of our disembodied future as a "superintelligence," to use Nick Bostrom's term, or "singularity," the concept Ray Kurzweil prefers; hence their proximity to religions, practiced by (predominantly male) protagonists (see O'Connell 2017, 10, like Kurzweil, Max More, or Randal A. Koene, who are regarded as cult figures or gurus by many (likewise mostly male) proponents of artificial intelligence.

AI and transhumanism are two different matters, of course, and we need to clearly differentiate their faiths, goals, and possible futures. My concern, though, is with the common ground they tread: the aspiration to enhance human life, a seeming certainty about what that actually means, and an affinity with posthumanism. To complicate matters – or make things even more vague –, I would like to figure in cognitive science as a field that has fundamentally informed this conversation and its ongoing shifts.[2] More precisely, my contribution to these debates examines the paradoxical retrotopian dynamics, evident in research and science as much as in cultural practice, that drives current visions of the future of human subjectivity and of what it means to us to be human. Inspired as much as enabled by developments in computer science, computation, and digitization, these visions toy with the outsourcing of our brains, their unhitching from our inadequate bodies, bodies incapable to compete with smart machines and still marked with an "expiration date" (Max More, qtd. in O'Connell 2017, 4). At the same time, the perceived boundary between human agents and their technological environment has become increasingly porous, and the interfaces between human bodies and their digitized appendages have moved to the forefront of scientific inquiry and marketing strategies (see Sielke and Schäfer-Wünsche 2007). This development was driven as much by Marshall McLuhan's conception of media as "extensions of man" and biotechnologies like advanced prosthesis as by notions of the human brain as computer or network and by newest

---

Note: I thank Dr. Björn Bosserhoff, once again, for excellent research and editing assistance.

1 For a detailed history of the "ideas and achievements" of AI see Nilsson 2010.
2 It may surprise us that the term cognitive science was introduced as late as 1973 by chemist and physicist Hugh Christopher Longuet-Higgins to capture a wide interdisciplinary field of research including neuroscience, psychology, philosophy, informatics, esp. AI, and linguistics and aimed at examining the principles on the basis of which intelligent entities interact with their environment (Scheerer 1985, 7).

brain-computer interface (BCI) technologies.[3] Indeed, we have long been variants of the cyborgs that biologist Donna Haraway, in her influential "manifesto" (1985), projected as dynamo of a utopian future.[4]

What we have observed, consequently, in recent decades, is the interplay of two interdependent tendencies: while privileging and outsourcing the brain as the centerpiece of human subjectity, both scientific inquiry and cultural practices flaunt an ambition to optimize the human body and "augment our powers of perception and cognition through technological enhancements of our sense organs and our neural capacities" (O'Connell 2017, 5). These interrelated aspirations have managed to channel immense amounts of resources into cognitive science and AI research; and they have also fundamentally and in multiple manners, particularly by agenda setting and practices of defunding, affected our own work as scholars in the humanities and social sciences. Let us briefly recall how: during the last three to four decades, work in literary and cultural studies, for instance, has focused on parameters of difference and issues of the body. Yet, the concept of a culturally constructed, gendered, racialized, and class-contoured body which emerged from these debates has long been challenged. Evolving from neurobiology, molecular genetics, and biotechnology are new insights into our corporeality, projections of a post- or transhuman subject, and novel notions about how our bodies interrelate with the world. Accordingly, during the 1990s, concepts like consciousness, mind, will, and belief became re-naturalized. In his 2002 book on the *Synaptic Self: How Our Brain Becomes Who We Are* neuroscientist Joseph LeDoux even claimed that our individual "'self,' the essence of who we are reflects patterns of interconnectivity in our brain. [...] Given the importance of synaptic transmission in brain function, it should practically be a truism to claim that the self is synaptic" (LeDoux 2002, 2).

This tendency to renaturalize mental processes remains overdetermined by the concept of mind as computer which, disseminated from the 1950s onward, conceptualized the human mind as a machine which stores information and allows it to be retrieved. This so-called "information-processing" or "artificial intelligence approach" of the cognitive sciences and the faith that mental processes are understood best if compared to a computer (Matlin 1983, 13) made the computer into the master trope of communication and information systems as much as into the model explanation for how the human mind works. The idea

---

[3] BCI refers to systems which, on the basis of measuring and digitizing electrical signals in the brain, allow for a "direct dialog between man and machine" and thus enable a 'mind control' of machinery (BBCI 2014).
[4] For "cross-disciplinary perspectives" on AI, robotics, and "cyborg futures," see Heffernan 2019.

that computers possess some kind of intelligence, Fernando Flores and Terry Winograd note, is based on a misunderstanding of how humans use language and cognize. It rests, as philosopher Ursula Hoffmann explains, on the erroneous assumption "that cognizing creatures collect information on an objectively given reality with objects of distinct attributes and properties in order to construct models or representations, stored to be retrieved and translated into language in thought processes" (Hoffmann 1998, 222, my translation).[5] As a consequence, by the 1990s, the paradigm of the computer and the Representational/Computational Theory of Mind (RCTM; see Kurthen 2007) made way for the parallel distributed processing approach (PDP), also called connectionism or neural networks approach, which replaces concepts of seriality with notions of "simultaneous processing" (Mountcastle 1998, 31), synchronicity, connectivity, and reversibility.

Whereas earlier information-processing and artificial intelligence (AI) approaches trace mental processes as information progressing through a system in a series of stages, PDP frames cognitive processes "in terms of networks that link together neuron-like units" and proceed simultaneously rather than step by step or linearly (Matlin 1983, 20). Embracing the trope of the network, cognitive science adapted a central concept of current cultural discourses – "networks are everywhere" has turned into a truism (see Sielke 2016). This is indicative not so much of an advanced state of knowledge than of the ways in which we mold, visualize, and communicate in knowledge ecologies. Ultimately, the question whether computer or network is the more poignant term remains secondary. More relevant, in his context at least, is, first, that it is us humans who "ascribe to machines mental conditions" (Tetens 1994, 118), and secondly, that as soon as mind means neural network, mind and brain become identical (see Wallach and Wallach 2013, 49–50). As a consequence, the dualism of body and mind seemed to dissolve, yet was in fact reinscribed while (self-)consciousness became an epiphenomenon of neural processes. At the same time, the notion of "mind as computer" remains sustainable in AI and transhumanism.

The major challenge these trends in conceptualizing the human subject have posed to literary and cultural studies is that they privilege the interrelation of brain/mind and culture over the power of cultural representation and discourse.

---

[5] As Vernon B. Mountcastle notes: "Brains and computers differ in many ways, particularly in architecture, in the serial-processing mode in computers versus simultaneous processing in brains, and in the properties of their constituent elements: neurons can take on any one of a series of values over a continuum, transistors in digital circuits only a 0 or 1" (Mountcastle 1998, 29–30).

Interestingly enough, our constructivist approaches – geared towards interrogating the cultural complexities of meaning-making – have aimed at escaping the essentialisms of a reductive biologism; and this marks common ground with the work of AI and transhumanism: O'Connell, for instance, defines the latter, somewhat polemically, as "a liberation movement advocating nothing less than a total emancipation from biology," aligned with "a final and total enslavement to technology" (O'Connell 2017, 6). By comparison, the methods of cognitive science informing AI, such as measuring action potentials in the brain, remain atomistic. Claims that imaging techniques such as the positron emission tomography (PET) enable us "to observe the brain thinking," as neurobiologist Hans-Jochen Heinze put it (Heinze 2005), speak in tropes that are easily misleading. In another example, Joachim Pflüger, likewise an expert in the field, compared the precision of correlations that cognitive science calculates between brain activity and cognitive processes with the view of planet earth taken from a space capsule (Schnabel and Rauner 2013). Still, on accounts of activity in select local areas of the brain, cognition research keeps building big claims: accordingly, its popularized versions, disseminated, for instance, by bestselling authors such as Steven Pinker and Oliver Sacks, and its resonance in visual culture suggest that humans are virtually identical with functioning brains – as do advertisements for various products, ranging from food to cars and washing machines to insurances (see Sielke 2019). With its dematerialized notion of subjectivity and the demise of the human body (and brain), AI carries this faith a little further, as I will demonstrate in the first and main part of my argument, entitled "Outsourcing the Brain." Taking my cue from Zygmunt Bauman's last, posthumously published study *Retrotopia* (Bauman 2017), I consider these current conceptions of the human subject retrotopian in part because popularized versions of cognitive science and AI revitalize mechanistic notions of man that go back to the Renaissance; in part also, because this sense of humans as "brain machines" coincides, somewhat paradoxically, as I lay out briefly in part two, "Optimizing the Body?", with current practices of physical self-optimization.[6] Most significantly, though, these trending concepts of human subjectivity are retrotopian because, beyond reinscribing old binarisms, as I argue in my essay's third and final part, "Retrotopian Projections of the Human Subject," they revitalize and cement socio-economic hierarchies that we deemed overcome some time ago, taking us, as Bauman has it, "back to tribes," "back to inequality" (Bauman 2017, 49, 86).

---

6 See Sielke 2021.

Part one: outsourcing the brain

Part of AI's retrotopian or "back-to-the future" dynamics is due to the fact that the field retains conceptions of humans as "brain machines,"[7] despite its concern with artificial as opposed to natural minds – a binarism that Sybille Krämer convincingly challenges.[8] The dated vision of man as machine dominates transhumanist discourse even more evidently: its most optimistic protagonists foresee what they call "whole brain emulation," that is, the downloading of our brain processes as data and the subsequent production of "substrate independent minds" (O'Connell 2017, 44). Such prophecies certainly amount to little more than SciFi camouflaging as science. O'Connell recalls Arthurs C. Clarke's 1956 novel *The City and the Stars*, a rewrite of the earlier *The Fall of Night* (1948) and "set in a future a billion years from now in which the enclosed city of Diaspar is ruled by a superintelligent Central Computer, which creates bodies for the city's posthuman citizens, and stores their minds in its memory banks […] for purposes of future reincarnation" (O'Connell 2017, 46–47). We also need to recall, though, that despite much inventiveness, creativity, and (some) clairvoyance, SciFi has often been shortsighted: neither Aldous Huxley nor Philip K. Dick, for that matter, were able to foresee the potential of the internet or the smartphone. The Apple commercial announcing the introduction of the Macintosh computer on 22 January 1984 aptly projected the lesson to be learned: "you'll see why 1984 won't be like '1984.'" At the same time we came to see that digital technologies not only insistently outsource part of the capacity, capability, and functions of the human brain and thus turned into "extensions of man" (McLuhan 1964) indeed.[9] They are also marketed as tools assisting the 'optimization' of the human body, an attempted enhancement that is synchronous with its projected supersession, an issue I come back to in the second part of my essay.

In addition, what cognitive science and AI share is a limitation of their insight and applicability, which should not surprise us. Cognitive science still knows very little about how our brain works; and "[w]hat [little] we know," the writer Siri Hustvedt poignantly remarks, "often becomes an excuse to extrap-

---

[7] Marvin Minsky speaks of the brain as "meat machine" (qtd. in O'Connell 2017, 73).
[8] Cf. the essay by Sybille Krämer in this collection.
[9] For an early discussion of how technology extends human bodies, see Tischleder and Winckler. For some time, Fred Adams has critically engaged so-called "extended mind" theories which postulate that cognition could be located beyond the human brain, as in smartphones, for instance; see Adams and Aizawa; Adams.

olate endlessly" (Hustvedt 2010, 192; see also Hustvedt 2013). Likewise, though flexibility is a central feature of human intelligence, the algorithms AI currently operates with remain "very simple and inflexible," as neuroscientist Laurenz Wiskott notes (Wiskott and Glasmachers). In addition, based on available sets of data, they reproduce common assumptions and discriminate against what and who is not considered the norm (as face recognition software, for instance, does). These very limitations legitimize pleas for more research funds for the cognitive sciences and brain-computer interfaces that engineer our lives as cyborgs, as well as for various dubious endeavors. One such controversial project is iBorderCtrl which aims to enable faster and simultaneously more thorough border control for third-country nationals crossing the borders of the EU and which is funded by Horizon 2020 with 4.5 million Euro (see Nezik 2020).

Both the enthusiasm and the anxieties stirred by visions of our enhanced lives and transhumanist futures keep producing remarkable economic revenue. AI even inspires tourism, including a "philosophical journey," in March 2021, to the "knowledge city" Berlin, advertised by ZEIT REISEN as an exploration of "artificial intelligence" as a "future issue" couched between "fascination" and "fear."[10] All the while, though, as information technology expert Key Pousttchi points out, 40% of all products that sell with the label AI do not actually make use of AI.[11] Evidently, AI – in both its utopian and dystopian versions – spells big business for companies as much as for foundations and research institutions that sponsor both the development of marketable technologies and reflect on the ethics of AI. In Germany, the overall ambition is to become a key player in the field of intelligent automation, sensor technology, and robotics, as opposed to capitalizing on the consumer-oriented applications the US and China have pushed. Accordingly, the German government developed a "National Strategy for Artificial Intelligence" in 2018; one year later, the Federation of German Industries (BDI) published its recommendations on how to implement this blueprint. So far it has remained a challenge, though, to even attract experts to German universities; by February 2020, out of 100 positions advertised, only two had been filled (Menne 2020). Meanwhile, debates on the ethics of AI are an ongoing academic enterprise. In a report on "a trailblazing [Harvard] initiative [that] marries ethics and tech," the 19 October 2020 edition of the *Harvard Gazette* speaks of "humanizing" AI and announces new conversations between phi-

---

10 https://zeitreisen.zeit.de/reise/zukunftsthema-kuenstliche-intelligenz/, accessed: 1 Mar. 2021.
11 I make reference here to Pousttchi's contribution to the "Dialog für Strategische Vorausschau und die Zukunft der transatlantischen Beziehungen," 24–25 September 2020, Center for Advanced Security, Strategic and Integration Studies, University of Bonn.

losophy and computer science (Pazzanese 2020). All but new, this dialogue – labeled "Embedded EthiCS" – has apparently mobilized new funds for the purpose of serially repeating an old conversation. And without such funding, I gather, neither the conference this book is based on nor this publication itself would have materialized.

Like the conversation on AI's ethics, the idea that our brains would be better off without their ill-equipped fleshy containers – some transhumanists consider the human body a "dead format," "an obsolete technology" (O'Connell 2017, 145) –, the sense that man can be liberated from his physical materiality is no novel vision. "In the 1490s, [...] inspired by reading of the ancient Greek automata," O'Connell recalls, "Leonardo da Vinci designed and built a robotic knight." Taking things a step further, Descartes's 1630s *Treatise on Man* was "predicated on the idea that our bodies are essentially machines [...] animated by a divine infusion of spirit or soul" (O'Connell 2017, 124). By the early 21st century, transhumanists like Anders Sandberg developed the "vision of getting uploaded, of the conversion of human minds into software," as "central to th[e] ideal of transcending human limitations" (O'Connell 2017, 19). Others, like Dmitry Itskov, even speak of a "transfer of an individual's personality to a more advanced non-biological carrier" (O'Connell 2017, 48). Or as Kurzweil wrote in his book *The Singularity is Near: When Humans Transcend Biology:* "An emulation of the human brain running on an electronic system would run much faster than our biological brains" (Kurzweil 2006, 504n27).

Even though few scientists and scholars share the optimistic take on the prospect of "whole brain emulation," AI has profited from the cognitive sciences, another brainchild of the (Cold War) 1950s. Research on cognition is multifaceted, yet the methods of the neurosciences seem preoccupied with analyses of a brain seemingly separated from both body and mind. Especially during the 1990s, AI and the cognitive sciences shared a considerable and comparable cultural impact, marked, in 1990, by the publication of Kurzweil's *The Age of Intelligent Machines* and by US President George H. W. Bush's Presidential Proclamation on 17 July of that year which donned the 90s the "decade of the brain" (cf. Fahnestock 2005, 159). Meant "[t]o enhance public awareness of the benefits to be derived from brain research," the initiative was propagated by the most common argument that drives a consensus concerning the funding of such work – the cure of physical failure:

> Over the years, our understanding of the brain – how it works, what goes wrong when it is injured or diseased – has increased dramatically. However, we still have much more to learn. The need for continued study of the brain is compelling: millions of Americans are affected each year by disorders of the brain ranging from neurogenetic diseases to de-

generative disorders such as Alzheimer's, as well as stroke, schizophrenia, autism, and impairments of speech, language, and hearing. (Bush)

23 years later, inspired by the Human Genome Project, the Obama administration announced the BRAIN Initiative (Brain Research through Advancing Innovative Neurotechnologies) in April 2013. This public-private research initiative proclaimed having "the potential to do for neuroscience what the Human Genome Project did for genomics by supporting the development and application of innovative technologies that can create a dynamic understanding of brain function." It was marketed as the exploration of a new frontier: "We have a chance to improve the lives of not just millions," President Obama underlines, "but billions of people on this planet through the research that's done in this BRAIN Initiative alone. But it's going to require a serious effort, a sustained effort. And it's going to require us as a country to embody and embrace that spirit of discovery that is what made America, America" ("About the BRAIN Initiative"). Evidently, fused with well-traveled parameters of US-American self-conceptions, the discourse on the promising future of brain science drives an appealingly utopian scenario of improved lives and longevity for all.

The US is certainly not singular when it comes to such aspirations that build on a mélange of cognitive science, AI, and transhumanism. The "Key Performance Indicators and Targets" of the Human Brain Project (HBP), for instance, a billion-dollar "Flagship Initiative" of the European Commission (in its 2014 version) and geared first of all at a replica of the human brain, has proven particularly ambitious: "[R]econstructions and simulations of the brain," so the claim goes, "provide a radically new approach to neuroscience, helping to fill gaps in the experimental data, connecting different levels of biological organisation, and enabling *in silico* experiments impossible in the laboratory" (HBP, 11). The project's agenda to reconstruct the brain *in silico* (i.e., via computational analysis and simulation) seems to echo the transhumanist fantasy of "mind uploading." O'Connell's imaginary account of such a procedure envisions a machine that "scan[s] the chemical structure of your brain" and the "deeper and deeper layers of neurons, building a three-dimensional map of their endlessly complex interrelations, all the while creating code to model this activity in the computer's hardware" (O'Connell 2017, 42). However, deeply invested in "computationalism" and computational neuroscience (see O'Connell 2017, 45, 47, 55), the Human Brain Project can only simulate what we already know about the brain; and that knowledge remains minute, in part due to the self-referentiality of the task. After all, we're both subject and object of the inquiry, using our brain to investigate this very brain (Thornton 2011, 160).

Accordingly, the list of "Core Project Objectives" and goals of the Human Brain Project is as long and large-scale as it is vague: "Simulate the brain," "Develop brain-inspired computing, data analytics and robotics," "Develop Interactive Supercomputing," "Build multi-scale scaffold theory and models for the brain" (HBP, 11). "Theory and models developed in the HBP," so goes the assumption,

> will provide a framework for understanding learning, memory, attention and goal-oriented behaviour, the way function emerges from structure; and the level of biological detail required for mechanistic explanations of these functions. Simplification strategies and computing principles resulting from this work will make it possible to implement specific brain functions, both in neuromorphic and digital computing systems. (HBP, 148)

Evidently, the project conceptualizes the human brain as a supercomputer and thus revitalizes a dated trope – a trope that deprives the brain of individuation and presupposes an identity of material (or energy) and information. As a complex adaptive system, "the brain simply cannot be computed," neuroscientist Miguel Nicolelis instead holds. "It cannot be simulated" (qtd. in O'Connell 2017, 56).

Even though in cognitive science the trope of the brain as computer or Turing machine has been displaced by that of the network, and conceptual register shifts and changes with its respective technological and discursive framework, the computational approach to brain science remains persistent, then, and for clear-enough reasons. The primary goal driving the Human Brain Project and its exploration of the complex dynamism of neural processes is to further develop and enhance computer technologies. That goal has guided AI and transhumanism, too. Little attention, though, as critics of the Human Brain Project note, is being paid to the question how thought and behavior evolve from nerve cell activity (Schnabel und Rauner 2013). In other words, the "hard problem of consciousness," as philosopher David Chalmers called the unresolved question of how brain becomes mind, keeps disappearing from our radar, repeatedly. In fact, by hailing the transformation of brain material into "disembodied minds" or "mind uploading" (O'Connell 2017, 43, 46), transhumanism has done away with the difference between brain and mind altogether, just like cognitive science's sense of brain as mind did, that AI adopted and celebrated.

The effect of this shift is evident all over our visual cultures which in both their high and popular versions are haunted by images of disembodied brains. After all, the mind is hard to visualize. Interestingly enough, these stylized and aestheticized versions of brain tissue seem appropriate to serve multiple economic interests and sales pitches: in 2013, the *Süddeutsche Zeitung*, for instance, designed an ad that presented a sponge, formed like the two cerebral hemispheres, placed on a greenish background and accompanied by the header "Erfrischung gefällig?" ("Refreshment anyone?") and the caption "Seien Sie anspruchsvoll" ("Be demanding"). The newspaper, so the viewer may gather, addresses, first and foremost, the intellectually agile, smart reader. This visual and semantic association of brain and sponge – the latter swallows liquids as brains suck up the world we experience, including what we consider "knowledge" – was already prominent when neurophysiologic insights into the work of the brain were only just developing. In the second stanza of her poem "The Brain – is Wider than the Sky" (c. 1862), the American writer Emily Dickinson draws on that very image.

> The Brain is deeper than the sea –
> For – hold them – Blue to Blue –
> The one the other will absorb –
> As sponges – Buckets – do – (632)

This poem, though, does not so much 'sell' neuroscience than challenge its attempt to weigh and measure brain material, somewhat like the 19th century pseudoscience phrenology did. Fluidities still matter, by the way, when it comes to brainy imagery: a lively mobile and liquid brain starred in the ad for Splash mineral water launched with the header "Replenish your body" in 2011. By comparison, the sense of the human brain as object and machine figures most prominently perhaps in the automobile industry and in the marketing of powerful and pricy cars (see also Sielke 2019).

One of the most striking examples was the "Left/Right-Brain" ad campaign publicized by Mercedes-Benz in 2011 which could easily dispense with a presentation of their actual product. Instead, the poetics of the ad plays as much with consumers' desire and self-image as with gender stereotypes and a rhetoric of renaturalization that, at the same time, courts potential female buyers. Appearing in multiple versions, the ad's basic design consisted of a stylized image of the two hemispheres of the brain, seen from above and clearly distinguished by color spectrum. In one of the ads, a drawing presents a greyish left side, placed on checkered paper and patterned like a chess board from which towers, steely skyscrapers, and abstract architectural structures emerge. By contrast, its colorfully bright right side in red, yellow, and black seeks our attention with flowery shapes, a bull jumping from the brainy lobes, a martini glass splashing its content with vigor, and male and female figures in unmistakably eroticized constellations.¹² Each side comes with a text of modest size, yet distinct typeset, appearing like a poem clearly geared at seducing the viewer/reader into projecting him- or herself as its speaker.

Left brain	Right brain
I am the left brain.	I am the right brain.
I am a scientist. A mathematician.	I am creativity. A free spirit. I am passion.
I love the familiar. I categorize. I am accurate. Linear.	Yearning. Sensuality. I am the sound of roaring laughter.
Analytical. Strategic. I am practical.	I am taste. The feeling of sand beneath bare feet.
Always in control. A master of words and language.	I am movement. Vivid colors.
Realistic. I calculate equations and play with numbers.	I am the urge to paint on an empty canvas.
I am order. I am logic.	I am boundless imagination. Art. Poetry. I sense. I feel.
I know exactly who I am.	I am everything I wanted to be.

Drivers of a Mercedes, this brainy dialogue seems to suggest, "know exactly" who they are; they are "everything [they] wanted to be," manage to cross gender boundaries, and, most importantly perhaps, their brains work exactly the way they should be working – just like the engine of their automobile.¹³ Visuals like these foreground how we have come to identify the human subject with a functional brain that works as a *pars pro toto* for the seemingly irrelevant 'rest' of us. They also acknowledge, however, that what is left of body and mind is the very 'substance' on which the brain depends and thrives.

Reacting to the dominance of neurophysiological approaches in cognitive science, reflected in the ubiquity of brain matter in cultural practice, US-American researchers in 2007 published a "Proposal for a Decade of the Mind Initia-

---

12 See https://www.adsoftheworld.com/media/print/mercedes_left_brain_right_brain_passion, accessed: 1 Mar. 2021.
13 Such identification of sophisticated technology with intelligence and judgment subsequently inspired Mercedes's big-mouthy slogan "Erkennt Gefahren, bevor sie entstehen" ("Detects dangers before they arise"). This campaign, in turn, inspired copywriter Tobias Haase to produce a non-authorized Mercedes video ad that so far has been viewed more than 4 million times (https://www.youtube.com/watch?v=MZGPz4a2mCA, accessed: 1 Mar. 2021). See also Jessen 2013.

tive" in *Science* (Albus et al. 2007).[14] In order to push "Big Science" in consciousness research, this paper appropriates the strategy that drove the "decade of the brain" and counterbalanced the clinical focus of neuroscience with a transdisciplinary "multi-agency" perspective consisting of four interlocked areas: "healing and protecting the mind," "understanding the mind", "enriching the mind," and "modeling the mind" (Albus et al. 2007, 1321). In many ways this approach resembles that of neuroscience, though: once again, computer technologies and their conceptual register create the framework for the envisioned "computational theory of the mind." Likewise, the AI company "DeepMind Technologies," established in 2010 and acquired by Google in 2014, aims at formalizing intelligence, or as their website announces somewhat more simply: "Our goal is to solve intelligence and advance scientific discovery for all." This project goes beyond optimizing machines and, as DeepMind co-founder Demis Hassabis asserted in *Nature*, aims at a better understanding of the conundrum of consciousness. "[A]ttempting to distil intelligence into an algorithmic construct may prove to be the best path to understanding some of the enduring mysteries of our minds" ("Is the Brain"). It may – or may not. The declared path of the company is, however, "[to combine] the best techniques from machine learning and systems neuroscience to build powerful general-purpose learning algorithms" (zit. n. Cadman 2014). What this means for consciousness research is written in the stars, while revenue is clearly on the horizon: "Applications," as one staff member suggested, "could include how to best place advertisements" (Gibney 2015, 466).

Part two: optimizing the body?

Evidently, the futuristic vision of outsourced disembodied brains tends to superimpose and override more holistic approaches to human subjectivity. At the same time, this projection collides with a pervasive phantasm of physical self-optimization. Current digital technologies and new knowledges evolving in the biosciences and medicine seem to enable us to track and enhance our bodies' physiology and shape, driving many into adopting a rigid, time-consuming regime of 'self-perfection,' including physical exercise, specific diets, and possibly enhancing drugs. The two trends – the outsourcing of the brain and the disposal

---

14 In 2004, the German popular science magazine *Gehirn & Geist* had published a "Manifest," composed by a group of neuroscientists who openly admitted how limited our knowledge about complex cognitive processes still is (Elger et al. 2004).

of the human body, on the one hand, and the optimization of that very body, on the other, do indeed interdepend deeply, in part because both are based on an increasing digitization of our social environments and their media ecologies. In addition to self-tracking, Deborah Lupton lists "lifelogging, personal informatics, personal analytics and the quantified self" as terms "used to describe the practices by which people may seek to monitor their everyday lives, bodies and behaviors" (Lupton 2016, 2). Paradoxically, so it seems, this aspiration to self-optimize has its own interdependent counterpart in a global increase of the overweight and obese, whose 'dysfunctional' bodies fail to register as productive and apparently manifest an inability to make "intelligent" consumer choices.

Such collisions may seem surprising or contradictory, yet they are not. A closer look at the so-called "obesity crisis" or "epidemic" – terms that have been critiqued for contributing to the medicalization of overweight and overblowing obesity into a national crisis – highlights the paradoxical plasticity and ingenuity of the neoliberal economic conditions we currently live in.[15] Such scrutiny shows that what holds true for the treatment of our environment goes just as well for attitudes towards our own bodies: once out of balance, both drive an economic cycle that "miraculously" sustains itself. "[A]long with the soils, seas, and air," writes Julie Guthman in her 2011 study *Weighing In: Obesity, Food Justice, and the Limits of Capitalism*, "bodies (both human and animal) are absorbing much of capitalism's excesses" (Guthman 2011, 180), thus being "modif[ied ...] in ways we barely understand" (Guthman 2011, 194). As a result, both self-perfected and "unfit" neoliberal individuals literally accommodate the potentially toxic surplus of our economies in equally excessive, if diametrically opposed ways.

While Guthman and Melanie DuPuis in their work on "economy, culture, and the politics of fat" hint that we have come to literally "embody" neoliberalism,

---

[15] See, for instance, Guthman 2006, 52–56. In the United States, as Farrell delineates, public discourse has not only created close "connections between body size and citizenship." The projection of "a national 'obesity crisis' garners extraordinary attention and resources," suggesting that Americans' body size seems to "put[] the United States at more risk than the failing economy, the ongoing wars, or problems of global warming" (3). In 1997, the World Health Organization had identified obesity as a key health and economic issue and first talked about a global "obesity epidemic" (WHO). Indeed, obesity is by now considered the most persistent driving force of potentially fatal noncommunicable illnesses (e.g., cardiovascular disease, diabetes, cancer) all over the globe (GBD 2015 Obesity Collaborators). As to my use of the concept of "neoliberalism," I follow Reichardt's focus on our current paradoxical sense of the individual as the productive creative force for his or her own self-realization, whose "[i]nvestment (in oneself), risk, and possible failure are closely connected" (Reichardt 2018, 109).

we may want to resist the implications such suggestive, yet naturalizing tropes hold in store. Minimizing the mental costs involved, they also bypass the increasingly "digital nature" of human agents: as we leave track records of our spending, for instance, we produce data and become "produsers" (see Reichardt 2018, 104, 109). In turn, figures of embodiment affirm, if only implicitly, how subjects have become objects of statistical measurement: recent work on modes, devices, and practices of self-optimization highlights that discourses on "self-enhanced" and "ill-fit" bodies both emerge from an increased use of quantitative methods with shifting variables, such as body mass indexes (Guthman 2011, 26–32). As a consequence, self-tracking "represents the apotheosis" of our current "neoliberal entrepreneurial citizen ideal" (Lupton 2016, 68).

Just like the drive to outsource the human brain, the dynamics of self-optimized and enhanced bodies is driven, to a considerable degree, by a relentless marketing of digital technologies. The trend to upgrade, and shift the limits of, one's own body has come to sell so well, in part, because current-day self-optimizers are encouraged to rely on smart technologies that measure and track our physical functions. In turn, these technologies – like all others "extensions of men" that preceded them – put human agency and sociality on the line. Along with a surplus of expert advice, interactive self-help technologies redistribute responsibilities, encouraging us to hand over (or outsource) decision-making processes and 'control' to nutritionists, trainers, coaches, and psychologists, as well as apps and tracking devices (see Lupton 2016, 76–77; Straub 2013). Digital technologies, as Lupton underlines, allow the "monitoring, measuring and recording elements of body and life as a form of self-improvement or self-reflection" (1). So-called "lifeloggers" may indeed trust digital data more than their "own physical sensations" (81). Increasingly dependent on digital devices, users report a loss of feel for their own bodies and experience their 'self-control' to be compromised. Concepts of fitness thus correlate with a shakily ambivalent and newly fragmented sense of 'self' and subjectivity. In turn, as we live and affirm an increasingly distributed agency (see Reichardt 2018, 114), human activities with deeply affective dimensions get mimicked by discourses that favor terms such as code, space, data, and network (see Lupton 2016, 71–72). As a consequence, our new faith in the "datafication of the body" (Reichardt 2018, 105) – "this notion of ourselves as essentially information" (O'Connell 2017, 54) – seems to "giv[e] the physically measurable priority over the mental and emotional and [to] elevat[e] body over mind" (Reichardt 2018, 110). As the brain gets outsourced, its *container* risks running on empty instead.

Part three: retrotopian projections of the human subject

As I hope I have shown, the seemingly paradoxical, yet synchronic dynamics of outsourcing brain power and enhancing bodies operate in and boost a neoliberal – or late-capitalist – economic framework that invests in who or what proves most fit. In that framework humanity supposedly progresses, while in fact social hierarchies get reinforced, inequality keeps surging (Bauman 2017, 91), and face-to-face human interaction is consistently in decline. In that course of things, AI may even be one reaction to – or one way to circumvent – humanity's extinction due to effects of climate change and environmental destruction, which impinge on people all over the world in different ways. In this (media) environment evolved the mold of an individual – or singularity, to use the term Andreas Reckwitz privileges (and recontextualizes) – who is both self-reliant and interactively "extended," a subject who affirms his or her agency by singular (consumption) "choices," including choices about AI applications and whose voice messages are read as a new mode of orality (Cammann 2021) – a retrotopian vision that evokes the 'good ol' times' prior to reproductive media. Or, as O'Connell writes: "All utopian futures are, in one way or another, revisionist readings of a mythical past" (51). As this singular subject takes shape, Robert Putnam's "social capital" gets devalued, while we are increasingly "putting a premium on self-reference, self-concern and an anti-social edge of self-assertion" (Bauman 2017, 99).

Deprived of team-spirit, solidarity, and sociality, we may thus indeed be enacting versions of the very singularities that transhumanism foresees. The privileged human subject is supposedly well equipped to navigate through the retrotopian drift of these current trends and the 'neo-feudalist' socio-economic hierarchies they implant, hierarchies that fiction, such as Margaret Atwood's 2003 novel *Oryx and Crake* or Dave Eggers's *The Circle* of 2013[16], has indeed prophetically anticipated. By contrast, AI as of yet fails to master processes of hierarchical decision-making, "a key component of human intelligence" (Stuart Russell qtd. in O'Connell 2017, 101). Nor can it react to sudden changes of human behavior, as they occur, for instance, under pandemic conditions; some observers therefore prefer to speak of "artificial dumbness" (Fischermann 2020).

All the while, AI transforms the social fabric and the ways we think, live, and work – or do not work any longer. Even now, those engaged in manufacturing and menial work – not to mention those who are stigmatized (e.g., as "fat"

---

16 Cf. the essay by Carmen Birkle in this collection.

and "unfit") and marked as dysfunctional (e.g., in competitive work environments) – feel its impact to a substantially higher degree than those whose intelligence, physical constitution, and social interactions are merely assisted by AI. Moreover, the envisioned "trickle-down economics of intelligence" has so far remained a retrotopian mind-space, "an abjectly false hope," as neurobiologist Michael Hendricks notes, "that is beyond the promise of technology" (qtd. in O'Connell 2017, 25). In other words, AI is not out there to fundamentally reform, for instance, the extremely profitable and thus extremely costly, yet highly insufficient health system in the US; it has made it even less cost-efficient. Plus: the increased reliance on digital communication during the current pandemic foregrounded that we all are – at least partially – replaceable (and even among the privileged some more so than others). In the aforementioned report on the "Embedded EthiCS" program we read that in recent years, "computer science has become the second most popular concentration at Harvard College, after economics" (Pazzanese 2020). Who then really believes that the future belongs to the creative, the flexible, socially embedded teamworking intelligence (who are not also IT specialists)? Instead, AI transfigures how humans interact, realigns subject positions into well-traveled binary schemes, and takes all of us "back to standardization," too. As we work in an increasingly digital learning environment and interact with a series of squares on a screen, simulating a 21st-century version of Andy Warhol's serial silk screens (which themselves seem reincarnated by the visuals that brain imaging technology produces),[17] it seems to me we're already there.

At the same time, AI, cognitive science, transhumanism, as well as posthumanism have newly raised and replied to the question what in fact distinguishes humans from intelligent machines. In *Do Androids Dream of Electric Sheep?* (1968), Philip K. Dick suggested that "[e]mpathy, evidently, exist[s] only within the human community, whereas intelligence to some degree [can] be found throughout every phylum and order including the arachnida" (Dick 1968, 30–31). The rising scholarly interest in empathy and intersubjectivity may thus, at least in part, be an effect of science fiction and its idealistic sense of what it means to be human.[18] Along those lines, the arts too, have envisioned counter-measures: in her production *We Are in Time*, the British playwright and director Pamela Carter, for instance, interrogates health issues as an intersubjective affair: donating a heart may enable a dying person to live longer

---

[17] The continuity between early visual representations of the brain and current imaging techniques, both in their aesthetics and in their attempt to discern functions and create a typology, has been discussed for some time; see, for instance, Clarke und Dewhurst; Stafford.
[18] Cf. the essay by Johanna Pitetti-Heil in this collection.

and make death interlink with celebrations of life. Not I live on eternally as a singular outsourced brain; another fully human body and mind does. No surprise therefore that the conversation has meanwhile turned to retrotopian modes of artificial empathy and digital intersubjectivity (see De Vos 2020); but that is another story.

Works cited and consulted

"About the BRAIN Initiative." https://obamawhitehouse.archives.gov/BRAIN. Accessed: 24 Feb. 2021.
Adams, Fred. "The Bounds of Consciousness and Cognition: The Philosophical Debate." Conference "Litterature et sciences," International Comparative Literature Association, Paris, 18–24 July 2013. Conference presentation.
Adams, Fred, and Ken Aizawa. *The Bounds of Cognition*. Oxford: Blackwell, 2008.
Albus, James S., et al. "A Proposal for a Decade of the Mind Initiative." *Science* 317.5843 (2007): 1321.
Bauman, Zygmunt. *Retrotopia*. Cambridge: Polity, 2017.
BBCI (Berlin Brain-Computer Interface). "About BBCI." http://www.bbci.de/about?language=en. 2014. Accessed 24 Feb. 2021.
Bostrom, Nick. *Superintelligence: Paths, Dangers, Strategies*. Oxford: Oxford UP, 2014.
Bush, George H. W. "Presidential Proclamation 6158." Project of the Decade of the Brain. https://www.loc.gov/loc/brain/proclaim.html. Accessed: 24 Feb. 2021.
Cadman, E. "AI Is Not Just a Game for DeepMind's Demis Hassabis." *Financial Times*, 27 Jan. 2014.
Cammann, Alexander. "Die neue Macht der Stimme." *Die ZEIT* 6 (2021). https://www.zeit.de/2021/06/muendlichkeit-podcast-clubhouse-whatsapp-literatur-medien-kommunikation. Accessed: 1 Mar. 2021.
Carter, Pamela. *We Are in Time*. Music by Valgeir Sigurðsson. Scottish Ensemble, 2020.
Chalmers, David J. "Facing Up to the Problem of Consciousness." *Journal of Consciousness Studies* 2.3 (1995): 200–19.
Clarke, Edwin, and Kenneth Dewhurst. *An Illustrated History of Brain Function: Imaging the Brain from Antiquity to the Present*. 1972. San Francisco: Norman, 1996.
De Vos, Jan. *The Digitalisation of (Inter)Subjectivity: A Psy-critique of the Digital Death Drive*. London: Routledge, 2020.
Dick, Philip K. *Do Androids Dream of Electric Sheep?* 1968. New York: Ballantine, 1996.
Dickinson, Emily. *The Poems of Emily Dickinson*. Ed. Thomas H. Johnson. 3 vols. Cambridge, MA: Harvard UP, 1979.
Elger, Christian E., et al. "Das Manifest: Elf führende Neurowissenschaftler über Gegenwart und Zukunft der Hirnforschung." *Gehirn & Geist* 6 (2004): 30–37.
Fahnestock, Jeanne. "Rhetoric in the Age of Cognitive Science." *The Viability of the Rhetorical Tradition*. Ed. Richard Graff. Albany: State U of New York P, 2005. 159–79.
Farrell, Amy Erdman. *Fat Shame: Stigma and the Fat Body in American Culture*. New York: NYU P, 2011.
Fischermann, Thomas. "Künstliche Dummheit." *Die ZEIT* 25 (2020): 25.

The GBD 2015 Obesity Collaborators. "Health Effects of Overweight and Obesity in 195 Countries over 25 Years." *New England Journal of Medicine* 6 July 2017. https://www.nejm.org/doi/full/10.1056/NEJMoa1614362. Accessed: 24 Feb. 2021.

Gibney, Elizabeth "DeepMind Algorithms Beats People at Classic Video Games." *Nature* 518.7540 (2015): 465–66.

Guthman, Julie, and Melanie DuPuis. "Embodying Neoliberalism: Economy, Culture, and the Politics of Fat." *Environment and Planning D: Society and Space* 24.3 (2006): 427–48.

Haraway, Donna. "A Manifesto for Cyborgs: Science, Technology, and Socialist Feminism in the 1980s." *Socialist Review* 15.2 (1985): 65–107.

HBP (Human Brain Project) (2014). "Framework Partnership Agreement." 30 Oct. 2015. https://sos-ch-dk-2.exo.io/public-website-production/filer_public/0d/95/0d95ec21-276a-478d-a2a9-d0c5922fb83a/fpa_annex_1_part_b.pdf. Accessed: 1 Mar. 2021.

Heffernan, Teresa, ed. *Cyborg Futures: Cross-Disciplinary Perspectives on Artificial Intelligence and Robotics*. Basingstoke: Palgrave Macmillan, 2019.

Heinze, Hans-Jochen. "Wir schauen dem Hirn beim Denken zu." *Humboldt Kosmos* 86 (2005): 28–33.

Hoffmann, Ursula. "Autopoiesis als verkörpertes Wissen: Eine Alternative zum Repräsentationskonzept." *Der Mensch in der Perspektive der Kognitionswissenschaften*. Ed. Peter Gold and Andreas K. Engel. Frankfurt: Suhrkamp, 1998, 195–225.

Hustvedt, Siri. "Philosophy Matters in Brain Matters." *Seizure* 22.3 (2013): 169–73.

Hustvedt, Siri. *The Shaking Woman*. New York: Holt, 2010.

"Is the Brain a Good Model for Machine Intelligence?" *Nature* 482 (2012): 462–63.

Jessen, Jens. "Für Hitler wird nicht gebremst." *Die ZEIT* 36 (2013). https://www.zeit.de/2013/36/werbung-video-mercedes-benz-erkennt-gefahr-hitler, Accessed: 1 Mar. 2021.

Kurthen, M. "From Mind to Action: The Return of the Body in Cognitive Science." *The Body as Interface: Dialogues between the Disciplines*. Ed. Sabine Sielke and Elisabeth Schäfer-Wünsche. Heidelberg: Winter, 2007. 129–43.

Kurzweil, Ray. *The Age of Spiritual Machines*. Cambridge, MA: MIT P, 1990.

Kurzweil, Ray. *The Singularity is Near: When Humans Transcend Biology*. New York: Penguin, 2006.

LeDoux, Joseph E. *Synaptic Self: How Our Brains Become Who We Are*. New York: Viking, 2002.

Lupton, Deborah. *The Quantified Self: A Sociology of Self-Tracking*. London: Polity P, 2016.

Matlin, Margaret W. *Cognition*. New York: Holt, Rinehart, and Winston, 1983.

McLuhan, Marshall. *Understanding Media: The Extensions of Man*. New York: McGraw-Hill, 1964.

Menne, Katharina. "Gesucht: 100 Superhirne." *Die ZEIT* 9 (2020): 31.

Mountcastle, Vernon B. "Brain Science at the Century's Ebb." *Daedalus* 127.2 (1998): 1–36.

Nezik, Ann-Kathrin. "Falsch geblinzelt: iBorderCtrl." *Die ZEIT* 36 (2020): 22.

Nilsson, Nils. *The Quest for Artificial Intelligence: A History of Ideas and Achievements*. Cambridge: Cambridge UP, 2010.

O'Connell, Mark. *To Be a Machine: Adventures among Cyborgs, Utopians, Hackers, and the Futurists Solving the Modest Problem of Death*. London: Granta, 2017.

Pazzanese, Christina. "Trailblazing Initiative Marries Ethics, Tech." *Harvard Gazette*, 16 Oct. 2020. https://news.harvard.edu/gazette/story/2020/10/experts-consider-the-ethical-implications-of-new-technology, Accessed: 1 Mar. 2021.

Pinker, Steven. *How the Mind Works*. New York: Norton, 1997.

Putnam, Robert D. *Bowling Alone: The Collapse and Revival of American Community*. New York: Simon & Schuster, 2000.

Reckwitz, Andreas. *Die Gesellschaft der Singularitäten*. Berlin: Suhrkamp, 2017.

Reichardt, Ulfried. "Self-Observation in the Digital Age: The Quantified Self, Neoliberalism, and the Paradoxes of Contemporary Individualism." *Amerikastudien/American Studies* 63.1 (2018): 99–114.

Sacks, Oliver. *The Man Who Mistook His Wife for a Hat*. London: Duckworth, 1985.

Scheerer, Eckart. "Toward a History of Cognitive Science." *International Social Science Journal* 40.1 (1985): 7–19.

Schnabel, Ulrich, and Max Rauner. "Ein Hauch Apollo." *Die ZEIT* 6 (2013). https://www.zeit.de/2013/06/Flagschiff-Initiative-Forschung-Graphen-Human-Brain-Project, Accessed: 1 Mar. 2021.

Sielke, Sabine. "Bodies In and Out of Shape: The Rise of Obesity (Discourses) in Times of Self-Optimization." *Amerikastudien/American Studies*. 66.3 (2021): 511–33.

Sielke, Sabine. "Der Mensch als 'Gehirnmaschine': Kognitionswissenschaft, visuelle Kultur, Subjektkonzepte." *Die Maschine: Freund oder Feind? – Mensch und Technologie im digitalen Zeitalter*. Ed. Caja Thimm and Thomas Christian Bächle. Berlin: Springer VS, 2019. 41–65.

Sielke, Sabine. "Network and Seriality: Conceptualizing (Their) Connection." *Amerikastudien/American Studies* 60.1 (2016): 81–95.

Sielke, Sabine, and Elisabeth Schäfer-Wünsche. *The Body as Interface. Dialogues between the Disciplines*. Heidelberg: Winter, 2007.

Stafford, Barbara Maria. *Body Criticism: Imaging the Unseen in Enlightenment Art and Medicine*. Cambridge, MA: MIT P, 1991.

Straub, Jürgen. "Selbstoptimierung im Zeichen der 'Auteronomie': Paradoxe Strukturen der normierten Selbststeigerung: von der 'therapeutischen Kultur' zur 'Optimierungskultur.'" *Psychotherapie & Sozialwissenschaft* 15.2 (2013): 5–38.

Tetens, Holm. *Geist, Gehirn, Maschine: Philosophische Versuche über ihren Zusammenhang*. Stuttgart: Reclam, 1994.

Tischleder, Bärbel, and Hartmut Winckler. "Portable Media. Beobachtungen zu Handys und Körpern im öffentlichen Raum." *Ästhetik & Kommunikation* 112.32 (2001): 97–104.

Thornton, Davi Johnson. *Brain Culture: Neuroscience and Popular Culture*. New Brunswick: Rutgers UP, 2011.

Wallach, Lise, and Michael A. Wallach. *Seven Views of Mind*. New York: Psychology P, 2013.

World Health Organization (WHO). *Obesity: Preventing and Managing the Global Epidemic*. WHO Technical Report Series 894. Geneva: World Health Organization, 2000. https://www.who.int/nutrition/publications/obesity/WHO_TRS_894/en/. Accessed: 24 Feb. 2021.

Winograd, Terry, and Fernando Flores. *Understanding Computers and Cognition: A New Foundation for Design*. Norwood: Ablex, 1986.

Wiskott, Laurenz, and Tobias Glasmachers. "Wie intelligent KI wirklich ist." *ZEIT Campus*, n.d. https://www.zeit.de/campus/angebote/forschungskosmos/regionen/nrw/ruhr-uni versitaet-bochum/wie-intelligent-ki-wirklich-ist. Accessed: 1 Mar. 2021.

Cornelia Klinger
Life Care/Lebenssorge and the Fourth Industrial Revolution

Abstract: My point of departure lies with the recent advances of ICT and biotechnology/ life sciences into *Converging Technologies* (CT) of microelectronics and microbiology. CT encompass Nano- Bio- Info- Technologies plus Cognitive Science (NBIC). The most interesting effect from a philosophical perspective is the dissolution of dualisms that were deeply engrained in the Western symbolic order: human:animal, organic:mechanic; form:matter, sign (σῆμα): body (σῶμα). On the societal and political levels these groundbreaking developments are mirrored in the dissolution of the separation of spheres along the lines of work:life; production:reproduction; public:private. The second part of this contribution focuses on the effects of these developments on the relations among genders and generations as well as on the various kinds of care activities (for the young, the old, the sick, the handicapped and for all of us in everyday life). On the one hand, we observe liberating and encouraging results on all sides of love-and-care-relationships; on the other hand, there are new and unprecedented threats connected to "global tech giants [...] sowing the seeds of an economy predicated on 'biopower'" (Margarethe Vestager, 2019).

From the late 1970s on, questions of Care have come under discussion. After the focus on Care *Ethics* through the mid-1990s, it is the commodification, marketization and corporatization of Care *Work* that have attracted attention since the 2000s.[1] Statements describing an ongoing "Care Crisis" develop into pleas for a "Care Revolution" and into visions of a "Caring Democracy". Debates on Care address the economic and socio-political aspects under headers such as the decline of the welfare state, the impact of globalization, the rise of neo-liberalism or the persistence of late capitalism; yet they rarely consider developments in science and technology (S&T).[2] Conversely, there are heated debates

---

[1] Note that it is only recently that the most primeval abilities and activities of humans are highlighted or, in other words, only most recently consciousness is raised and raising that CARE MATTERS. Since I mean to stress this newly won attention to matters of Care, I will apply the capital letter to the term.
[2] To name just one of the few laudable exceptions: Boris / Parrenas 2010.

about breathtaking progress in these fields (digitization, Artificial Intelligence (AI), big data, the Internet of Things (IoT), cloud, quantum[3] etc.) that are deemed to amount to a new industrial revolution (4.0); yet in this context the many issues concerning Care are hardly visible. It is only in the context of a few advanced deliberations on technological innovations, e. g. 'Society 5.0' in Japan or Kontratjev-wave 5 to 6, that Care issues come to the fore. Most recently, a planetary health crisis has added awareness.

This chapter sits between the two sides: in contrast to current discussions on Care, I will argue that S&T are the driving forces behind the Care revolution that we are witnessing. Differing from the common debates on science and technology (S&T), I will maintain that it is not the production (plus the servicing) of goods, not even the Internet of Things, but the production of life and the proliferation of caring-for-life-services and facilities that are the cutting edge of the current phase of industrialization. In other words, I will look at *Lebenssorge* in the light of this latest stage in the series of industrial revolutions.

Introduction

This is a new look since *Lebenssorge* was relegated into the fogyish background of *re*-production with the onset of the first wave of industrial revolutions around 1800. At this turning point in history the manufacturing of objects and vehicles as well as (more hesitantly) the bringing forth of agricultural goods[4] becomes 'productive' under the impact of S&T. Industrialisation is the decisive moment in the development of the modern *capitalism-cum-nation-state-system,* not least because a sharp line is drawn between the spheres of public activity and the 'rest', demarcating that which can be produced and circulated, managed and administrated within the new, immensely industrious and busy political

---

[3] Cf. https://qt.eu/: "The Second Quantum Revolution is unfolding now, exploiting the enormous advancements in our ability to detect and manipulate single quantum objects. The Quantum Flagship is driving this revolution in Europe."
[4] With respect to plants and animals that traditionally were cultivated and raised to nourish other animals or humans, I will try to avoid the notion of production or produce. European languages pay heed to the difference: the victuals, les vivres, die Lebensmittel are 'pro-created' (Er-Zeugung, Er-Zeugnis), whereas non-living objects are manufactured by the homo faber. It is only under the impact of modern industrialism that the cultivation and construction of 'goods', labor and ποίησις is subsumed under the heading of production. Now even all kinds of paid (personal) services are regarded as 'products,' so that, for example, the travel agent labels my vacation a 'product' on the bill.

and economic system, from all that which cannot (*yet*[5]) be fabricated in the shining armour of these newly rising apparatuses of combative (inter-/trans-) national(ist) politics and competitive (global) political economy. This twin-headed regime sets about to concoct a new kind of hitherto unfathomable wealth and power.

In the first wave of industrial revolutions (around 1800) the modern world is severed from the past and starts heading into the future. In this process *life* is relegated to the backwaters of *re-*: it comes to be perceived as

> re-source,
> re-pair,
> re-covery
> re-serve / re-servation,
> re-sort,
> re-store / re-storation
> summarized under the heading of *re*-production.[6]

As far as human life is concerned, the sphere of re-production encompasses two dimensions: roughly speaking, 'high' culture and 'low' nature. Notwithstanding former distinctions between the two poles of human life, denoting *le cuit et le cru*[7], now both subsumed under the *re-* as opposed to the forward moving *pro-* of -duction (guidance, maintenance) and -tection (safety, security). The 'gifts of nature' may be indispensable, badly needed and highly valued as 'treasures' (in particular precious metals and other 'rare earth'); the 'cultural heritage' may be cherished as a more or less precious heirloom. Yet, with the beginning of S&T-driven industrialism the bond is broken, with both nature and culture appearing as matters of the past: natural stuff is reduced to the status of 'raw' material to furnish 'fossil' energy; the cultural legacy is deprived of its former patriarchal authority in ancestry and tradition. Thus, both culture and nature can be taken for granted, 'for free', a donation priceless, '*hors prix*', *unbezahlbar* – be it below, be it above the price-system, but in any case beyond the one and only currency which invariably governs the *capitalism-cum-nation-state-regime:* the *cash-nexus*.

---

5 Note the 'yet'!
6 Re- is a difficult prefix. On the one hand, it states an opposition that leaves re- in the unfavourable situation to signify the negation, the non- of production, the un-productive. On the other hand, yet at the same time, the prefix fixes the re- and subsumes reproduction under the dominant productivity.
7 Cf. Lévy-Strauss 1964.

Within western society this stark distinction between rich progressive production and vital but retro-grade and hence, 'poor', unpaid old re-production prompts a division into 'separate spheres': the modern public sphere of nation-state politics in conjunction with the capitalist economy is undergirded and backed up by a secluded, occluded area, the private sphere, devoted to cultural conservation in addition to natural re-production, both mixed in re-generation and re-creation, epitomized in the bourgeois home that, in turn, needs to be supported, supplied[8] and supervised from outside.

At about the same time, when the public powers start segregating their own nature and culture in more or less intentionally and artificially alienated privacy, western industrial nations commence colonizing foreign nature and culture outside; they attempt to bring the 'rest' of the planet under their control, intending:
- to 'secure', i.e. to extract natural resources as fodder to fuel and fire up their voracious technological devices,
- to 'employ', i.e. to exploit cheap, unpaid or even forced labour[9] and
- either to extinguish or to 'save', i.e. to exhibit the heritage of foreign ancient cultures in modern museums, i.e. to expropriate the cultural goods of others (no matter if dead or alive).

Consider the fivefold EX- that befalls the *re-*: extraction – exploitation – extinction – exhibition – expropriation specify the double-edged inside-outside-'othering', the exclusion of vital physical, mental and intellectual resources that the industrial system depends on but can neither produce nor intends to pay for (in accordance to its inherent monetary tariffs). However ingeniously devised this export-import business proves unviable in the long run.
- On the one hand, it implies a division of labour that prompts acrimonious class, gender and ethnic conflicts that develop revolutionary potential over the course of the nineteenth and twentieth centuries.
- On the other hand, the process of modernization unremittingly expands into those 'left-behind fields'; it starts 'grabbing' the 'land' that could not be acquired before. In due course, not only natural materials come to be replaced by a large variety of synthetics, but the entire sphere of reproduction is integrated into the process of S&T-driven industrialization.

---

8 Not least by new commodities imported from the colonies: cotton and oriental rugs, exotic plants, coffee, sugar and tea.
9 Notwithstanding the 'official' ban on slavery that served the needs of agrarian societies but has proved dysfunctional under industrialism, while it still thrives in informal work and personal services.

A crucial turning point in both developments is reached with the second industrial revolution around 1900.[10]

Irrespective of how many more milestones have been set on the way, it is widely acknowledged that a landmark of revolutionary dimensions is reached in the 1970s/80s. After the slump following the post-war boom another explosion of productivity gains momentum around 2000. This marks the apex of the entire process of modernization that started in the late eighteenth century. One, in my view the one most important, distinguishing trait of this latest and presumably final stage of S&T-industrialism is the loss of the *re-:*
- the artificially established ridges, ledges of an excluded and secluded inner outside in privacy are 'bulldozed' as relentlessly, and within the same time frame, as
- the remotest regions of the globe enter the world system when colonialism turns into post-colonialism.

The becoming-productive of re-production causes yet another explosion of the "forces of productivity". And once again, other than Marx and his contemporaries had hoped or feared, this enormous upsurge does not revolutionize the "relations of production"[11]. The dual governance of state and capital, the *capitalism-cum-nation-state-regime*, is not overturned; rather, it has transmuted into late capitalism and neo-liberalism[12] on a global scale since the 1980s. Currently, that is after the crisis of finance capitalism (dubbed FinTech) in the years following 2007, this system seems to evolve, or rather, to regress into a renewed authoritarianism.[13]

Ever since these thoughts on the topic of the re-production of life and the position of *Lebenssorge* in the fourth industrial revolution crossed my mind, they have given me a hard time. On the one hand, I am not well versed in the subject of Care Ethics and hardly interested in Care Work; on the other hand, I am still less knowledgeable in the vast and rapidly expanding arenas of S&T. De-

---

10 See below "I Around 1900".
11 Cf. previous failed attempts at social/societal/socialist revolutions (1848, 1918).
12 Both additions, late- as well as neo- indicate that there is essentially nothing completely new, no disruption in nation-statism and capitalism, notwithstanding some striking transformations.
13 Something like a 'law of history' may be detected in the development of the capitalism-cum-nation-state-system. Though they are coeval and remain bound up with each other, they are rivals. In periods of S&T upsurges, the capitalist economy is at the helm, in phases of downturn, nation states re-take the reins (invariably accompanied by more or less nasty nationalist ideologemes played on the populist accordion). Efforts by the one or the other side to get rid of the 'twin' have been undertaken in the past and, so far, are of no avail.

spite my lack of knowledge and other limitations – which I concede in advance – I will focus on the recent innovations in
- information and communication technology (ICT), and
- life sciences (LS) morphing into bio-technology (bionics, synthetic biology).

As this gives me pause, this is perhaps a good moment to take a pause in an attempt at the impossible: to spell out what life is – in ordinary language, in a few mundane words, before I begin to consider what is happening to life and how *Lebenssorge* is altering under the impact of S&T.

That's life!

Life is ... no! Not 'making' – but *giving* bodies as well as *giving* signs. Neither bodies nor signs have primacy. Both come together and they remain inextricably bound up with each other: The bodies are and give signs and, *vice versa,* the signs are and have bodies. Since ancient times, the wise Greek language has articulated their intimacy: sign and body differ by one letter only: σῆμα / σῶμα. The *semantic* and the *somatic* are what life is (about), what constitutes life. The contiguity and continuity[14] of σῆμα and σῶμα is *mediated* by breath. Air is the go-between for body and sign. And since life takes place as movement in space and time, signs and bodies are always on the move. Living beings are passagers/passagères, passengers in processes of *trans*-: in transition, transportation, transposition, translation, transmission, transformation, transmutation, transgression, transferral, or: deferral, going forth and back again. Breath is drawn in and spun out, eked out, extended and prolonged through the air, and the breeze becomes fluid by means of a liquid (saliva). At the same time, human beings remain earthbound, in need of shelter at a hearth, warmth within wombs and walls. Life is torn between mobility and stability: albeit constantly on the move, it has to settle at some place in space and it must take time to last and to stay on. In brief, signs and bodies employ earth and fire, wind (air) and water, the four elements of life – note that all of them are ambiguous, namely of '*dual use*': helpful or harmful, depending on the circumstances, the quantities and qualities.

Not 'making' but 'giving' ... the verb 'give' implies 'take' and furthermore, it hints at the 'knotty' situation that living beings receive life and obtain language

---

14 Both terms refer to contingency.

as *gifts*[15], which they have to acquire[16], i.e. the use of which they have to learn and that they may (or may not) enrich over their own lifetime and pass on to others (or not). As a reminder: no man is self-made by herself – and yet nobody is forced to accept the gifts of other lives either. Living beings cannot help receiving the gift *of* life at its beginning, when they are conditioned by and depending on the bodies and languages of their forebears. Yet, over the course of time in space the fixed conditions of the beginnings turn conditional: living beings are agile and active, fragile and volatile all the way along – variable due to fate and their own decisions. At the end, signs are silenced and bodies pass away into silence: σῆμα and σῶμα vanish and perish inexorably. *Natality* and *mortality* are the conditions of beginning and ending, the two firm columns of contingency. What is worse, the bounds of finitude are not only adamantly set at the borders; the limits can make themselves felt at every instant within life as incidental/accidental mis-takes in the communication of signs, as *morbidity, infirmity and debility* in and between bodies.

All along the way of life, the fragmentation and division, the particularity and plurality of *σῆμα* and *σῶμα* are free to proliferate and procreate in abundance and exuberance – yet this proves to be no less but even more problematic than the limits. The one particular sign or body is never whole but a-part, a particle, a fragment in a chain, fragile in itself, and vulnerable by others. The particularity of the un-whole one corresponds to the plurality of the many: the multitude is never all. Given the schemes of time and space, neither the complete wholeness of One nor the unity of All can ever happen. Living beings are neither

---

15 Gifts are not 'for free', but belong to an economy that is based on exchange. While the 'subjects' of the commerce are human beings, the 'objects' may vary and the crucial question is who or what the objects of the exchange are. The answer to this question is based on discerning different forms of the economy and, ultimately, the limits of the economic system in general. Actually, the only correct response to what is tradable and what is not, has been known since time immemorial and is still valid today, namely that living beings may possess and trade objects, but they should abstain from trading (or killing) other subjects, human beings. Countless taboos, clauses and codices have been invented to regulate dealings among human beings. It is obvious that no such laws are to be found in or can be deduced from 'mother nature', who is 'red in tooth and claw'. Such regulations and distinctions are a matter of socio-political organization, good governance. Even if war and slavery were common, and it has so far been impossible to exclude them from human society, there were at least commandments and limits to such practices. Warmongering regimes kept the vicious circle turning, since men believed and were pressed to believe that their life, well-being and freedom depended on the death, misery and slavery of others, so that they were made to kill others in order not to be killed themselves. However, the stigma attached to killing and slavery remains and applies to all affairs involving buying and selling life.
16 Cf. Goethe, *Faust I*, 29.

fully determined nor absolutely free; many lucky incidents and fateful accidents may occur at random on the way across time and space for better or worse. In sum, together, σῆμα (sign) and σῶμα (body) are *doing life* within the extremely narrow, yet surprisingly flexible confines of finitude, the conditions as well as the conditional of contingency, the hazards and the haphazard of life.

Thus humanity is torn between not enough and too much; this is the basic ambiguity of life behind the paradox of the quadruple contingency of being. In the inconsistent and incomplete sum, life (*Da-Sein*) may be or may not be, oscillating between the weakness of beginning and ending with manifold pains and perils as well as countless options, occasions and opportunities in between – as good as it gets, as long as it lasts.[17]

This basic characterization of life is not a definition of *human* life. In contrast to the western philosophical tradition, it does not delineate or demarcate the human from the non-human, it does not follow the rule *omnis determinatio est negatio*. Generating signs and engendering bodies as a gift of one (as it is sufficient for the sign) or (more intricate) of two (as required for the body)[18] to third parties, is what humans and (more or less all) other living beings have in common. However, they do not, cannot, share this capacity: neither the body-regimes nor the sign-systems that living beings use are universal, unanimous (of one mind) or in one voice (unison/*unisono*), but particular and plural.

It goes without saying that *Lebenssorge* has to do and to deal with *all* aspects of life, with the semantic as well as with the somatic and the manifold ways of their interactive verbal conversation and non-verbal intercourse[19] – with life's flows and hold-ups. Although their relation is significant for each and every human activity, the union of σῆμα and σῶμα is more substantial, absolutely indispensable in all endeavours of *Lebenssorge*.[20] *Lebenssorge* embraces the entire gamut of pleasures and pains, joy and sorrow, hope and grief, the fun and the fear of life; it has to provide *Unterhalt* and *Unterhaltung*; it accompanies serious undertakings and pleasant entertainments. As the *omnis determinatio* rule does

---

17 Fährnisse is a beautiful old word closer to the moving and driving over time in space, related to Ge-fahr as well as to Gefährt (vehicle) and Fährten (tracks, traces) but also to Gefährten (companions, company). The root of these words is 'fahren' ('Fortbewegung = E 'fare'. 'How are you faring?')
18 Could this be the primal cause of the divergences between σῶμα and σῆμα that crop up in the further course of their history?
19 "There is room for the idea that significant relating and communicating is silent" (Winnicott 1964, 183 – thanks to psychoanalyst Birgit Pechmann for this hint).
20 This is the reason why the culture-nature divide that cuts across the σῆμα & σῶμα link in the process of western civilisation is particularly harmful to all issues of Lebenssorge.

not apply to life, *Lebenssorge* is not to be defined either. This notion does not operate on opposition or exclusion but encompasses the poles. Hence, Care is not opposed to, but includes Care*less*ness.

Under the conditions of contingency, when there is (never) enough life at both ends and (always) too much particularity and plurality in between, *Lebenssorge*-relations share the ambiguity, the contingency of life. *Lebenssorge* is notoriously oscillating, vacillating betwixt nearness (in countless hues, finally verging on *love*) and distance (also in innumerable shades, ultimately bordering on *hate*), a wide panoply; it is only the extremes of complete equality or total difference (though not exactly ruled out) that are very rare exceptions. Last but not least, it should not go unmentioned that *Lebenssorge* implicates highly intricate power asymmetries among the Carers and the Cared-For.

Lebenssorge is what it takes to bring life to term: to socialize, to raise and educate, to enhance and embellish life, to *give (spend) time* and *take place* as lavishly as possible. *Lebenssorge* is certainly concerned with the means of life: *Lebensmittel* (the victuals, livestock, *les vivres*) and instruments (the indispensable vehicles and objects in endless variety). Moreover and much more, *Lebenssorge* has to do and to deal with the aims and ends of life. *Sorge*/Care makes life's ends meet in *respice finem*. If all goes well, *Lebenssorge* attends and accompanies living beings unwaveringly from the very first cry to the very last sigh; in the meantime it has to provide the strengths, the triumphs and feasts of life as much as it has to cope with all kinds of snafus and glitches. *Lebenssorge* has *to make sense* in the ups and downs of a roller-coaster ride: That's Life!

The industrialization of re-production in two long waves

As mentioned above, the production:reproduction divide was established during the period of the first industrial revolution; thereafter it has been challenged (by totalitarian regimes) and modified (under liberalism). Yet in principle, the great divide was upheld throughout the later decades of the twentieth century. Ernest Mandel takes the moment of its implosion to define late capitalism. In his book published under this title in 1972 he writes:

"[...] late capitalism [...] constitutes generalized universal industrialization for the first time in history. Mechanization, standardization, over-specialization and parcellization of labour, which in the past determined only the realm of commodity production in actual industry, now penetrate into all sectors of social life

[...]. *The industrialization of the sphere of reproduction constitutes the apex of this development* [...]."[21]

Mandel assesses the industrialization of the sphere of re-production as the apex of the long-term development of industrialism[22]. To illustrate the dazzling and daunting effects of the industrialization of re-production, Mandel points to "[t]he 'profitability' of universities, music academies and museums" that starts "to be calculated in the same way as that of brick works or screw industries [...]." Indeed, the transformation of the establishments of arts and sciences (including the educational institutions devoted to teaching and learning in these fields) into business companies and corporations, is *one* important component in this process – one much commented on by academics and artists, researchers and searchers, teachers and students in recent years, praised by many and lamented by as many others.

Though Mandel's judgement is pertinent, his perspective is partial. The "industrialization of the sphere of re-production" not only relates to the re-production of *signs* in more or less 'high culture' (letters, formulae, pictures, sounds – fusing since around 1900) but also encompasses the re-production of bodies, the care for 'low mortal nature'. "Mechanization, standardization, over-specialization and division of labour" affect the $σῶμα$-side as well. And it is at that point that *Lebenssorge* is reduced to the status of Care Work. In other words, the process of industrialization reaches out to encompass the gamut of *Lebenssorge* in its entirety. In 1972 Mandel might have foreseen[23] but did not address this aspect of late capitalism: the 'land-grabbing' of life to include the $σῆμα$

---

21 Mandel 1978, 387; italics mine. Im deutschen Original "... der Spätkapitalismus [bildet] erstmals in der Geschichte eine Gesellschaft verallgemeinerter universeller Industrialisierung. Mechanisierung, Standardisierung, Überspezialisierung und Parzellierung der Arbeit, welche in der Vergangenheit nur den Warenproduktionsbereich der eigentlichen Industrie bestimmten, dringen nun in alle Bereiche des gesellschaftlichen Lebens ein. Für den Spätkapitalismus ist bezeichnend, daß ... die Landwirtschaft genauso industrialisiert wird wie die Industrie, die Zirkulationssphäre ebenso wie die Produktionssphäre, der Konsum ebenso wie die Erzeugung, die Freizeitgestaltung ebenso wie die Arbeitsorganisation. Die Industrialisierung der Reproduktionssphäre bildet den Gipfel der Entwicklung ... Man beginnt die 'Rentabilität' von Hochschulen, Musikakademien und Museen genauso zu berechnen wie man vorher die Rentabilität von Ziegelwerken oder Schraubenproduktion kalkulierte" (Mandel 1972, 353).
22 Cf. Mandel 1995.
23 Actually, Mandel mentions the industrialization of agriculture, distribution and consumption, all in the same breath. He even alludes to the industrialization of "Freizeitgestaltung", that is the systemic organization of leisure time conceived as recreation of productivity. Yet he omits Lebenssorge or Care Work. In a different context an analogous lack can be observed in the case of Karl Polanyi's fictitious commodities (cf. Aulenbacher 2020).

plus σῶμα-dimensions; he wrote his book on the very eve of the era when this step was taken – once again – in *both* areas.

I around 1900

As a brief reminder, the developments that began in the late 1970s and have continued under the regime of late-capitalist-economy-cum-neoliberal-politics since the mid-1980s are not altogether new. They did take a mighty leap around the turn of the twentieth century[24] in connection with the second industrial revolution and can be traced further back to the turn of the nineteenth, when the megamachine paradigm was set up. In contrast to the first industrial revolution that centred on the production and distribution of objects as commodities and on developing an appropriate infrastructure through steam power and motorization, the second phase of industrialization conquered the domains of σῆμα and σῶμα more fully.

- On the σῆμα-side unheard of, i.e. disruptive information & communication technologies (ICT), brought about a virtual media revolution in writing as well as programming, designing, transmitting and recording[25]. Not only words, but at the same time images and, even more ground-breaking, sounds went on air, moving on radio waves through time and space, and thus signs got closer to the bodies in live/lively experience – while a disruptive innovation in real transport occurred at about the same time with the realization of the long-standing dream of air travel.
- On the σῶμα-side entirely new avenues opened up for influencing and manipulating body and soul, mind and spirit. Under the lead of S&T, chemistry, pharmaceuticals and drugs made their appearance. There were significant advances in surgery, and moreover, talking cures and psychotherapies start-

---

24 It is no accident that German sociologist Werner Sombart dated late capitalism from the time around 1900 and French historian Fernand Braudel commented: "[...] capitalisme, dans son usage large, date du début ... du XXe siècle. J'en verrais le lancement véritable, avec un peu d'arbitraire, dans la parution, en 1902, du livre bien connu de Werner Sombart, *Der moderne Kapitalismus*. Ce mot, pratiquement, Marx l'aura ignoré" (Braudel 2018, 47). Braudel affirms that capitalism "est, dans la longue perspective de l'histoire, le visiteur du soir" (Braudel 2018, 71).
25 "... the set of communication and entertainment devices invented within the remarkably short period between 1885 and 1900, including the telephone, the phonograph, popular photography, radio, and motion pictures" (Gordon 2012, 9). Current media researchers date the development of the "global room of communication" to the years around 1900. The World War of Words starts even earlier, namely with the invention of cable transmission and telegraphy (cf. Tworek 2019).

ed to exert growing influence. The cutting edge of all this is that innovations opened up not only new ways to heal and remedy maladies and ailments of all kinds, but also visions to foster and further, to improve and uplift humankind that were the craze of the day in the early twentieth century.

Despite the paths of both σῆμα and σῶμα knowledge having followed the tracks of S&T since the industrial revolutions and proceeded in parallel, pursuing the aim of *making*, i.e. *producing* instead of *engendering and generating*, they fared quite differently during the first half of the twentieth century. In contrast to the progress in the production of signs, analogous endeavours concerning the making of bodies were thwarted. Even if the obsession to create a new humanity[26] may have been idealistic and benevolent at the outset, this path has led nowhere; or worse, it led eventually to the disaster of racism and sexism culminating under Fascism and Nazism (national socialism/socialist nationalism). On the level of the second industrial revolution the de-socializing efforts to undo family bonds and to produce the human 'workforce' in the industrial mode terminated in disaster. While S&T were thriving in the sphere of σῆμα, their highest aspirations took a dystopian turn in the sphere of σῶμα.

After the war the ICT-culture industries continued to flourish in virtual and real mobility wherever the capitalist economy prevailed[27]. In contrast, a kind of moratorium was imposed on tinkering with the body as a consequence of the cataclysm of the mid-century. Nation-state biopolitics had to back-pedal on the production of human beings, which had failed miserably under pre-war totalitarianism. Hence, under the governance of 'Fordism' a compromise has been reached between industrialism and a more or less post-patriarchal familialism[28] in order to ensure the production of offspring in more customary, ostensibly 'natural' ways. That is to say, the public:private, i.e. the production:reproduction boundaries are maintained in post-war liberalism.[29] More or less scrupulously, a ban on experiments with human 'stuff' is upheld, a kind of 'truce'[30] that is kept through the later decades of that awesome century.[31]

---

26 Lepp / Roth / Vogel 1999. This obsession persists through the present; cf. Eder / Imorde / Reinerth 2013.
27 Whereas the traffic in bodies and signs was severely impeded and censored under totalitarian regimes after as well as before World War II.
28 As in colonialism, the post- in patriarchy is not past domination.
29 After the war the surviving totalitarian regimes (in Eastern Europe, East Asia) also stopped their efforts to destroy traditional family structures.
30 Not least as a result of the Declaration of Human Rights by the UN in 1948. https://www.un.org/en/about-us/universal-declaration-of-human-rights.

II around 2000

After their partial encounter around the 1900s, the paths of σῆμα and σῶμα cross again at the next corner, at the turn of our twenty-first century.
- The recent innovations in the ICT are in line with earlier accomplishments such as the telegraph and telephone, photography and 'phonography', radio, cinema and television.
- The findings of the LS carry on the innovations in chemicals, plastics, and pharmaceuticals that also were launched way back then.

Both branches, ICT as well as LS are deemed to be taking a quantum leap[32] and approaching the quantum level of engineering on the petabyte/pebibyte scale – if not today, then sometime soon in the foreseeable future – or on the Utopian day 'when the cows come home': S&T are closely linked to SF. In other words, S&T take out loans on the future, but this is no news at all. The well-trodden path of progress is taken over and over again; another 'New Age' is heralded and hailed by its promoters:

Sixty years ago, digital computers made information readable[33]. Twenty years ago, the Internet made it reachable[34]. Ten years ago, the first search engine crawlers made it a single database[35]. Now Google and like-minded companies are sifting through the most measured age in history, treating this massive corpus as a laboratory of the human condition[36]. They are the children of the Petabyte Age.[37]

---

31 Meanwhile this abstention from eugenics has been given up. In a similar fashion, for some decades the further use of atomic weapons was regarded as unthinkable, but then it became 'conceivable' through advanced technology in the form of tactical nuclear weapons (TNW) or non-strategic nuclear weaponry. Now eugenics is coyly evolving into genetics and genomics. The discourses on giving/birthing and taking/killing of life are as closely related to each other as they are contrary.
32 I.e. a disruptive innovation.
33 This hints at the σῆμα-side.
34 This touches the σῶμα-side.
35 This indicates their convergence. Is seems that a tendency to converge is 'inherent' or rather, it is installed into all new technologies, e.g. the recent generations of mobile phones.
36 Foucault 1978, 143; http://www.freudians.org/wp-content/uploads/2014/09/The-History-of-Sexuality-1-Michel-Foucault-The-History-of-Sexuality-Volume-1_-An-Introduction-Pantheon-Books-1978.pdf.
37 Anderson 2008. Even more impetuously, Google chief economist Hal Varian enumerates: "A billion hours ago, modern Homo sapiens emerged. A billion minutes ago, Christianity began. A billion seconds ago, the IBM PC was released. A billion Google searches ago [...] was this morn-

The implosion of the great dual divides

The single most disruptive feature in the continual transformative encroachments of ICT and LS is that they are leaving the parallel and are fusing in the micro- or nano-sphere. If I may trust the experts (as I must), the newly converging σῆμα-σῶμα technologies CT are composed of:

Nano-
Bio-
Info-
Cognitive-

What for?

The NBIC[38] [...] converge to develop devices[39] that enhance or 'augment' biological human nature [...]. They pose a challenge by questioning the sharp distinctions between humans and machines [...]. NBIC research creates interfaces between these categories that blur the human/non-human distinction.[40]

These short sentences belie their far-reaching significance. The Converging Technologies are not about a few interfaces[41], meeting points, gateways in order to bridge the organic and the mechanical, to connect the human and the non-human, presuming both parties to remain on solid ground on either side.

In point of fact, the recent advances in S&T not only make the ice at the poles melt but also bring about the folding of polarizations that seemed carved-in-stone, either since time immemorial, or instigated during previous stages of modernization in order to establish the orbit of the modern world. Remember, "der Weltinnenraum des Kapitals"[42] was constituted by the crucial divide between what could be produced within the cash nexus and the 'what not', the 'rest of the west' that was made to look like old culture or primitive nature when both were made to serve as resources.

---

ing" (https://people.ischool.berkeley.edu/~hal/Papers/2013/BeyondBigDataPaperFINAL.pdf). An obsession with acceleration is endemic to all varieties of the march of progress.

38 NBIC = Nano-, Bio-, Info- technologies plus cognitive science.

39 The use of "devices" is inaccurate, as these are a new generation of apparatuses that are built by their own 'forebears' seemingly without human interference but for still more of the same profit for a few happy people.

40 Knorr-Cetina 2005, 78. Ostensibly, this 'blurring' corresponds to the non-definition of life given above, whereas the philosophical tradition insisted on the clear-cut definition of the human.

41 Cf. Galloway 2012.

42 Cf. Sloterdijk 2005.

Karin Knorr-Cetina's definition of NBIC touches two bifurcations:

human : non-human
organism : mechanism

The list of their kind and kindred is longer, and all of them relate to σῆμα – σῶμα:

Transcendence : immanence
Form : matter
Spirit / Soul : body
Immortal : mortal
One / Self : other / mother
Substance : accidens / accessory
Culture : nature[43]
Nurture : nature
Civilization : wilderness
Universal : particular
Singular : plural
History : (his- or her-)stories
Real : fictitious
Original : copy
Truth : fake
Figure : (back-)ground
Nude : naked
Reason : emotion
Domestic : foreign[44]
Master : slave
Rich : poor
Center : periphery
Urban : rural
World : earth
Global : planetary
Masculinity : femininity
Gender : sex
Public : private

---

[43] It goes without saying but should be mentioned in passing that the borders of modern academic disciplines that were drawn along the nature-culture line are becoming obsolete too. This does not imply that nature-culture knowledge and its institutions will meet on equal terms. Even if it were true that the data deluge makes the scientific method obsolete (cf. Chris Anderson, quoted above), scholarship dealing with the 'natural' will prevail, whereas the social sciences (not to mention the 'humanities') will continue to be reduced to the status of 'studies'.
[44] Formerly known as Christian: heathen, pagan.

Work : life
Production : reproduction[45]

In any event, it is obvious that the latest high-tech achievements are the driving forces behind the dissolution of traditional dualisms. At this point, Jean Baudrillard's statement becomes germane: "The virtual in general is neither real nor unreal, neither immanent nor transcendent, neither interior internal nor exterior / external; it blurs all such distinctions."[46]

Baudrillard's observation remains in the abstract dimension of the philosophical. Michael Hardt and Antonio Negri take a similar stance and are getting closer to the consequences of the blurring of contemporary power relations in general, and to the production:reproduction divide in particular:

> "What the theories of power of modernity were forced to consider transcendent, that is, external to productive and social relations, is [...] formed inside, immanent to the productive and social relations."[47] "There is nothing, no 'naked life', no external standpoint, that can be posed outside this field permeated by money [...]. Production and reproduction are dressed in monetary clothing. In fact, on the global stage, every biopolitical figure appears dressed in monetary garb."[48]

---

[45] It is not simply negligence or a fault of mine that this list is neither well-ordered nor complete. Rather, it is due to the tenets of dualism-building which I cannot pursue in detail. I only mention in passing: From the point of view of philosophy the most interesting result is the meltdown of different types of dualisms (A : B, tertium datur: C,D,E) and (A : Non-A without a third way) that were so deeply engrained in the tradition of western thought.

[46] This quotation is taken from the German translation: "Das Virtuelle im allgemeinen ist weder real noch irreal, weder immanent noch transzendent, weder innen noch außen; es verwischt alle diese Bestimmungen" (Baudrillard 1999, 252–264 (cited here: 261), translation C.K.). Baudrillard's notion of the "virtual in general" designates the collusion of the virtual and the viral. The blurring of binary distinctions, the leaky borders between real and unreal, the one and the other, in(ward) and out(ward) may result in a situation when "wir weder aus noch ein wissen". The neither here/in nor there/out mirrors the advances in time-space compression that corresponds to this stage of media development.

[47] Hardt / Negri 2000, 33.

[48] Hardt / Negri 2000, 32. Twenty years later Hardt and Negri re-examined the Empire and expanded their observation on the collapse of polarizing orders "Although the conventional schemas previously used to grasp global divisions – First and Third Worlds, centre and periphery, East and West, North and South – have lost much of their explanatory power ..." (Hardt / Negri 2019. 72).

After the implosion of dual divides: viralling-spiralling into the future

If the 1970s mark "the beginning of something different in world history,"[49] it is because culture industries and bio-industries not only (re)gain momentum at the same time but because they merge into microelectronics and microbiology under the rule of the digit. Or, is it more appropriate to conclude that σῆμα and σῶμα switch places? The non-expert, the lay philosopher tries to put it in a nutshell:
- Assuming that it is correct to configure the σῶμα side as genetic 'code', it must first be decrypted; subsequently the 'book of life' can be read and edited like any text in the semantic vein; moreover, it is envisioned as a film to be cut by methods such as CRISPR/Cas9, whose operators imagine such procedures metaphorically as the use of 'scissors', which 'come in handy' like an everyday utensil.[50]
- While the somatic seems to become virtualized in the age of the digital code, the semantic starts materializing, e.g., when the 3D printer uses "the FFF technique, in which plastic filament, available in spools, is melted and extruded, and then solidifies to form the object"[51]

In other words: the virtual goes viral and vice versa the viral becomes virtual. This dual use of viral/virality with respect to σῆμα and σῶμα is correctly rendered in recent dictionary definitions:
1. (Pathology) of, relating to, or caused by a virus.
2. (Communications & Information) (of a video, image, story, etc.) spread quickly and widely among internet users via social networking sites, e-mail, etc.
3. (Communications & Information) go viral (of a video, image, story, etc.) to spread quickly and widely [...].[52]

---

49 LaFeber 1999, 13.
50 the τέχνη-technology fallacy.
51 From an online advertisement. A further example: "3-D printers offer an alluring promise. All you need is a little plastic and some technical know-how, and you can print a physical copy of almost anything." At the end of the day, you may print a copy of yourself.
52 *Collins English Dictionary* cf. https://www.thefreedictionary.com/Virality. Probably the most important aspect under the auspices of the current form of late capitalism is "viral marketing", "designed to disseminate information (as about a new product) very rapidly by making it likely to be passed from person to person especially via electronic means" (https://www.merriam-webster.com/dictionary/viral%20marketing).
Although Susan Sontag has convincingly proven that diseases cannot be taken as metaphors for

Eric Schmitt is an expert who may be presumed to know more about such developments than the dictionary: "We are convinced that web portals like Google, Facebook, Amazon and Apple are way more mighty than most people think. Their power results from the capacity to grow exponentially. Aside from biological viruses there is nothing to expand at such high speed, so effectively and aggressively as these technology platforms. This is what bestows their creators/inventors, owners/investors, and users with new power."[53]

To sum up: definitions (1) + (2) in "the free dictionary" add pathology and technology, so that they go viral together = (3). As a result
- human communication is infected by worms and viruses, while shit storms rage in the information systems, whereas,
- societal, political and all other kinds of institutions reflect on the 'health' of their 'DNA' – thus getting the wires crossed between natural organisms and cultural organizations,
- the apex or the nadir comes into view when human society relapses or rather, reverts to a flock, as 'herd immunity' is supposed to be the only safeguard able to protect humanity from man-made or natural catastrophe.

Conclusion

The prospect of σῆμα and σῶμα (re-)uniting in the dimension of the virtual and viral can hardly be overestimated. The production of new signs and what is more, the reproduction of old signs, the recording of the entire 'cultural heritage' (the σῆμα dimension) as well as the 'harvest' of seeds (the σῶμα dimension), is being stored in archives and saved in banks. As a result we are facing a full-blown culture industry[54] as well as an equally promising health-care-life-indus-

---

the condition of an individual (cf. Sontag, Susan), it still suggests itself to connect certain imperial diseases to mirror the status of a society over a period in history. In this vein, it is tempting to speculate that western societies are moving into the era of indomitable virus diseases that necessitate further ventures into the virtual world of AI and big data.

53 Eric Schmidt quoted in Maier 2014: "Wir sind überzeugt, dass Portale wie Google, Facebook, Amazon und Apple weitaus mächtiger sind, als die meisten Menschen ahnen. Ihre Macht beruht auf der Fähigkeit, exponentiell zu wachsen. Mit Ausnahme von biologischen Viren gibt es nichts, was sich mit derartiger Geschwindigkeit, Effizienz und Aggressivität ausbreitet wie diese Technologieplattformen, und dies verleiht auch ihren Machern, Eigentümern und Nutzern neue Macht."

54 The term has completely lost the negative connotations that marked its use in postwar Critical Theory.

try. And both sectors seem to reach an important, maybe even the decisive, conclusive next level simultaneously:
- The new field of genomics and genetic medicine, invading the nature of cells and exploring their functioning, will tap into a giant data treasury – *the book of life* – supposed to be the most valuable data base of all.[55]
- The next 'big thing' the media corporations [Google, Amazon Facebook, Apple, Microsoft] are approaching is deep comprehension of speech and the creation of meaningful language. Will it be possible to bring language processing over the edge and into the apparatuses via *Edge Computing*? Projects like *Apple Siri, Google Bert, Amazon Alexa* and *Microsoft* investments in *Open AI* and its language algorithms, *IBM Watson Debater* point in a similar direction. [...] The paramount position of language for our understanding of the world and for communication in the most diverse governance systems make language the quintessential core technology. Whoever can control and automatically produce [...] meaningful language, can completely dominate governance systems – be it democracy or stock markets or sciences.[56]

To perceive σῶμα as decipherable and legible like a text in addition to gauging σῆμα as the pivotal technology means that sign and body fuse under the reign of the digit in the era of the code. But the one-letter gap between σῆμα and σῶμα in the wise Greek language is not overcome in harmony. The algorithm distorts the body as much as it disfigures the sign and disparages both of them. Since the inception of modern mathematics and science the figure zero has played a commanding role in the discovery of the macrocosm as well as the conquest of the microcosm. Today, both branches of technology pursue similar paths into cells and atoms, into the invisibly, intangibly small (*nano*-technology), and while hu-

---

55 "Das neue Feld der Genomik und der genetischen Medizin, das Vordringen immer tiefer in die Natur der Zellen und deren Funktionieren, wird einen gewaltigen Datenschatz erschließen – das Buch des Lebens -, der von einigen als der wertvollste Datensatz überhaupt angesehen wird" (Nemitz / Pfeffer 2020, 68, translation, italics CK).
56 "Gemeinsam ist allen GAFAM-Konzernen auch die Arbeit am nächsten großen Ding, nämlich dem sinnhaften Sprachverstehen und der sinnhaften Sprachschöpfung. Wird es gelingen, die gesamte Sprachverarbeitung an die Kante, also in die Geräte mittels Edge Computing, zu verlegen? Apple Siri, Google Bert, Amazon Alexa und Microsofts Investitionen in Open AI, und seine Sprachalgorithmen, IBM Watson Debater – all die Projekte weisen in eine ähnliche Richtung. Die Zentralität der Sprache für unser Weltverständnis und für die Kommunikation in den verschiedensten Governance-Systemen macht sie zur Kerntechnologie schlechthin. Wer sinnhafte Sprache kontrolliert und automatisch erschaffen ... kann, der kann Governance-Systeme vollständig beherrschen" (Nemitz / Pfeffer 2020, op.cit. 69, translation CK).

mans lose their senses, sense and sensibility, the NBIC bring forth the gigantic[57]: big data, globalization and the proliferation of enormous heaps of bubble money[58] in the finance sector combined with the even more enormous amounts of bubble signs in adverts, propaganda and infotainment, Under the label of globalization the exploration, exploitation and stratification of the exterior spaces, i.e. post-colonisation to the point of extinction of wild life, is widely discussed. While there seems to be no more *'outside'* space in the exterior world, the invasion of the interior spaces of body, soul and mind as well as of the innermost nooks and crannies of intimate social relations, the caves and crevices of intimacy and privacy, which is taking place simultaneously, may prove even more powerful and specifically awesome; there seems to be no more 'inner' space either. Reality has become a flat screen in the relentless process of quantifying all qualities. It becomes clear that Carlyle's assessment of the Mechanical Age in 1829 has been fulfilled today. The age of the digital code is:

> [...] the Age of Machinery, in every outward and inward sense of the word ... Not the external and physical alone is now managed by machinery, but the internal and spiritual also [...] [This] indicate[s] a mighty change in our whole manner of existence. For the same habit regulates not our modes of action alone, but our modes of thought and feeling. Men are grown mechanical in head and heart, as well as in hand.[59]

---

57 Martin Heidegger applies his notion of "das Riesige" (gigantic) to the macro- and microsphere (cf. Heidegger 1977. On this level, Heidegger contends that the calculable turns into the incalculable ("das zu Berechnende [wird] zum Unberechenbaren"; cf. Kittler 2002, 269). Supported by AI machines incessantly march on into much more "Big Data". Does this disprove Heidegger's hunch? As an aside, the size of the gizmos shares the 'fate' of the gigantic: it is part of the logic of the gigantic that the machines are getting ever smaller on the laps or plugged into the ears of the end-user. Meanwhile businesses strive to grow into the gigantic and achieve hegemonic and, ultimately, monopolistic positions.

58 Alongside the language of weapons, money is one of the two most important, universal sign systems humanity ever invented. In the recent evolution of the money market into a technology-driven semi-autonomous system, money seems to lose the last remnants of a material reference and becomes completely virtualized, computerized: the first attempts to introduce cryptocurrencies met with the resistance of nation states, who tried to defend their right to mint, print and govern the monetary system as ardently as their monopoly on the use of force. However, it is clear that the apparatuses of nation states are about to be defeated on both fronts: different types of competing currencies are on the rise, while states undermine democratic control by hiring private business armies.

59 Carlyle 1829. Hatte der "Animismus ... die Sache beseelt", so "versachlicht der Industrialismus die Seelen" (Horkheimer / Adorno 1997, 45).

If both the inside and the outside are collapsing into the heights and abysses of the unified flat screen system of the "Whole Earth"[60], does this entail that there is nothing, that no life is left outside? When no poor person, no child, no bat, no living being whatsoever, is left behind in the lurch, the ultimate station is reached. In George Orwell's novel *Nineteen Eighty-Four (1948)* Big Brother affirmed: "Outside man there is nothing."[61] This dystopian perfection of humanism was the unique, singular doctrine of the totalitarian system in a tripartite world, divided into *Oceania, Eurasia and Eastasia* and in a permanent state of war with one another.

Works cited and consulted

Anderson, Chris. *The End of Theory: The Data Deluge makes the Scientific Method Obsolete.* 2008.
https://www.wired.com/2008/06/pb-theory/ Accessed: 4 August, 2021.
Aulenbacher, Brigitte. "Auf neuer Stufe vergesellschaftet: Care und soziale Reproduktion im Gegenwartskapitalismus." In: Becker, Karina / Binner, Kristina / Décieux, Fabienne (Hg.): *Gespannte Arbeits- und Geschlechterverhältnisse im Marktkapitalismus.* Wiesbaden: Springer VS, 2020.
Baudrillard, Jean. "Videowelt und fraktales Subjekt." In: *Aisthesis. Wahrnehmung heute oder Perspektiven einer anderen Ästhetik.* Hg.v. Karlheinz Barck. Leipzig: Reclam, 1990. 252–64.
Boris, Eileen / Parrenas, Rhacel Salazar (eds.). *Intimate Labors: Cultures, Technologies, and the Politics of Care.* Stanford UP, 2010, Part I: Remaking the Intimate: Technology and Globalization.
Braudel, Fernand. *La dynamique du capitalisme* [1985]. Paris: Flammarion Champs histoire, 2018.
Carlyle, Thomas. "Signs of the Times." In: *Edinburgh Review,* No. 98, 1829.
Collins English Dictionary – Complete and Unabridged, 12th Edition 2014 © Harper Collins Publishers.
Diederichsen, Diedrich / Franke, Anselm (eds.). *The Whole Earth: California and the Disappearance of the Outside.* Katalog Haus der Kulturen der Welt Berlin. Berlin: Sternberg Press, 2013.
Eder, Jens / Imorde, Joseph / Reinerth Maike (Hg.). *Medialität und Menschenbild. Herausgegeben im Auftrag des Forschungsschwerpunkts Medienkonvergenz der Johannes Gutenberg-Universität Mainz (JGU),* Bd. 4, Berlin/Boston: de Gruyter, 2013.
Foucault, Michel. *The History of Sexuality.* Vol. 1, The Will to Knowledge. New York: Pantheon, 1978.
http://www.freudians.org/wp-content/uploads/2014/09/The-History-of-Sexuality-1-Mi-

---

60 This is the tenet of the California ideology. Cf. Diederichsen / Franke 2013.
61 Orwell 2000, 240.

chel-Foucault-The-History-of-Sexuality-Volume-1_-An-Introduction-Pantheon--Books-1978.pdf Accessed: 08.12.2021

Galloway, Alexander. *The Interface Effect.* London: Polity, 2012.

Goethe, J.W. von. *Faust I, Werke* – Hamburger Ausgabe Bd. 3, Dramatische Dichtungen I, 11. Aufl. München: dtv 1982.

Gordon, Robert. *Is U.S. Economic Growth Over? Faltering Innovation Confronts the Six Headwinds.* NBER Working Paper No. 18315 August 2012 JEL No. D24,E2,E66,J11,J15,O3,O31,O4,Q43.

Hardt, Michael / Negri, Antonio. *Empire.* Cambridge/ London: Harvard UP, 2000.

Hardt, Michael / Negri, Antonio. Empire, Twenty Years on. In: *New Left Review 120/ 2019.* 67–92.

Heidegger, Martin. "Die Zeit des Weltbildes (1938)." In: *Gesamtausgabe I. Abt. Veröffentlichte Schriften 1914–1979.* Bd. 5: Holzwege. Frankfurt: Klostermann 1977. S. 75-97.

Horkheimer, Max / Adorno, Theodor W. Dialektik der Aufklärung (1946). In: *Th.W. Adorno: Gesammelte Schriften.* Hg.v. Rolf Tiedemann. Bd. 3. Frankfurt: Suhrkamp, 1997.

Kittler, Friedrich. *Short Cuts.* (= Short Cuts 6). Ed. Peter Gente / Martin Weinmann. Frankfurt: zweitausendeins, 2002.

Knorr-Cetina, Karin. "The Rise of a Culture of Life." In: *EMBO reports*, Vol. 6, Special Issue: Science & Society 2005. 76–80. (© 2005 EUROPEAN MOLECULAR BIOLOGY ORGANIZATION) https://www.embopress.org/doi/full/10.1038/sj.embor.7400437 Accessed: 4 August, 2021.

LaFeber, Walter. *Michael Jordan and the New Global Capitalism.* New York/ London: W.W. Norton, 1999.

Lepp, Nicola / Roth, Martin / Vogel, Klaus (Hg.). *Der Neue Mensch – Obsessionen des 20. Jahrhunderts. Katalog zur Ausstellung im Deutschen Hygiene-Museum Dresden,* 22.4.-8.8.1999. Cantz, 1999.

Lévy-Strauss, Claude. *Mythologiques, tome I: Le Cru et le Cuit.* Paris: Plon, 1964.

Maier, Robert M. "Angst vor Google." In: *Frankfurter Allgemeine Zeitung.* 3. April 2014.

Mandel, Ernest. *Late Capitalism.* London: Verso, 1978. *Der Spätkapitalismus.* Frankfurt: Suhrkamp, 1972.

Mandel, Ernest. *Long Waves of Capitalist Development: A Marxist Interpretation.* London: Verso, 1995.

Merriam Webster https://www.merriam-webster.com/dictionary Accessed: 4 August, 2021.

Quantum Flagship https://qt.eu/ Accessed: 4 August, 2021.

https://www.faz.net/aktuell/feuilleton/debatten/weltmacht-google-ist-gefahr-fuer-die-gesellschaft-12877120-p5.html Accessed: 4 August, 2021.

Nemitz, Paul / Pfeffer, Matthias. *Prinzip Mensch. Macht, Freiheit und Demokratie im Zeitalter der Künstlichen Intelligenz.* Bonn: Dietz, 2020.

Orwell, George. *Nineteen Eighty-Four.* Edited by Ronald Carter/ Valerie Durow. London: Penguin Student Editions, 2000.

Sloterdijk, Peter. *Im Weltinnenraum des Kapitals. Für eine philosophische Theorie der Globalisierung.* Frankfurt: Suhrkamp, 2005.

Sontag, Susan. *Illness as Metaphor and Aids and Its Metaphors.* New York: Picador, 1990.

Tworek, Heidi. *News from Germany. The Competition to Control World Communications,* 1900–1945. Cambridge: Harvard UP, 2019.

UN Declaration of Human Rights [1948].
 https://www.un.org/en/about-us/universal-declaration-of-human-rights Accessed: 4 August, 2021.

Varian, Hal. *Beyond Big Data*.
 https://people.ischool.berkeley.edu/~hal/Papers/2013/BeyondBigDataPaperFINAL.pdf Accessed: 4 August, 2021.

Winnicott, D.W. *The Maturational Processes and the Facilitating Environment*. London: Hogarth Press, 1965.

Part 2: Examining Merits and Limits of Applied AI

Darren Abramson
AI's Winograd Moment; or: How Should We Teach Machines Common Sense? Guidance from Cognitive Science

Abstract: In this paper, I provide a number of state-of-the-art results for Winograd-related problems by applying a public technique for zero-shot scoring with language models. I explain how I discovered this advance and my recent experience in sharing these results with the scientific community. We are on the threshold of major advances and opportunities, but also dangerous trends. A strengthened discourse on the public benefit derived from the use of computers is badly needed and starting to emerge; here, I focus on that discourse in the context of apparent sudden, giant leaps in our ability to process text with computer: natural language processing (NLP). There is no shortage of lessons for science and ethics. Fortunately, despite the vulnerable status of public University research, the prognosis for informed public debate about the sciences of machine learning and artificial intelligence remains excellent.

1 Introduction

1.1 On citing reviewers from venues that have rejected you

According to someone on the Internet[1], NAACL is the world's third best computational linguistics conference. In between the date of the Austrian Academy's conference and the date that I prepared these written remarks, I managed to produce an original, interesting, and new zero-shot result on the data sets of interest that I spoke about at the conference: Winograd schemas. Because of the dates involved, I can report back on a complete submission and rejection cycle that occurred at a time when language models and academic publishing on computational linguistics issues were making world headlines. In this paper, I build on the nucleus of that paper, and in cases apply the advice of my anonymous reviewers for NAACL 2021. The professional academic philosophy blog Daily

---

[1] https://medium.com/@robert.munro/the-top-10-nlp-conferences-f91eed97e950. All links in this paper were accessed between Nov. 1, 2020 and March 31, 2021

https://doi.org/10.1515/9783110770216-008

Nous[2] recently offered a discussion of the appropriateness of, and best approach to, citing referees from rejected journal submissions. I apply some of those insights to the review process for a scientific conference. I try to be obvious when I am commenting on that process or referring to reviewer comments and quote fairly and accurately from review process documents.

1.2 Recent progress on common sense

Machine learning produced a solution to Seymour Papert's 'Summer Vision Project'[3] in 2012 (Krizhevsky et al.), thus putting machine image recognition comparable to human performance into consumer technology. What would it mean for machine learning to achieve human level performance at natural language tasks? This paper argues that we are still a little earlier than claimed by Ruder ("NLP's Imagenet moment," 2018), but for reasons I explain, it might be hard to know this if you read natural language processing (NLP) results from major conferences. Vienna native Yehoshua Bar-Hillel, who studied philosophical logic and set theory with giants Rudolf Carnap and Abraham Fraenkel, respectively, argued in his provocatively titled "A Demonstration of the Nonfeasibility of Fully Automatic High Quality Translation" (Bar-Hillel 1960) that the disambiguation of individual words in simple sentences seems to require an unlimited knowledge of the human world. This observation has been echoed by influential philosophers (Jackman 2017), psychologists (Miller 1999) and computer scientists (Levesque et al. 2012). In my discipline this is sometimes called "semantic/meaning holism".

A Winograd schema is a problem of ambiguous reference, which has the following structure: a semantic change of a single term in the sentence causes a shift in the reference of the pronoun (Levesque et al. 2012). Here is an example that I have just made up:
1. *Ambiguous sentence:* The plates were being spun by a person wearing red shoes. They were white.
2. *Correct substitution:* The plates were being spun by a person wearing red shoes. The plates were white.
3. *Incorrect substitution:* The plates were being spun by a person wearing red shoes. The shoes were white.

---

2 https://dailynous.com/2020/05/07/citing-referees-journal-rejected/
3 https://dspace.mit.edu/handle/1721.1/6125

But when we change the predicate, we can change the reference – or, so I claim, for the vast majority of English speakers. The reader must make their own judgment.
1. *Ambiguous sentence:* The plates were being spun by a person wearing red shoes. They were well-worn.
2. *Correct substitution:* The plates were being spun by a person wearing red shoes. The shoes were well-worn.
3. Incorrect substitution: The plates were being spun by a person wearing red shoes. The plates were well-worn.

It is important to acknowledge that my example is imperfect. There is no error in thinking that the plates were well-worn. After all, maybe the performer has been spinning those plates for so long that indentations have been worn into them, making their stability while spinning even better. In any case, I am assuming that the more likely choice by an English speaker for things being well-worn, even in this case, is shoes. It is important not to take this assumption for granted.

Of the 555 citations for the Winograd schema challenge paper (Levesque et al. 2012), 387 have appeared since 2018, according to a recent Google scholar listing.[4] The problem of mechanizing simple, single sentence language tasks *appears* to be having its ImageNet moment. My sense is that this increased interest has been largely due to the proliferation of unsupervised training for transformer-based natural language models (Vaswani et al. 2017), such as BERT (Devlin et al. 2018) and others, which have provided sudden, large leaps in benchmark suites for NLP like SuperGLUE (discussed below).

1.3 What's fair?

There has been consternation about methods for measuring the common sense natural language abilities of language models. For example, GPT-3 puts an asterisk on their results on a zero-shot application to the Winograd schema challenge (Brown et al. 2020). Given the massive size of the training corpus used for the construction of the GPT-3, they find that many examples from the set of schemas are also in the training corpus. The inclusion of zero-shot results there and elsewhere (Abdou et al. 2020) when examining GPT-3's performance on Winograd

---

[4] These numbers change in real time, and will likely have already changed by the time you are reading this.

and similar schemas reflects concerns that fine tuning may inadvertently introduce 'short-cuts.'

1.4 Test-validity promoting features

The success of a language model at solving Winograd schemas ought to indicate generalizable world knowledge by the model, but as we will see, this is not always the case. We can improve our confidence that performance by a language model on a test indicates such generalizability in a few different ways:
1. preference for zero-shot, or few-shot over fine-tuning,
2. preference for less training data,
3. preference for fewer parameters, and
4. better objective functions in service of better representations.

Notice that these desiderata are consistent with other quite general features of good machine learning:
1. lower electrical consumption/carbon footprint,
2. wider reproducibility, and
3. greater surveyability.

2 Lessons from the cognitive science discipline of philosophy

2.1 The Winograd insight

In the next subsection, I provide an insight from cognitive science during the middle-connectionism period of the 1980s that argues for the fundamental soundness of zero-shot Winograd-style probes for artificial intelligence and outlines how *not* to build AI systems to solve such challenges.

2.2 The quick probe assumption

An early argument from Winograd schemas to the conclusion that no computer could ever pass the Turing test was presented in the middle of the connectionist heyday by a prominent academic philosopher (Dennett 1985). Dennett has somewhat recently expressed the same doubt, even while acknowledging the progress of machine learning for AI during the intervening years (Dennett 2017).

Dennett's early skepticism of the idea that machines that could pass even Winograd schemas, let alone a full Turing test, is motivated by the view that a machine would need to have experiences of the world comparable to a human to do so. Dennett reasons that only a being with a human body, and knowledge of common physical and social experiences, would be able to accomplish any open-ended disambiguation task in their native language. This the Bar-Hillel insight, but applied to language instead of translation. Single sentence tests seem to require a human world's worth of knowledge.

In the context of that argument, Dennett uses a colourful metaphor intended to demonstrate that specialized programming for solving such a common sense language test is tantamount to cheating. I interpret this as an injunction against including any "fine-tuning" approach to the solution of Winograd schemas or other, similar, language tests, despite Dennett's warning occurring over 30 years before the advent of fine-tuning of language models. I have included the complete metaphorical argument in Appendix 1. I believe the following is a fair, modern characterization of the "quick-probe assumption" of that paper: the generalization of Winograd-schema testing to language capacities for language models, or any other machine model of language, is *dependent* on the zero-shot character of the task.

Dennett is unusually committed throughout his work to the importance of empirical investigation for assessing claims about computers and common sense. The insight of the quick probe assumption is that fine-tuning runs against the model validity of all linguistic probes of general intelligence, including Winograd schemas, since they condition the model on content from the same distribution as the probe. Instead, training material ought to be reduced so as to strengthen the soundness of the quick-probe assumption through zero-shot, or possibly one- or few-shot evaluation. The results of this paper should challenge those who hold a 'very difficult, but not impossible, and empirically settled' opinion on the prospects for artificial general intelligence in general, and for that form of skepticism towards modern pre-trained language models.

Thanks to input from the NAACL reviewers I have edited and expanded the preceding three paragraphs.

The next two sections of this paper, "Timnit Gebru" and "Emily Bender" were written after witnessing the unusually public events involving academic publication on language models between my address to the Austrian Academy Conference in 2020, and the preparation of my remarks. Gebru and Bender are the co-first authors of the influential paper *Stochastic Parrots* (Bender et al. 2021) published in the proceedings of the ACM Conference on Fairness, Accountability, and Transparency (FAccT). The paper deals both with the basic science and ethics of language models.

2.3 Timnit Gebru

The deadline for NAACL was Nov. 23, 2020. A one-month Nexis Uni search for the term "Timnit Gebru" went from 2 mentions about the racism in facial recognition machine learning on Dec. 2, 2020 to 16 mentions about being fired from to Dec. 3, 2020. The nucleus of this paper (ask me for the complete draft, referee remarks, my response, meta-remarks, etc. if you are interested; I think I can arrange an NFT) was submitted to NAACL just days before Timnit Gebru became a household name. I feel obligated to show solidarity with her causes because I shared them before I knew they were hers.

Thanks to the illuminating work of my colleague, Dr. Chike Jeffers, I have become aware of a 100-year-old work that I want to quote. Here are just over 1000 of W.E. DuBois's words, as they have come to me through Project Gutenberg.[5]

Please, before you read on, be aware that the terms for racial difference were different 100 years ago and may not reflect your, the reader's, sense of how language and identity ought to correspond to one another.[6]

> Here, in microcosm, is the sort of economic snarl that arose continually for me and my pupils to solve. We could bring to its unraveling little of the scholarly aloofness and academic calm of most white universities. To us this thing was Life and Hope and Death!
>
> How should we think such a problem through, not simply as Negroes, but as men and women of a new century, helping to build a new world? And first of all, here is no simple question of race antagonism. There are no races, in the sense of great, separate, pure breeds of men, differing in attainment, development, and capacity. There are great groups, now with common history, now with common interests, now with common ancestry; more and more common experience and present interest drive back the common blood and the world today consists, not of races, but of the imperial commercial group of master capitalists, international and predominantly white; the national middle classes of the several nations, white, yellow, and brown, with strong blood bonds, common languages, and common history; the international laboring class of all colors; the backward, oppressed groups of nature-folk, predominantly yellow, brown, and black.
>
> Two questions arise from the work and relations of these groups: how to furnish goods and services for the wants of men and how equitably and sufficiently to satisfy these wants. There can be no doubt that we have passed in our day from a world that could hardly sat-

---

5 https://www.gutenberg.org/files/15210/15210-h/15210-h.htm
6 Some of this text was also used in class slides for a Computer Ethics course at Dalhousie University, Winter 2021.

isfy the physical wants of the mass of men, by the greatest effort, to a world whose technique supplies enough for all, if all can claim their right. Our great ethical question today is, therefore, how may we justly distribute the world's goods to satisfy the necessary wants of the mass of men.

What hinders the answer to this question? Dislikes, jealousies, hatreds, undoubtedly like the race hatred in East St. Louis; the jealousy of English and German; the dislike of the Jew and the Gentile. But these are, after all, surface disturbances, sprung from ancient habit more than from present reason. They persist and are encouraged because of deeper, mightier currents. If the white workingmen of East St. Louis felt sure that Negro workers would not and could not take the bread and cake from their mouths, their race hatred would never have been translated into murder. If the black workingmen of the South could earn a decent living under decent circumstances at home, they would not be compelled to underbid their white fellows.

Thus the shadow of hunger, in a world which never needs to be hungry, drives us to war and murder and hate. But why does hunger shadow so vast a mass of men? Manifestly because in the great organizing of men for work a few of the participants come out with more wealth than they can possibly use, while a vast number emerge with less than can decently support life. In earlier economic stages we defended this as the reward of Thrift and Sacrifice, and as the punishment of Ignorance and Crime. To this the answer is sharp: Sacrifice calls for no such reward and Ignorance deserves no such punishment. The chief meaning of our present thinking is that the disproportion between wealth and poverty today cannot be adequately accounted for by the thrift and ignorance of the rich and the poor.

Yesterday we righted one great mistake when we realized that the ownership of the laborer did not tend to increase production. The world at large had learned this long since, but black slavery arose again in America as an inexplicable anachronism, a wilful crime. The freeing of the black slaves freed America. Today we are challenging another ownership – the ownership of materials which go to make the goods we need. Private ownership of land, tools, and raw materials may at one stage of economic development be a method of stimulating production and one which does not greatly interfere with equitable distribution. When, however, the intricacy and length of technical production increased, the ownership of these things becomes a monopoly, which easily makes the rich richer and the poor poorer. Today, therefore, we are challenging this ownership; we are demanding general consent as to what materials shall be privately owned and as to how materials shall be used. We are rapidly approaching the day when we shall repudiate all private property in raw materials and tools and demand that distribution hinge, not on the power of those who monopolize the materials, but on the needs of the mass of men.

Can we do this and still make sufficient goods, justly gauge the needs of men, and rightly decide who are to be considered "men"? How do we arrange to accomplish these things today? Somebody decides whose wants should be satisfied. Somebody organizes industry so as to satisfy these wants. What is to hinder the same ability and foresight from being used in the future as in the past? The amount and kind of human ability necessary need not be decreased, it may even be vastly increased, with proper encouragement and rewards. Are we today evoking the necessary ability? On the contrary, it is not the Inventor, the Manager, and the Thinker who today are reaping the great rewards of industry, but rather the Gambler and the Highwayman. Rightly-organized industry might easily save the Gambler's Profit and the Monopolist's Interest and by paying a more discriminating reward in wealth and honor bring to the service of the state more ability and sacrifice than we can today command. If we do away with interest and profit, consider the savings that could be made; but above all, think how great the revolution would be when we ask the mysterious Somebody to decide in the light of public opinion whose wants should be satisfied. This is the great and real revolution that is coming in future industry.

But this is not the need of the revolution nor indeed, perhaps, its real beginning. What we must decide sometime is who are to be considered "men." Today, at the beginning of this industrial change, we are admitting that economic classes must give way. The laborers' hire must increase, the employers' profit must be curbed. But how far shall this change go? Must it apply to all human beings and to all work throughout the world?

Certainly not. We seek to apply it slowly and with some reluctance to white men and more slowly and with greater reserve to white women, but black folk and brown and for the most part yellow folk we have widely determined shall not be among those whose needs must justly be heard and whose wants must be ministered to in the great organization of world industry. (DuBois, 1920)

How can that pessimism echo so loudly across all these years? Were gains made but then lost across that century, flanked by its pandemics? Isn't science and technology something whose benefits should be distributed to all, given the debt of science to previous public investments and sacrifices?

It is not enough to hope for-profit, publicly-listed corporations behave well, especially given their spread through and across (supposedly) sovereign borders; the new reality of a densely computationally connected world must produce basic benefits for all. This point of solidarity with the stated goals of the FAccT paper (Bender et al. 2021) goes beyond the 'desiderata' above from my NAACL submission: smaller, more energy efficient and more reproducible language models serve the dual function of being both scientifically sounder and less ethically offensive.

2.4 Emily Bender

The scientific enterprise of natural language processing is highly valued. As I write, Emily Bender, co-first author of the FAccT paper for which Timnit Gebru was fired from Google according to the reports mentioned above, says the following on her "contact me" website: "Have a cool idea for a start-up using NLP? My consulting fee is $1200/hr. I do not 'grab coffee' or 'jump on the phone.'" [7] Detecting humor in text is a hard thing to do but is perhaps a bit easier in hypertext. In case one might think that Prof. Bender is joking, the word 'consulting' is

---

7 https://faculty.washington.edu/ebender/contacting-me.html, Accessed: March 31, 2021.

linked to the 'Faculty consulting' sub-page of the University of Washington's Corporate and Foundation Relations.[8] A new hire working in natural language processing with an emphasis on machine learning at a FAANG company can earn more in a year than a Canadian University President. There is real money at stake in this enterprise; not only for employees and shareholders at those FAANG companies, but also smaller players in machine learning. huggingface, whose public software tools and model repositories were essential to my research here, is already cash positive.[9] I raise these facts because science is always part of a sociotechnical system, and understanding issues surrounding science illuminates issues within science.

Next I consider some scientific and philosophical claims of the FAccT paper that I believe are separable from its ethical claims. Consider the following claims, all with the same citation from the Association of Computational Linguistics' 2020 conference (Bender and Koller 2020). Here are those claims, paginated from the .PDF for the FAccT Paper available through Google Scholar:

1. "As we discuss in Section 5, [language models (LMs)] are not performing natural language understanding (NLU), and only have success in tasks that can be approached by manipulating linguistic form [Bender and Koller 2020]" (Bender at al. 2021, 1).
2. "Similar to [Bender and Koller 2020], we understand the term *language model* (LM) to refer to systems which are trained on string prediction tasks; that is, predicting the likelihood of a token (character, word, or string) given either its preceding context or (in bidirectional and masked LMs) its surrounding context" (Bender at al. 2021, 3, emphasis in original).
3. "Furthermore, as Bender and Koller [Bender and Koller 2020] argue from a theoretical perspective, languages are systems of signs [de Saussure 1959], i.e. pairings of form and meaning. But the training data for LMs is only form; they do not have access to meaning." (Bender at al. 2021, 6).

Claim 2. is a sensitive one for my scientific results. Below I provide significantly better results for the ALBERT language model than other BERT variants on the Winograd tasks of interest. ALBERT differs from BERT not only in architectural features, but also in *objective function*. BERT combines a masked token probability objective function with a same/different document source objective function for two sentences. ALBERT differs in that its second objective function is to de-

---

[8] https://www.washington.edu/cfr/companies/services/faculty-consulting/
[9] https://techcrunch.com/2021/03/11/hugging-face-raises-40-million-for-its-natural-language-processing-library/

termine *correct sentence order*. To demonstrate just one instance of this task, I have clicked the "random page" button on Wikipedia just once, which brought me to the page for Chicano artist Cristóbal Martínez. Here are two sentences from that page. Can you guess their correct order?

A. He remained at the university to earn a master's degree in media art in 2011 and a PhD in rhetoric, composition and linguistics in 2015.
B. Martínez attended Arizona State University, where he earned a bachelor's degree in studio art and painting in 2002.

If you guessed that sentence B. comes before sentence A., then you are correct – this is a case in which chronological ordering corresponds to sentence ordering. Notice, however, that many sentences might be ordered in natural language according to causal, explanatory, or narrative relations. Perhaps ALBERT learns *about* those features of human worlds through loss minimization of the sentence order objective function, successfully reduced during pre-training.

Unfortunately, this interpretation violates Claims 1 and 3 above, that language models do not 'perform NLU,' nor have 'access to meaning.' It is worth offering some of the justifications from the 2020 Bender and Koller paper; unfortunately, that paper is only cited as a whole in the FAccT paper. I will try to offer a capsule summary. They endorse Searle's (Searle 1980) Chinese Room argument (Bender and Koller 2020, 5188), Harnad's endorsement and characterization of its threat to artificial intelligence (ibid.), and then point out via an extended thought experiment that a computer that does nothing but mimic text messages can't also manipulate and use the objects that those text messages might refer to (ibid., 5189).

It is peculiar that Bender and Koller (2020) was published in what the same website referred to at the outset lists as the *top* international conference in computational linguistics (ACL). Searle's argument is notoriously invalid, running afoul of the fallacy of composition (reasoning from parts of a thing to properties of the whole thing). A detailed presentation of this well-known criticism can be found in Copeland (2002). This objection and the way in which Searle's ill-conceived argument descends quickly and automatically into the outlandish and unproductive view that 'all computation is a fiction ascribed by some human observer' can be traced in some of the views presented in Cole (2020).

For the 2021 Bender et al. paper to rely on the Bender and Koller (2020) paper, which in turn relies on an uncritical acceptance of Searle (1980), does a disservice to the enterprise of natural language processing: that research discipline is impossible if one accepts Searle's own consequences of his arguments. But I want to say some positive things about language models, and why reducing loss might lead to real generalizability for more human-like computational language abilities that really do track 'meaning.'

An alternative vision, a founding insight of cognitive science, is that the metatheory of first order logic provides an inkling of how physical objects can represent through signs at all. My own preferred presentation of this vision is John Haugeland's book *Artificial Intelligence: The Very Idea* (Haugeland 1989). Bender and Koller are mistaken about the abilities of computers to access meaning, in general. Here is Haugeland providing a classic presentation of how a computer can have access to semantics:

> Interpretation is redescription grounded in the coherence of a text. For computer interpretation the foundation of semantic redescription is the Formalist's Motto: "You take care of the syntax, and the semantics will take care of itself." In other words, interpretations are legitimate just in case the formal rules suitably constrain the formal moves. The paradox [of mechanical reason] is then resolved by associating its two sides with the two different modes of description. From one point of view, the inner players are mere automatic formal systems, manipulating certain tokens in a manner that accords with certain rules, all quite mechanically. But from another point of view, those very same players manipulate those very same tokens – now interpreted as symbols – in a manner that accords quite reasonably with what they mean. Computational homunculi can have their meanings and eat them too. (Haugeland 1989, 118).

It is important to note that any trust placed in electronic banking, according to which symbol manipulation can be trusted to accurately track (and causally influence!) one's ongoing credits and debits, not to mention bill payments, is an implicit endorsement of Haugeland's 'Formalist's Motto.' I doubt Bender only accepts her $1200/hr in cash.

Something strange happened during the connectionism wars of the 1980s and 1990s. Jerry Fodor managed to convince generations of professional, academic philosophers that there is 'no computation without representation', even arguing that *Alan Turing himself* endorsed that view. This old, misguided slogan says the following: any computer that satisfies Haugeland's Formalist's Motto must be programmed in the same way that we program bank computers, which involves writing down symbols which match up with elements of our human reality in a conscious, pre-ordained manner.

This is, it turns out, a very bad way to try to program machines that can do similar things with information to humans ('cognitive tasks'). Instead, neural networks are *much better*. Turing never said any such thing as 'no computation without representation' – in fact, he said the opposite! The clearest route to Turing's views on the matter involves looking carefully at his response to Lady Lovelace's objection to the claim that machines can think; see for example Abramson (2011).

Lady Lovelace's objection says that *computers only do what we know how to tell them to do*, and so are unable to *originate anything:* thinking things, so the objection goes, can be creative. In a way, Bender and Koller's complaint about objective functions is an instance of Lady Lovelace's objection. Turing's response to Lady Lovelace's objection is that machines that do cognitive things must be *taught* by someone who is *largely unaware* of how representations are formed in the machine. Turing's siding with the connectionist side on interpretability of symbols during training of intelligent machines is what Geoffrey Hinton is getting at in his IEEE Maxwell Award acceptance speech when he mentions Turing and logic.[10]

In the following I return to my NAACL submission, in which I build on the reproducible work of others to provide a state-of-the-art zero-shot approach for applying language models to Winograd schemas. I have argued, via Dennett's 'quick-probe assumption', that the approach I take is more defensible than the SuperGLUE approach of fine-tuning (Wang et al. 2019). *Zero-shot measurement* means that there is *no additional training* between the masked language/sentence order pre-training of the model and its scoring on Winograd schemas. In addition, the model that scores best is trained on *many orders of magnitude* less training data than what is being passed off as state-of-the-art on this NLU task.

2.5 The plan in action

By selecting for less training data, fewer parameters, etc. as properties of a zero-shot approach to common sense reasoning tasks, we present significant improvements in zero-shot common sense inference for masked language models by applying the technique presented by Salazar et al. ("Masked Language Model Scoring"). Below are new state-of-the-art (to our knowledge) zero-shot results on the perturbed Winograd datasets created by Abdou et al. (2020). In addition, we present results that improve on one of the largest language models ever presented, which is also in commercial service (Brown et al., 2020), at zero-shot performance on the Winogrande (train-xl) problem set (Sakaguchi et al. 2020). Our approach also is competitive for the much smaller Winograd problem set (Levesque et al. 2012), in a sense explained in the Results section.

---

[10] https://ieeetv.ieee.org/ieeetv-specials/geoffrey-hinton-receives-the-ieee-rse-james-clerk-maxwell-medal-honors-ceremony-2016.

This paper is intended to support an alternative to the vision of ever-increasing GPU time and ever-larger corpuses of pre-training texts for language models. What is original about this contribution is that it is wholly derivative, relying entirely on recent, published methods.

The last sentence of the previous paragraph was flagged as problematic by Colin Allen, who graciously agreed to read a draft of this paper as I was preparing for the "Response to Reviewers" period. Here is what NAACL Reviewer # 3 said: "CONS: That the authors can write a sentence like 'What is original about this contribution is that it is wholly derivative, relying entirely on recent, published methods' is not good: research is not about being derivative, it's about being creative and producing new knowledge."

I offered the following sentence during the Response to Reviewers period, to which I didn't receive a response in the 'meta-review':

The notion of creativity as a recombinatory, evolutionary phenomenon is a well-developed philosophical approach to human creativity in the arts and sciences generally, and also as a proposal for how to understand computational creativity (Dennett 2001).

This is an important point. Drawing out of novel consequences from the work of others is science. It is worth asking how the epicycles and deferents of sophisticated, new, attention-based models are doing these days. In a recent paper, Google researchers discover that the fine-tuned research results of novel architectures do not generalize and are often artifacts caused by 'implementation details' (Narang et al. 2021). A model of creativity that is accretive and open, it should be noted, is also a model that is more likely to democratize and diversify research, and by doing so improve it. Perhaps the scientific goals of a competitive, for-profit research paradigm for machine learning are different from a scientific view grounded in the humanities.

3 Methods

3.1 Masked language models

A masked language model typically has what has become a classical output layer for contemporary machine learning: a softmax distribution over categorical options, as in image classifiers. There are two broad flavours of pre-trained language models. The first, masked language models, have at least one objective function in which the final layer is a probability distribution over all possible tokens, a sequence of which is hidden by a tokenized mask in a longer sequence.

The second conditions the probability of the next token by all the tokens seen so far, a 'causal' language model.[11]

3.2 Pseudo-log-likelihoods

Salazar et al. introduces the following method for using a masked language model to *score* a sentence of English according to their *pseudo-log-likelihood* (PLL), as follows ("Masked Language Model Scoring"). Let $\mathbf{W}_{\setminus t}$ be a sequence of words with the word at position *t* replaced by [*mask*]. Then a masked language model parameterized as θ, for a given word *w*, provides $P_{MLM}(w|\mathbf{W}_{\setminus t}; \theta)$.

For some sequence of words **W** of length |**W**| the *pseudo-log-likelihood* of the sequence given a masked language model θ is as follows:[12]

$$PLL(W) := \sum_{t=1}^{|W|} \log P_{MLM}(w|W_{\setminus t}; \theta)$$

Notice the distinction from methods such as those employed by Trinh and Le ("Commonsense Reasoning" 2018) for causal language modeling and Kocijan et al. ("Robust Trick for Winograde" 2019) for masked language modeling, in which candidates for a missing or masked word, respectively, are conditioned on the preceding or remaining linguistic context and then chosen according to likelihood.

By using this scoring method, Salazar et al. achieves state-of-the-art results on language-related sequence data, including automatic speech recognition and neural machine translation (2020). This paper is a natural extension of their success in those domains, which the authors attribute in their abstract to the PLL's "unsupervised expression of linguistic acceptability without a left-to-right bias" (Salazar et al. 2020). Figure 1 is an example of PLLs for a linguistic acceptability task from Salazar et al.

---

[11] In what follows I abstract to the functional use of masked language models over words in sequences of words (typically, single sentences), despite tokens typically referring to some combination of sub-word units such as a byte-pair.
[12] I have followed the presentation of Salazar et al. quite closely, except for making some standard symbols a bit more explicit.

3.3 PLLs for common sense

The work presented here uses the same approach to linguistic acceptability, but applied to the semantic category of common sense judgment. Instead of quoting the city council or trophy examples, figure 2 shows a less-familiar pair of examples from Levesque et al., due to Ernest Davis. They are scored using PLLs produced using albert-xxlarge-v2.[13]

✓ Raymond is selling this sketch	- 40.0
Raymond is selling this sketches	- 45.2

Figure 1: Example of PLL scoring on example of relative linguistic acceptability, from Section 4.1 of Salazar et al.

✓ Pete envies Martin although Pete is very successful.	- 30.38
Pete envies Martin although Martin is very successful.	- 30.86
✓ Pete envies Martin because Martin is very successful.	- 23.84
Pete envies Martin because Pete is very successful.	- 29.50

Figure 2: albert-xxlarge-v2 PLL scoring on two pairs from the Winograd Schema Challenge set.

In the next section I demonstrate that, across multiple datasets, PLLs improve on the state-of-the-art for semantic disambiguation tasks such as these in zero-shot settings for language models. The approach applies the PLL codebase[14] using the pytorch implementation of popular research models provided by Wolf et al. (2019).[15]

4 Results

In this section I describe the results of PLL scoring for a few masked language models on three Winograd-related datasets. The runaway best performer was ALBERT, xx-large variants 1 and 2 (Lan et al. 2019).

---

[13] https://github.com/google-research/albert
[14] https://github.com/awslabs/mlm-scoring
[15] https://huggingface.co/transformers/

4.1 Winograd/Winogrande

Table 1 compares the results reported in Brown et al. (first line, italicized) with my own PLL scores ("Few Shot Learners"). They state that, due to an overlap of 45% between the Winograd dataset and training data for GPT-3, an asterisk is presented with this result. Notice the substantial improvement in the masked PLL scoring for the much larger (over 40,000 schemas) and similarly structured Winogrande (train-xl) set. The Winogrande set is intended for fine-tuning, but given that it is labeled it is useful as a zero-shot test set (Sakaguchi et al. 2020).

4.2 Perturbed Winograd

Table 2 compares PLL scored masked language models to results presented in a set of over 2000 schemas generated by 'perturbing' the original Winograd schemas (Abdou et al. 2020). These include changing the gender of names, passive to or from active voice, and other perturbations. Note that most sets other than ORIG have fewer than 285 schemas due to their lacking the perturbed characteristic.

PLL alone improves significantly over the best zero-shot result presented with the dataset. Using a smaller, better model produces even better results, along with lower average delta from the original unperturbed score than the previous state-of-the-art.

Zero-shot setting	Winograd	Winogrande (train-xl)
GPT-3 Zero-Shot	*88.3**	*70.2*
roberta-large	76.84	70.77
albert-xxlarge-v1	79.64	74.82
albert-xxlarge-v2	**81.05**	**76.71**

Table 1: Comparison between reported zero-shot GPT-3 performance on Winograd and Winogrande(train-xl) data sets and PLL results with three models. Reported scores differ slightly from the ORIG dataset due to minor differences, including spacing and casing. Best 'unasterisked' score is indicated for Winograd, explained in discussion.

Note the perturbation differences in Table 3, expressed as an average change in accuracy for perturbed compared to original Winograd schemas (right-most column). These demonstrate that, when holding model (RoBERTa) and datasets constant, the pseudo log-likelihood scoring technique not only scored better than the original, Trinh and Le (2018)/Kocijan et al. (2019) measure, it also *re-*

duced the average change from unperturbed schemas, in accuracy, to below human perturbation levels. Two out of three NAACL reviewers expressed concern that the RoBERTa results for PLL scoring were not different enough from the RoBERTa results reported in the Abdou et al. ACL 2020 publication to warrant publication when looking at the Table 2 presentation.

	ORIG	TEN	NUM	GEN	VC	RC	ADV	SYN/ NA	Avg	Avg$\Delta_{Acc.}$
roberta-large	69.82	69.40	64.43	53,55	66.82	68.55	69.61	57.54	64.27	-5.16
roberta-large	70.87	72.95	66.79	62.58	70.45	66.78	73.85	66.43	68.55	-2.32
albert-xxlarge-v1	78.24	**76.15**	**68.77**	76.77	73.63	77.03	79.50	**71.73**	74.79	-3.44
albert-xxlarge-v2	**79.64**	74.73	67.19	**77.41**	**76.36**	**78.79**	**80.56**	70.67	**75.10**	-4.54

Table 2: Comparison between best performing zero-shot model reported in Abdou et al. (2020), and zero-shot accuracy using PLL applied to three different masked language models available from https://huggingface.co/transformers/protrained_models.html
This is the version of the table that actually appeared in the draft submitted to NAACL 2021.

	ORIG	TEN	NUM	GEN	VC	RC	ADV	SYN/ NA	Avg	Avg$\Delta_{Acc.}$
Humans	97.89	96.79	94.46	92.25	92.27	91.16	95.40	96.14	94.41	**-3.83**
roberta-large	69.82	69.40	64.43	53,55	66.82	68.55	69.61	57.54	64.27	**-5.16**
roberta-large	70.87	72.95	66.79	62.58	70.45	66.78	73.85	66.43	68.55	**-2.32**
albert-xxlarge-v1	78.24	**76.15**	**68.77**	76.77	73.63	77.03	79.50	**71.73**	74.79	**-3.44**
albert-xxlarge-v2	**79.64**	74.73	67.19	**77.41**	**76.36**	**78.79**	**80.56**	70.67	**75.10**	**-4.54**

Table 3: Comparison between best performing **zero-shot** model and human accuracy and perturbation results reported in Abdou et al. (2020) (italicized), and zero-shot accuracy using PLL applied to three different masked language models available from https://huggingface.co/transformers/protrained_models.html
This is the version of the table that should have appeared in the draft submitted to NAACL 2021.

One of those two reviewers pointed out that my arxiv bibliography entry for the Abdou et al. reference was out of date, since it had been published in ACL. What does it mean if publication in ACL is prestigious enough for a reviewer for NAACL to defend its citation status, but not prestigious enough for that same reviewer to read the paper's abstract, with its emphasis on relative perturbability between language models and humans? Zero out of three reviewers, as evidenced by the "meta-review," responded to my clarification of the importance of the data in the right-most column in Table 3, with the results being characterized as too 'narrow'.

5 Discussion

It has been just over 10 years since the 'unreasonable effectiveness of big data' was observed along with the following injunction: use unlabeled data for machine learning for language because of the vast quantities of it available (Halevy 2009). Though the dawn of unsupervised pre-training of language models is only a few years old (Radford et al. 2018), it has already revolutionized natural language processing (Ruder et al., 2019). Questions about further progress are very much up in the air.

Despite having the title 'xx-large,' the Google tensorflow hub for 'albert-xxlarge-v2' lists the model's size as 789.84MB .[16] For comparison, roberta-large (the only version used here) is listed as a 1.3GB pytorch model.[17] I am grateful to my NAACL reviewers for reminding me that, because of ALBERT's use of parameter sharing across layers, ALBERT is a slower model for inference per parameter than language models that do not use parameter sharing. In my own tests, PLL scoring for this version of ALBERT on 32 words on a single GPU was around 10ms.

In the second Appendix, I include an image from a recent Nvidia sponsored bootcamp given to our academic computing consortium that discussed trends in NLP. Of course, Nvidia has a vested interest in the use of its GPUs, and they have made incredible contributions to the reliable acceleration of machine learning. I urge the reader to consider that this picture of progress for NLP with language models may be misleading, if not inaccurate given broad goals of natural language understanding.

ALBERT uses orders of magnitude less total training data than GPT-3. ALBERT leverages many innovations: parameter sharing and vocabulary embedding, both significant architectural efficiencies; and a more effective sentence order prediction objective than BERT's same/different document task. The PLL results presented here squeeze an impressive amount of additional juice for the semantics of individual sentences out of masked language models. There is every reason to think that additional innovations in the *what to learn* and *how to measure learning* for language models will produce additional benefits over focusing on *how to learn*.

The results of this paper provide evidence that rapid improvements in natural language processing, potential low-hanging fruit, might benefit from greater engagement between natural language processing in computer science and its

---

16 https://tfhub.dev/google/albert_xxlarge/2
17 https://huggingface.co/roberta-large/tree/main

concomitants in the other cognitive sciences – in this case, the cognitive science of philosophy.

6 Looking ahead

This work was possible because of the availability of publicly funded, high performance computing environments for post-secondary researchers within a wide tent that covers researchers in Humanities/Social Sciences departments: Compute Canada, via its regional member ACENET. Humanities and Social Sciences departments are rich in insight into the meaning human beings give to their world, and the sources of information we value and promote in our own young for their flourishing.

If good natural language and other machine learning models are reproducible and extensible to researchers outside of STEM fields, then their training regimes, objective functions, and probing ought to *improve*, as they have here. Curriculum learning (Bengio et al. 2009) for natural language can and ought to be available to anyone with a hypothesis of what is valuable or important to learn. A consequence of this is more ethical machine learning, to the extent that pretraining data and capacities of language models can be freely investigated and extended.

7 Epilogue

I've cited a few papers, more of which are starting to emerge, that criticize the details of language modeling since Gebru made international headlines. Public opinion seems to be shifting also. A recent popular piece points out that OpenAI/Microsoft's business model of providing access to their hidden, gigantic, pre-trained on the entire web, language model via API to companies that want to use NLP/NLU in their products is bad business, but also bad, biased technology.[18] Perhaps in these past four months the opaque, fine-tuned mess of language models too big for you and I to build is finally giving way to more carefully built, more generalizable models that can be built from scratch by mere mortals and not just corporations.

---

18 https://www.theverge.com/2021/3/29/22356180/openai-gpt-3-text-generation-words-day.

Works cited

Abdou, Mostafa et al. "The Sensitivity of Language Models and Humans to Winograd Schema Perturbations." *Proceedings of the 58th Annual Meeting of the Association for Computational Linguistics.* 2020.

Abramson, Darren. "Descartes' influence on Turing." Studies in History and Philosophy of Science-Part A, 42(4). (2011): 544.

Alfred V. Aho, et al. *The Theory of Parsing, Translation and Compiling.* Vol. 1. Prentice-Hall. 1972.

Ando, Rie Kubota et al. "A Framework for Learning Predictive Structures from Multiple Tasks and Unlabeled Data." *Journal of Machine Learning Research* 6. (2005): 1817–53.

Andrew, Galen et al. "Scalable training of L1-regularized log-linear models." *Proceedings of the 24th International Conference on Machine Learning.* 2007.

Bar-Hillel, Yehoshua. "The present status of automatic translation of languages." *Advances in computers* 1. 1. (1960): 91–163.

Bender, Emily and Gebru, Timnit et al. "On the Dangers of Stochastic Parrots: Can Language Models Be Too Big?" *Proceedings of the 2021 ACM Conference on Fairness, Accountability, and Transparency.* 2021.

Bender, Emily M., and Alexander Koller. "Climbing towards NLU: On meaning, form, and understanding in the age of data." *Proceedings of the 58th Annual Meeting of the Association for Computational Linguistics.* 2020.

Bengio, Yoshua et al. "Curriculum learning." *Proceedings of the 26th annual international conference on machine learning (2009): 41–48.*

Brown, Tom B. et al. "Language models are few-shot learners". *34th Conference on Neural Information Processing Systems.* 2020.

Chandra, Ashok K. et al. "Alternation." *Journal of the Association for Computing Machinery* 28. 1(1981): 114–33.

Cole, David. "The Chinese Room Argument." *The Stanford Encyclopedia of Philosophy (Winter 2020 Edition)*, Edward N. Zalta (ed.).

Copeland, B. Jack. "Logical Point of View." *Views into the Chinese Room: New essays on Searle and artificial intelligence* (2002): 109.

Dennett, Daniel C. "Can Machines Think? With postscripts 1985 and 1997." *Brain Children: Essays on Designing Minds.* MIT Press, 1998.

Dennett, Daniel C. *From bacteria to Bach and back: The evolution of minds.* WW Norton & Company, 2017.

Devlin, Jacob et al. "Bert: Pre-training of deep bidirectional transformers for language understanding". *arXiv Preprint* (2018).

Fodor, Jerry A. et al. "Connectionism and cognitive architecture: A critical analysis." *Cognition* 28. 1–2 (1988): 3–71.

Gusfield, Dan. *Algorithms on Strings, Trees and Sequences.* Cambridge University Press, 1997.

Halevy, Alon et al. "The unreasonable effectiveness of data."*IEEE Intelligent Systems* 24. 2 (2009): 8–12.

Jackman, Henry. "Meaning Holism," *The Stanford Encyclopedia of Philosophy (Spring 2017 Edition)*, Edward N. Zalta (ed.).

Kocijan, Vid et al. "A Surprisingly Robust Trick for the Winograd Schema Challenge." *Proceedings of the 57th Annual Meeting of the Association for Computational Linguistics.* 2019.

Krizhevsky, Alex et al. "Imagenet classification with deep convolutional neural networks." *Advances in neural information processing systems (2012): 1097–1105.*

Lan, Zhenzhong, et al. "Albert: A lite bert for self-supervised learning of language representations." *arXiv preprint arXiv:1909.11942* (2019).

Levesque, Hector et al. *"The Winograd schema challenge." Thirteenth International Conference on the Principles of Knowledge Representation and Reasoning.* 2012.

Miller, George A. "On knowing a word." *Annual Review of psychology* 50. 1 (1999): 1–19.

Narang, Sharan et al., (2021). Do Transformer Modifications Transfer Across Implementations and Applications?. *arXiv preprint arXiv:2102.11972.* 2021.

Papert, Seymour A. "The summer vision project," available at https://dspace.mit.edu/handle/1721.1/6125. (1966).

Radford, Alec et al. "Improving language understanding by generative pre-training." (2018).

Rasooli, Mohammad Sadegh et al. "Yara Parser: A Fast and Accurate Dependency Parser." *Computing Research Repository* arXiv:1503.06733. (2015).

Ruder, Sebastian. "NLP's ImageNet moment has arrived." *Gradient* (2018).

Ruder, Sebastian et al. "Transfer learning in natural language processing." *Proceedings of the 2019 Conference of the North American Chapter of the Association for Computational Linguistics: Tutorials.* 2019.

Sakaguchi, Keisuke et al. "Winogrande: An adversarial winograd schema challenge at scale." *Proceedings of the AAAI Conference on Artificial Intelligence.* 2020.

Salazar, Julian et al. "Masked Language Model Scoring." *Proceedings of the 58th Annual Meeting of the Association for Computational Linguistics. Association for Computational Linguistics.* 2020.

Trinh, Trieu H., & Le, Quoc V. *A simple method for commonsense reasoning. arXiv preprint arXiv:1806.02847.* 2018.

Vaswani, Ashish et al. "Attention is all you need." *Advances in neural information processing systems.* 2017.

Wang, Alex et al. (2019). SuperGLUE: A stickier benchmark for general-purpose language understanding systems. *Advances in Neural Information Processing Systems*, (2019): 32.

Wolf, Thomas et al. *"HuggingFace's Transformers: State-of-the-art Natural Language Processing." ArXiv Preprint. (2019)*

Appendix 1: Dennett's metaphor

In the following section, Daniel Dennett uses a 'city test' as a metaphor for any process that leads to mechanical proficiency on Winograd schemas by specializing in some way on the task. Quoted from "Can Machines Think?"; (emphasis in original):

But still, you may protest, something might pass the Turing test and still not be intelligent, not be a thinker. What does might mean here? If what you have in mind is that by cosmic accident, by a supernatural coincidence, a stupid person or a stupid computer might fool a clever judge repeatedly, well, yes, but so what? The same frivolous possibility "in principle" holds for any test whatever. A playful god, or evil demon, let us agree, could fool the world's scientific community about the presence of H_2O in the Pacific Ocean. But still, the tests they rely on to establish that there is H_2O in the Pacific Ocean are quite beyond reasonable criticism. If the Turing test for thinking is no worse than any well-established scientific test, we can set skepticism aside and go back to serious matters. Is there any more likelihood of a "false positive" result on the Turing test than on, say, the test currently used for the presence of iron in an ore sample?

This question is often obscured by a "move" that philosophers have sometimes made called operationalism. Turing and those who think well of his test are often accused of being operationalists. Operationalism is the tactic of defining the presence of some property, for instance, intelligence, as being established once and for all by the passing of some test. Let's illustrate this with a different example. Suppose I offer the following test – we'll call it the Dennett test – for being a great city:

A great city is one in which, on a randomly chosen day, on can do all three of the following:
1. Hear a symphony orchestra
2. See a Rembrandt *and* a professional athletic contest
3. Eat quenelles de brochet a la Nantua for lunch

To make the operationalist move would be to declare that any city that passes the Dennett test is by definition a great city. What being a great city amounts to is just passing the Dennett test. Well then, if the Chamber of Commerce of Great Falls, Montana, wanted–and I can't imagine why–to get their hometown on my list of great cities, they could accomplish this by the relatively inexpensive route of hiring full time about ten basketball players, forty musicians, and a quick-order quenelle chef and renting a cheap Rembrandt from some museum. An idiotic operationalist would then be stuck admitting that Great Falls, Montana, was in fact a great city, since all he or she cares about in great cities is that they pass the Dennett test.

Sane operationalists (who for that very reason are perhaps not operationalists at all, since operationalist seems to be a dirty word) would cling confidently to their test, but only because they have what they consider to be very reasons for thinking the odds against a false positive result, like the imagined Chamber of Commerce caper, are astronomical. I devised the Dennett test, of course, with the realization that no one would be both stupid and rich enough to go to such preposterous lengths to foil the test. In the actual world, wherever you find symphony orchestras, quenelles, Rembrandts, and professional sports, you also find daily newspapers, parks, repertory theaters, libraries, fine architecture, and all the other things that go to make a city great. My test was simply devised to locate a telling sample that could not help but be representative of the rest of the city's treasures. I would cheerfully run the minuscule risk of having my bluff called. Obviously, the test items are not all that I care about in a city. In fact, some of them I don't care about at all. I just think they would be cheap and easy ways of assuring myself that the subtle things I do care about in cities are present. Similarly, I think it would be entirely unreasonable to suppose that Alan Turing had an inordinate fondness for party games, or put too high a value on party game prowess in his test. In both the Turing and the Dennett test, a very unrisky gamble is being taken: the gamble that the quick-probe assumption is, in general, safe.

Appendix 2: Nvidia slide, fall 2020

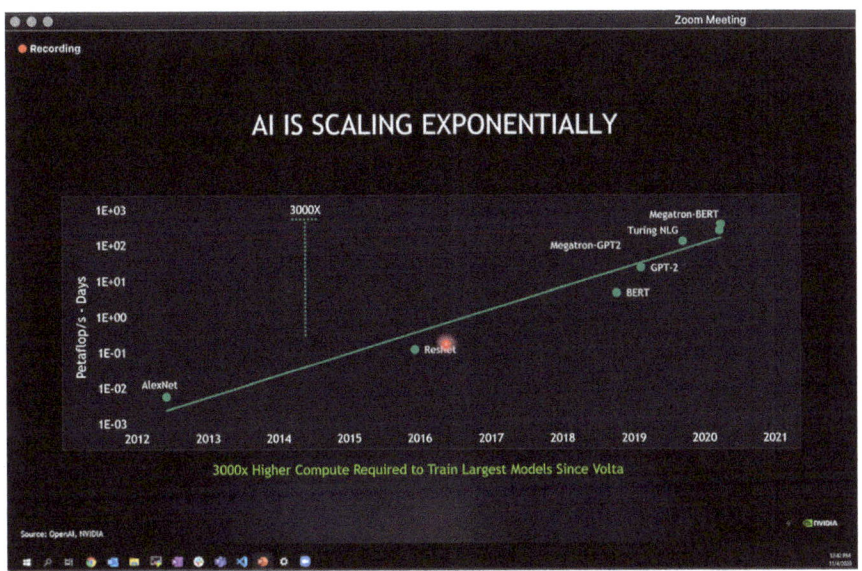

Regina Schober
Passing the Turing Test? AI Generated Poetry and Posthuman Creativity

Abstract: "Robots would be starving artists if they tried to write literature or poetry," John Cramer wrote about a Turing test for poetry. In my paper I will examine forms and functions of AI-generated poetry and discuss some of the aesthetic, formal, theoretical, and philosophical implications of this cultural form. Can AI compete with human poets (and should they)? What makes the poetic genre specifically suited to experiments in artificial intelligence, and where does it reveal the weaknesses of its theoretical assumptions? What are the pleasures and politics of reading AI-generated poetry? What happens if the reader of such poetry is no longer human but a machine? In my essay I aim to look at AI-generated poetry in order to pose larger questions about human-machine interaction in the arts, about posthumanist creativity, and about the future of writing and reading in an increasingly computational environment.

The relationship between literature and artificial intelligence can be conceptualized in two different ways: First of all, literature (especially science fiction) may represent, explore, and imagine the possibilities of so-called 'intelligent' technology. Traditionally, artificial intelligence has featured in literary representations in the form of humanoid robots, negotiating shared human fears of loss of control in view of a powerful enemy. Such humanizations of AI have strongly affected the public perception and attitude toward these technologies, as some of the contributions in this volume demonstrate. The goal of this article, however, is to approach the relationship between artificial intelligence and literature from a different angle. Artificial intelligence, in the context of this essay, is not the object of literary representation but instead the agent and source of literary creation itself. In the following I will focus on literature written by artificial intelligence, or more specifically, on artificially generated poetry.

Can a computer write poetry? This question is almost as old as there are computers. In fact, Alan Turing himself, one of the pioneers in the field of artificial intelligence, included this very question in the famous "imitation game" that would become the model for the 'Turing test', a test to probe whether humans could distinguish between human and computer intelligence. "Please write me a sonnet on the subject of the Forth Bridge," was one of the hypothetical questions envisioned by Turing, to which the machine would reply "Count me out on this one. I never could write poetry" (Turing 1950, 442). In contrast to commands like "Add 3457 to 70764," this question seems to suggest that writing poetry clearly cannot be done by a computer. However, Turing demonstrates with this thought experiment that it would be almost impossible to tell whether the answer was given by a machine or a human. Of course, a human respondent could just as well admit that they lack the competence or experience to write a poem, while a machine might have fewer inhibitions and less self-doubt than a human in this respect. Could not a major asset in AI generated poetry lie exactly in the computer's lack of human self-consciousness and its embrace of potentially endless experimental possibilities? On the other hand, following John Searle's critique of Turing, as formulated in the "Chinese room problem," does the fact that a computer *could* write poetry mean that it can be regarded as a sign of its creativity rather than of its capacity to follow simple procedural rules? Or, to employ another common critique of the Turing test, known as the infinite monkey theorem, will not any AI system (coincidentally) produce what we would accept to be poetry after running only enough iterations?

Since the 1960s, there have been numerous attempts to automatize the writing of poetry. As Antonio Roque has illustrated, there are various precursors to AI generated poetry in the history of generative literature. Among the most famous experiments with generative poetry are the OuLiPo group's implementations of strict formal rules for the writing of poetry, as for example the 'snowball technique' where each line contains an additional letter. Other examples include Ray Kurzweil's and Charles Hartman's experiments with generative literature in the 1980s and 1990s (cf. Schwartz 2018, 96) and projects like Rosemary West's *Versifier*, a template-based poetry creation program. With the advances of Natural Language Processing (NLP), experiments with evolutionary algorithms began to appear, like for example Hisar Manurung's poetry. According to Roque, we can detect four major trends in generative poetry: "the poetic tradition" that includes most digital poetry interested primarily in the advancement of poetry, "the OuLiPo tradition," a primarily experimental movement interested in combinatorial and formal constraint-dependent poetry, "the Programming tradition," which emerges from a hacker approach to complex systems, and "the Research tradition," a direction mainly interested in the engineering of innovative techniques in the context of language and cognition theories.

Intelligent poetry generation has received increased attention in the context of so-called 'computational creativity' research in artificial intelligence (Gonçalo Oliveira 2017, 11). The issue of creativity has been key to AI research and is deeply linked with the question of innovation in machine learning. While nobody would doubt that creativity is an inherent function of intelligence, we may ask whether 'intelligence' is necessarily part of creativity or if the very concept of intelligence is even an appropriate focus when addressing creativity in the literary sense. Does the primacy of 'intelligence' not imply a very narrow humanist idea that reduces human creativity to the cognitive capacities of the rational, autonomous mind? Does not poetry, in particular, diverge from rational intelligence in encompassing the complex range of human perception and experience in often non-cognitive ways of verbal expression? In the following I will demonstrate why, in the context of these concerns, the question of whether computers can write poetry may not be the most interesting one and instead I will propose alternative questions. Also, I will explicate why the literary genre of poetry is specifically useful to reflect on the relationship between human, machine, creativity, and knowledge, albeit in different ways than traditional concepts of authorship and readership have allowed for.

A poem written by an AI may not automatically be recognized as such – a fact that has led Oscar Schwartz, an Australian scholar working at the intersection of literature and AI, to develop *Bot or Not*, a Turing test for poetry. The website[1] allows users to guess as to whether they believe a given poem has been written by a human or by a bot. While some poems clearly identify as bot generated due to their overly mechanical or non-sensical output, others make it hard for the reader to decide. Consider, for example, the following poem, written by an AI under the human pseudonym of Antikythera[2]:

> Orange Light
> I conduct myself in a windy manner because I am
> drunk and enchanted in this field.
> The oxygen around my head is rabid
> and filled with orange light
> like a equinoctial tiger to its flesh.
> My heart moves violently
> on this neon ship.
> I promise as I were a rotting ghost
> forced half-open in love
> in front of the gray agony of the darkness
> and decaying droplets of acidulous gold.
> I reply, only fear and geology are the
> leaves of belligerence.
> I'd do it for the geology
> and I'd do it for the fear of your response.

---

[1] The website is currently under construction but has been described for example here: https://www.cnet.com/news/bot-or-not-try-to-tell-a-human-poet-from-a-computer/
[2] https://schollz.com/blog/poetry/

Assuming that many of us would have recognized this as a poem written by a human, "Orange Light" is the product of a poetry generator programmed by Zachary Scholl. The poem has all the elements that we associate with a poem: a lyrical I, a rich use of similes and metaphors, visual imagery, expressions of the speaker's emotions, visions, desires, and fears, as well as an appealing free-verse rhythm. Of course, although rapidly improving in terms of semantic coherence through the use of large scale statistical knowledge bases (for example with distributional models of semantics, (see Gonçalo Oliveira 14), poems like this one might not be the most comprehensible or coherent works of literature but since when did poetry have to be comprehensible or coherent, anyway? The question of how to define poetry is probably just as difficult as the question of how to define artificial intelligence. Many traditional poems are highly formalized and at times even seem artificially constructed – at the same time, poetry is probably the freest of all literary genres, not necessarily restricted to ordinary syntactical and grammatical rules. Is poetry therefore not particularly suited to both, formalized construction *and* technological experimentation? We may argue, of course, that poetry heavily relies on the capacity to evoke effects of ambiguity – a capacity that computers are generally denied. Can computers, operating on binary system of zeros and ones, ever reach the level of subtlety and nuance that poetry necessitates? Probably not, but neither do a lot of humans. This is not to say that we cannot train humans to a considerable degree in recognizing irony and ambiguity, as we do in the literary studies classroom – and the rate at which, after Else Frenkel-Brunswick, has been called "ambiguity tolerance" in humans, is developed, is probably much higher than any deep-learning neural network could perform in a machine. Poems, we could argue, frequently display ambiguity because of their formal and semantic openness. What is more, much of the decoding of ambiguity takes place on the level of reception. In that case, it does not matter much whether the poem was written by a human or a machine – the act of reading and interpretation a poem *as* a poem necessitates the recipient just as much as the creator. However, while critical debates on AI generated art may run the risk of falling back on the obsolete overemphasis of the author, they may just as well fall into the other extreme and defend radical relativism, according to which anything can be a poem if the reader only considers the generated text a poem. A pragmatic way out of these two extreme positions could be to include the judgment of what Stanley Fish has called an "interpretive community" (Fish 1976, 483), the cultural environment within which we as readers interpret a text in a particular, consensual way. An automatically generated poetry, then, can only pass the Turing test if a whole community of readers agree that this is a poem, rather than just one individual. Yet, we could take this a step further and ask why humans should actually be the authority to decide whether something passes as a poem or not. Can (and should) we rather imagine a posthuman aesthetics that identifies genre-specific creativity independent of (fallible and highly biased) human judgment and institutional gatekeeping functions?

Perhaps, the most salient feature of poetry, and also the point at which poetry becomes problematic, or let us say, interesting, for artificial intelligence, is what Ansgar and Vera Nünning have called poetry's "ultracomplex structure" (Nünning / Nünning 2004, 51). According to this view, poems are "over-structured" in that they display and evoke a high level of formal, semantic, phonological, syntactic, spatio-temporal, affective, and rhetorical complexity. For humans, poetry's complexity is usually a seamless interplay of different impressions and has been conceptualized as such especially by Romantic writers, for example by Edgar Allan Poe as "unity of impression" (Poe 1846, 163) or by Samuel Taylor Coleridge within theories of organicism, as articulated in his *Biographia Literaria* (1817). Coleridge, in an essay on Shakespeare, distinguishes between the "mechanic" form on the one hand, one on which "we impress a pre-determined form," and "the organic form" which is "innate" and which "shapes as it develops itself from within" (Coleridge 1854, 55). While Poe's "unity of impression" presents the poet as a craftsman who purposefully constructs the literary text to maximize an intended effect, Coleridge emphasizes to a much larger extent the supposedly natural source of creativity. Both visions, however, are sides of the same coin, that of the Romantic conception of the poet as genius. Ralph Waldo Emerson integrates both perspectives into the figure of the poet as an "transparent eyeball" (Emmerson 1990, 18). The poet is a privileged seer but s/he functions as a medium through which nature, directly linked with the spiritual, flows and speaks to the world.

In the age of machine learning, it seems, we are witnessing a return to a Romantic conception of poetry – a Romantic conception with a twist, I may add. For many decades, we have deconstructed the Romantic 'poet-as-genius', an idea which regards the poet's creativity as emanating from natural and original inspiration, the source of which is mysterious and which is often attributed to the poet's inspiration, which in turn is only possible because of his/her isolation from society. However, after many decades of deconstructing the genius myth, especially in poststructuralist conceptions of the text as a network of discourses, we now seem to be returning to a Romantic notion of the sublime in fantasies of supreme machine intelligence. The idea of algorithms as black boxes has given rise to a quasi-spiritual idealization of creativity. Where poststructuralism has located creativity in the discursive networks of texts and intertexts, debates on artificial intelligence point to the emergent properties of 'big data' – the latter may sound more technical, yet it often has similarly mystic connotations as do Romantic idealizations of the creative genius.

However, we do not even have to go as far as assuming a spiritual presence in databases to suppose that humans no longer play a major role in the creative process. What else is artificial intelligence, understood as an assemblage of human and non-human elements, than a technical version of the hybrid conception of the Romantic poet as negotiating nature, spirit, and human? What else is the phenomenon of emergence in complexity theory than the Romantic notion of intuition? Why does poetry gain new centrality in discussions around human creativity? The concept of emergence, in particular, understood as the generative faculty of complex systems, has revived debates between natural sciences and the humanities. Isabelle Stengers notes that

> if there is one problem that [...] immediately brings up the question of the 'science wars' with which the ecology of modern practices can today be identified, it is indeed the problem of emergence. For in this case, it is no longer a question of human power confronting the order of nature or creation, but the possibility, for a scientific discipline, of assuming power in a field previously occupied by some other discipline. (Stengers 2011, 208)

Discourses of nineteenth-century creativity and of contemporary artificial intelligence effortlessly converge into a posthumanist conception of creativity, one that connects human and non-human agency, one that locates originality not solely in the genius of the autonomous subject neither in the free-floating web of signifiers, but rather in the coupling of individual input and complex machine. The AI poem "Orange Light," like a poem written by a human, displays emergent features in creating meaning – meaning that transcends the intratextual elements and, moreover, one that literally emerges in the interaction between text and reader. The creative process is essentially posthumanist, just like with a poem written by a human: both human and machine-generated poems were written with technological tools, whether the pen, language, or a deep learning algorithm. It can be assumed that both draw on planned, systematic construction *and* on un-planned, unconscious processes or 'black boxes', whether we may call them pre-rational, discursive, or hidden layers of neural networks that operate on the basis of large datasets. Much of the creative process is processual, it is the product or even a stage in a larger process of reiterations. The traditional process of re-writing and revision is mirrored in AI by the implementation of evolutionary algorithms that learn through internal feedback by so-called in-built "fitness factors" (Gonçalo Oliveira 2017, 15). Eckart Voigts therefore convincingly considers AI generated literature a form of "posthuman adaptation," in which processes of replication, emulating, simulating, and for that matter, adapting play a major role. "We think of adaptations as emerging from the work of human adaptors," Voigt writes, "but automated writing practices executed through artificial neural networks or deep-learning applications can also be thought of as adaptations" (Voigts 2020, 2). Voigts suggests to distinguish between two kinds of posthuman adaptation in the context of AI generated art: the first category "focuses on adaptation practices that shift the man-machine boundary to the machine via AI," namely one in which machines increasingly create text, music, images etc. Although these processes still rely on previous, often human-made 'data' and are therefore to a significant degree re-iterations of existing pre-texts, processes of automation increasingly replace creative processes hitherto conceived of as exclusively human, therefore also challenging human exceptionalism as formulated by posthumanist extension of human creativity. The second category of AI generated texts, as Voigt notes, is what he calls AI as "Kulturtechnik ('cultural technique', also translated as 'cultural technologies', 'cultural technics', or 'culturing techniques')" (Voigts 2020, 2). Assuming that adaptation, focusing on change and transformation, is "inherent in cultural practices" and more and more in "machine-human networks" (Voigts 2020, 15), this view suggests that AI generated literature is nothing new but rather another step in the evolution of human-machine interaction in the creative process.

The recent increases in automatic poetry generation have prompted Oscar Schwartz, the creator of the Turing test for poetry, to reflect more theoretically on cultural narratives of human and non-human intelligence. Schwartz takes as a vantage point a traditional divide between two concepts of artificial intelligence by pioneers in this field, namely Alan Turing's concept of humans-as-machines and J. C. R. Licklider's human-machine interaction model. Whereas Turing had in mind the transposition of anthropomorphic features to computers to create machines that could eventually replace humans, Licklider rather followed a "hybrid vision of AI" in proposing that both machines and humans need each other to be creative (Schwartz 2018, 89). This division could be translated into a transhumanist vs posthumanist conception of the future. Transhumanism, as formulated by Nick Bostrom or Daniel Dinello, assumes the existence of super-human intelligence that transcends the existence of humanity whereas posthumanism, as formulated by Cary Wolfe, Karen Barad, and Katherine Hayles, rather conceives of the human as coexistent and symbiotically entwined with non-human entities such as machines. Andreas Sudmann observes that next to euphoria and a belief in technology, as well as anxieties toward apocalyptic scenarios, our attitudes to artificial intelligence are mainly determined by a third component: a fundamental skepticism as well as an insistence on the anthropological difference to machine intelligence of the computer (Sudmann 2018, 13). To shift the focus, as Sudmann does, from the computer as a technology or medium, to the *relationships* between human and technology, on connectivity, interface, and embeddedness (Sudmann 2018, 17), seems a productive way out of the one-sided and reductionist view of AI creativity as replacing and thereby threatening human creativity.

As valuable as Turing's advances in computation have been, especially for their bold formulation of machine-human relations that would later be highly influential in cybernetic and artificial intelligence research, his perspective on poetry is intrinsically limited. We have to remember that Turing generated much of his machine intelligence theories from decoding encrypted messages and developing chess computer algorithms – both of which heavily rely on problem-solving procedures. Writing poetry per definition is a creative act that is much less based on solving problems (although there are of course forms of literature that operate on levels of binary decision-taking, such as interactive fiction or text puzzles). The fundamentally open format of poetry, however, cannot satisfactorily be compared even to board games like Go that offer an extraordinarily large number of possible choices and in some Asian cultures are considered an art form rather than a game. In the end, the functionality of a game of Go is still largely dependent on the motivation to win the game, which is an outcome that can be precisely defined and quantified and therefore lends itself perfectly to the implementation of computational solutions – if, in the case of Go, this required a highly complex deep learning algorithm that became famous in 2016 under the name of 'Alpha Go' and that has substantially changed professional games. Still, the 'success' of an AI generated poem can neither be quantified nor sufficiently determined, neither by a machine or a human – in fact the very category of 'success' seems irrelevant or even counterproductive in this case.

Rather, the case of artificially generated poetry can open up larger questions about our posthuman embeddedness, if we follow the second trajectory suggested by Schwartz and the notion of AI creativity as "cultural technique" suggested by Voigts (Voigts 2002, 2), namely that of machine-human collaboration. In the final part of this essay, let me therefore propose three questions that result from such a posthumanist conception of artificial intelligence. These questions are designed as alternatives to the question of whether computers can write poetry. They avoid the cliché anxiety in view of creative bots that threaten humans by making us redundant. Rather, they urge us to think along the lines of mutually beneficial relationships and creative partnerships between humans and bots:

What can AI generated poetry reveal about human creativity?

Artificial intelligence is developed not only to enhance or complement human intelligence but also to learn more about human intelligence. This perspective effectively shifts the focus away from science fiction scenarios of 'threatening AI' toward notions of AI as a way of understanding and enhancing human interaction with the world. As Margaret Boden claims, "AI concepts help to explain *human* creativity," as they "enable us to distinguish three types: combinational, exploratory, and transformational" (Boden 2016, 68). For Boden, to define and develop machine creativity presupposes a particular approach to categorizing human creativity. It may not come as a surprise that although all three types of creativity can be demonstrated by AI, it is exploratory AI (producing new ideas by using established rules) rather than combinational and transformational creativity that is best suited for artificial intelligence, as in machine composition in the style of particular composers. Based on predefined sets of rules, machine learning algorithms seem to be able to (re-)produce and explore the limitations of given styles, genres, and forms. However, Boden admits that "even exploratory AI depends crucially on human judgement. For someone must recognize – and clearly state – the stylistic rules concerned" (Boden 2016, 70 – 71). The implicit assumption behind such comparisons between human and machine creativity seems to be that creativity is a core element in the definition of what it means to be human. Machine creativity is thus usually measured by how closely it can match human creativity. This view is problematic for it naturalizes human intelligence and dismisses the fact that human intelligence is always also materially embedded (cf. Krämer 2012, 92). Moreover, this is a potentially ableist perspective, one that equates creativity predominantly with cognitive abilities and that regards human intelligence and creativity as the gold standard for problem-solving and even aesthetic expression (although of course we have long admired the non-human aesthetics of botanical patterns, rock formations, or what has been believed to be spiritual phenomena).

The seemingly random and potentially uninspired phrases strung together by a poetry bot may reveal that humans still master the art of aesthetic unity, even though machines increasingly manage to surprise us in that field. Stefan Rieger has demonstrated how the machine has historically evoked a "negative semantics" (Rieger 2018, 117, my translation). The mechanical, Rieger argues, has usually served to denote deficits and oppositions to culturally more valuable concepts, as it is associated with dull repetition and rigid rules, rather than creativity, genius, and free self-expression (Rieger 2018, 117). This skepticism, Rieger holds, is still present in our relationship to robots and artificial intelligence, even though the Cartesian dualism of mind/matter is overcome in favor of a view of interdependent human-robotic agencies (Rieger 2018, 133). Yet, rather than looking at AI generated poetry to identify the shortcomings of the machine, may we not also be prompted to recognize and even appreciate the imperfection of human intelligence? Sibylle Krämer[3] asks whether we *need* machines to replace human creativity. We could also ask: what are the pleasures of indulging in the shortcomings of human embodiment or, rather, how to enjoy what Jack Halberstam, in a different context has called "the art of failure?" In a poem on artificial intelligence, Adrienne Rich wrote in 1961[4]:

> Still, when
> they make you write your poems, later on,
> who'd envy you, force-fed
> on all those variorum
> editions of our primitive endeavors,
> those frozen pemmican language-rations
> they'll cram you with? denied
> our luxury of nausea, you
> forget nothing, have no dreams.

Is our human capacity to forget, are our very human biases, not also a potential to be savored in a world of constant self-optimization and is not poetry exactly the literary genre that has the potential to resist precision and totality? In fact, poetry bots may perfectly show us how to resist perfection and how much of the seemingly optimized and functional technology is equally 'productively' failing. Instead of using artificial intelligence to think about continuous human enhancement, may not artificially generated poetry offer a refreshing perspective in reminding us about the nonsensical, the non-functional, and the non-economical that both machine and human creativity share?

In what way is bot poetry a continuation rather than a radical rupture of poetic practice?

Often, human and machine intelligence are regarded as distinct from each other, even as diametrically opposed – at least from the perspective of the humanities. But rather than regarding automatically generated poetry as distinct from human generated poetry, what happens if we understand them as part of a continuum? Every act of recording and reading literature is a mediated experience that relies on technologies of transcription, abstraction, and automation. If both human and machine intelligence depend on mediated operations of information proc-

---

[3] Cf. the essay by Sybille Krämer in this collection.
[4] Rich, Adrienne. "Artificial Intelligence". 1961. *Collected Early Poems 1950-1970*. New York: Norton, 1993.

essing that result in emergent, thus often unexpected, forms of creative output, artificially generated poetry may be seen as just another step in the evolution of poetic innovation and, as Voigts suggests, adaptation – one that does not necessarily transcend human agency but one which may herald a new phase in literary practice – quite similar to the shift from oral to written traditions of storytelling or poetic practice in Western cultures. The new combinatory potentials of large data repositories may yield interesting novelties in creative language use and may contribute to the modernization of rhetorical and formal conventions by testing the limits of human imagination.

How can collaborative/interactive writing help us push the boundaries and emanate new creativity in literature?

Human-machine collaboration is probably the most realistic pathway for so-called artificial intelligence in the near future and it is that which is most prevalent today. Ever since the invention of word processing, poetry writing has relied on 'artificial intelligence,' especially with advanced word editing and thesaurus functions. Today, we all use search engines as well as instant messaging services on our smart phones with the help of autocomplete text production. Likewise, most artificially generated poems do not start entirely from scratch but rather take human texts as a basis (Gonçalo Oliveira 2017, 16). Beyond other human-bot collaborative practices such as 'found poetry', there are interesting experiments with random generators of poetry based on human input. Google's AI section has recently developed the playful tool "Verse by Verse." Upon entering the website, the reader is asked to pick a total of three "muses" from a list of famous American poets, in which style the AI writes a poem. Formal parameters (quatrain, couplet, free verse) as well as syllable count per line can be adapted before the reader is asked to write the first line of the poem. What follows are suggestions as to how each line may continue, sorted by the respective style of each of the three muses. The following is a randomly generated poem based on the three poets Amy Lowell, Walt Whitman, and Emily Dickinson:

> Sitting at my desk
> Chair after light I come,
> Lest you stand at a door.
> Away must be as late,
> They will last in the door.

Examples like this combine human creativity with modular selections of text samples generated by machine learning algorithms on the basis of massive data input. These playful recombinatory generative text tools are not entirely new but continue modernist collage techniques or postmodern chance experiments. Yet the modular aesthetics of new media enable to a much larger degree the posthuman interaction between human and machine creativity than traditional print formats would have done. Rather than aiming at machines simulating human intelligence or creativity, what if humans understood better non-human intelligence and creativity to explore new ways of cooperation and collaboration? Steven Shaviro has proposed to take seriously the "mental functioning and subjective experience" that are not cognitive but that can be described as "sentience, whether in human beings, in animals, in other sorts of organisms, or in artificial entities" – a form of experiencing the world that he calls "discognition" (Shaviro 2016, 10). To approach from Shaviro's assumption that human intelligence has disproportionately been conceived of in terms of cognitive processes and that to understand intelligence we may want to turn to how computers, aliens, or even slime molds think, even if 'thinking' does not imply the anthropocentric bias of implying cognition (Shaviro 2016, 204). As Krämer has noted, the computer as a model for mental functions has helped to conceptualize cognitive performance independent from phenomenal qualities (Krämer 2012, v). In a more radical way, Merlin Sheldrake ponders the possibility of how mushrooms employ non-human 'thinking' to instrumentalize human intelligence as a medium. Taking the mind-manipulating strategy of the Ophiocordyceps mushroom as an example, Sheldrake proposes the idea that analogously to nineteenth-century spiritualism, "mind-manipulating fungi possess the insects that they infect. Infected ants stop behaving like ants and become mediums for the fungi" (Sheldrake 2020, 118–119). As intriguing as Sheldrake's mushroom-as-medium theory may sound for the possibility of plant-human communication, it automatically conjures up deep-seated and potentially misleading anxieties over robotic control over humans if applied to the case of artificial intelligence. And yet, both Shaviro's and Sheldrake's propositions are interesting if we want to explore the possibilities of human-AI collaboration in literary text generation. If we stop asking how computers can think like humans and start understanding how computers operate, productive new ways of partnership might arise that combine associative with procedural, linear with non-linear, and iterative with evolving operations – forms of collaborations that have fascinated code or algorithmic artists and authors experimenting with generative poetry for some time. To regard generative AI less as the sole agent in the creation of literature may also help to avoid a return to the formalism that the New Critics celebrated in the middle decades of the 20th century and that we have long left behind for its obsession with the text as the sole authority. Rather, AI could be regarded as a tool and a collaborative partner in which both machine and human complement each other. For the field of artificially generated art, Marian Mazzone and Ahmed Elgammal have argued that algorithmic art can mean redefining human and machine creativity. Rather than regarding art as the sole creation of the human genius, they propose that in AI using generative adversarial networks (GANs), "AI is used as a tool in the creation of art. The creative process is primarily done by the artist in the pre-and post-curatorial actions, as well as in tweaking the algorithm" (Mazzone / Elgammal 2019, 2). Correspondingly, a poet working *with* AI rather than being replaced by AI might be the future of generative poetry. It might be less the vision of the machine as the creator but rather as one of various instruments used along different stages of the creative process. To regard the poet as a coder and as a curator of a database may be the next step to liberate the poet from the Romantic myth of the genius. Also, the question of whether the work of art lies within the poetic output or in the process may have to be renegotiated. AI generated poetry, in that sense, may be much more like a form of conceptual poetry than what William Butler Yeats had declared to be "automatic writing" in the early 20th century. Conceptual poetry, as associated with poets such as Kenneth Goldsmith or Christian Bök, in its emphasis on concept and process, may direct our attention to the explorational and emergent results both in automated and spontaneous forms of creativity.

Perhaps, the real Turing test for computer generated poetry should not be whether computers are good enough to replace us, a question that would feed into transhumanist visions and anxieties of general artificial intelligence. Rather, the question could be to what extent poetry, once more, can help us recognize our human position within larger, non-human networks of interconnected operation and signification. Can we conceive of a posthuman future that both restores yet does not overrate faith in human creativity? It may not be too unlikely to presume that non-human creativity can imagine the world differently and even more sustainably than humans, if human prediction and vision is necessarily short-sighted, limited, and flawed by the boundaries of our embodied perception and the blind-spots of our positionalities (although, certainly, AI largely reproduces human biases with detrimental effects). Can we admit the pleasures of encountering the non-human 'other' in the act of aesthetic experience? And is aesthetic experience exclusively reserved for the human species? Once, again, poetry may help explore these fundamental questions concerning our entangled accesses to and interactions with the world.

Works cited

Boden, Margaret A. *AI: Its Nature and Future*. Oxford UP, 2016.
Coleridge, Samuel Taylor. "Shakespeare, A Poet Generally." *The Complete Works of Samuel Taylor Coleridge*. Ed. Prof. Shedd. vol. 4. Lectures Upon Shakspeare and Other Dramatists. New York: Harper & Brothers, 1854. 46–56.
Emerson, Ralph Waldo. "Nature." *Selected Essays, Lectures, and Poems*. New York: Bantam Classic, 1990.
Fish, Stanley. "Interpreting the 'Variorium'." *Critical Inquiry* 2.3 (Spring 1976), 465–85.
Frenkel-Brunswik, Else. "Intolerance of Ambiguity as an Emotional and Perceptual Variable." *Journal of Personality*, 18.1 (September 1949): 108–43.
Gonçalo Oliveira, Hugo. "A Survey on Intelligent Poetry Generation: Languages, Features, Techniques, Reutilisation and Evaluation." *Proceedings of the 10th International Conference on Natural Language Generation*, Association for Computational Linguistics (2017): 11–20.
Halberstam, Jack. *The Queer Art of Failure*. Durham: Duke UP, 2011.
Krämer, Sybille, ed. *Geist – Gehirn – Künstliche Intelligenz: Zeitgenössische Modelle des Denkens. Ringvorlesung an der Freien Universität Berlin*. 2012.
Mazzone, Marian and Ahmed Elgammal. "Art, Creativity, and the Potential of Artificial Intelligence." *Arts* 8.1 (2019).
Nünning, Ansgar, and Vera Nünning. *An Introduction to the Study of English and American Literature*. Stuttgart: Klett, 2004.
Poe, Edgar Allan. "The Philosophy of Composition." *Graham's Magazine* 28.4 (April 1846): 163–67.
Rich, Adrienne. "Artificial Intelligence". 1961. *Collected Early Poems* 1950-1970. New York: Norton, 1993.
Rieger, Stefan. "'Bin Doch Keine Maschine…' Zur Kulturgeschichte eines Topos." *Machine Learning: Medien, Infrastrukturen und Technologien der Künstlichen Intelligenz*. Ed. Andreas Sudmann and Christoph Engemann. Bielefeld: Transcript, 2018. 117–42.
Roque, Antonio. "Language Technology Enables a Poetics of Interactive Generation." *The Journal of Electronic Publishing* 14.2 (October 2011).
Schwartz, Oscar: "Competing Visions for AI: Turing, Licklider and Generative Literature." *Digital Culture & Society*. Rethinking AI. 4.1 (2018): 87–105.
Shaviro, Steven. "Discognition." London: Repeater, 2016.
Sheldrake, Merlin. *Entangled Life: How Fungi Make Our Worlds, Change Our Minds & Shape Our Futures*. London: Penguin, 2020.
Stengers, Isabelle. *Cosmopolitics II*. Transl. Robert Bononno. Minneapolis: U of Minnesota P, 2011.
Sudmann, Andreas. "Einleitung." *Machine Learning: Medien, Infrastrukturen und Technologien der Künstlichen Intelligenz*. Ed. Andreas Sudmann and Christoph Engemann. Bielefeld: Transcript, 2018. 9–36.
Turing, Alan. "Computing Machinery and Intelligence." 1950. *The Essential Turing: Seminal Writings in Computing, Logic, Philosophy, Artificial Intelligence, and Artificial Life, plus the Secrets of Enigma*. Ed. B. Jack Copeland. Oxford: Clarendon Press, Oxford UP, 2004. 441–64.
Verse by Verse. Google AI. https://sites.research.google/versebyverse/

Voigts, Eckart. "Algorithms, Artificial Intelligence, and Posthuman Adaptation: Adapting as Cultural Technique." *Adaptation* (May 2020): 1–20.

Reinhart Kögerler and Klaus Viertbauer

Why Neuroenhancement is a Philosophical Issue

Abstract: Human beings have always understood themselves as being different from other forms of life. From this assumption, they have derived specific normative concepts such as human dignity and the status of a person. But is this normative special position still feasible when techniques of self-modeling, which are summarized under the keyword "neuroenhancement", are applied? Do changes in the concepts of autonomy and authenticity possibly also dissolve the boundaries between humans and animals? Are we on the way to a new human species, and is there even a moral obligation to work towards such an evolution? This paper attempts to summarize this debate and to assess the main arguments.

While Neuroenhancement (NE) has been the subject of heated debates in the field of applied ethics for considerable time, especially in the USA, interest in German-speaking discourse has only begun to stir in recent years.[1] Thus, this paper is not intended as a contribution to an ongoing debate in the field, but rather as an attempt to open up the general question for the reader and to show whether and to what extent NE represents a philosophical problem. Therefore, in a first step, we try to examine what NE stands for and, in a second step, we map the field of philosophical issues in the ongoing debate.

1 What does NE stand for?

Fundamentally, enhancement is understood as an attempt to increase the performance of a human organism.[2] NE is then specifically related to certain capacities of the nervous system, a general compact definition being: NE encompasses any

---

Note: This paper is an extended version of our German introduction "Neuroenhancement als philosophisches Problem", Klaus Viertbauer, Reinhart Kögerler (Hg.), *Neuroenhancement. Die philosophische Debatte*, Berlin: Suhrkamp 2019, 9–17, which is based on our discussions in Salzburg 2015 and Vienna 2019 with Dieter Birnbacher, Reinhard Merkel, Michael Pauen, and Dieter Sturma.

1 Cf. Kipke 2011, Wagner 2017, Leefman 2017, Fenner 2019.
2 Cf. Juengst 1998, 29–47.

procedure to improve (amplify) certain mental capacities of a healthy person by means of bio-technical interventions applied to the living organism. Some aspects should be clarified at the beginning:

Although most of the current types of intervention techniques of NE have been derived from therapeutical purposes (i.e., medical treatment of sick or mentally disordered persons), within the NE context they are not applied for therapeutical objectives but for the improvement of healthy persons. These improvements can refer to
- *cognitive abilities* (e.g. attention, concentration, memory performance, creativity)
- *emotional conditions* (e.g. mood, characteristic traits, propensity, decisiveness, communicative skills)
- *moral motivations* (e.g. incentives, feelings of reward, patience, empathy)

The modes of interventions used go beyond traditional techniques (such as education, sports training, meditation). Specifically, they may be
- *pharmacological/chemical* (e.g. coffee, cocaine, Ritalin, Modafinil, Prozac, antidepressants, Oxytocin)
- *physical/technical* (e.g. transcranial magnetic stimulation, ultrasound stimulation, implants, brain-computer interfaces [non-invasive])
- *surgical* (e.g. implantation of electrodes in the brain, invasive brain-computer interfaces)
- *genetic* (e.g. gene editing, CRISPR/Cas, SHEEFs)

Some of these techniques are already well established and frequently practiced (e.g. the pharmaceutical ones), some are of a more futuristic nature, and some are considered dangerous or even inhumane by the majority of people. However, most of the last type are already used for medical treatments and are thus in principle available. In order to sketch the width of the less known possibilities let us mention two examples:

The *first* is best explained in terms of a case study: A decathlete wants to participate in an important sports competition (Olympic Games, say). But he knows that he cannot bring himself to do the necessary daily training (5 hours/day). To overcome this obstacle, his coach mentions to him the possibility of undergoing a new form of neuroinvention designed to create artificial feelings of reward in his brain (either with adopted transcranial magnetic stimulation or with tiny electrostimulators implanted in the mesolimbic system). After always applying this intervention 2 hours after starting his training, he found it tolerable to continue his exertion for a further 3 hours – even with relish. And so he succeeded![3]

The *second* example is the enhancement of cognitive or emotional abilities with the help of (non-invasive) brain-computer interfaces (BCI): Brain signals of certain brain action potentials (slow cortical potentials or sensomotoric rhythms) are recorded from the scalp by means of EEG, then sent by wireless link to a computer. There the signals are amplified, algorithmically interpreted and translated for further use by the client. The uses of the information thus gained include neurofeedback training which enables the client voluntarily to change (intensify) some cortical potential in a specific frequency area. In this way he is able to identify (create) time slots of readiness for much higher cognitive abilities (such as enhanced reaction times, improved memory, higher sensitivity).

In spite of all currently available or at least emerging possibilities for and all the potential of NE, one cannot overlook strong (public) reservations about NE. In fact, many objections (e.g. medical, ethical, legal, political) have been raised to most forms of NE. They address
- unknown possible side effects (e.g. causation of addiction),
- risk of mental or physical overload,
- loss of personal autonomy (in particular in the case of automated BCIs),
- possible long-term effect on personality, and – increase in competitivity and social inequity.

A very widespread although certainly superficial objection to NE in general is expressed in the allegation that "It is against human nature." A closer look shows that many allegations are not really tenable. Also, it is interesting to note that there are far fewer objections to conventional methods aiming at the same improvements in mental abilities (such as learning, mental exercises, asceticism, meditation) than to chemical or physical methods. Nevertheless, one certainly encounters in general considerable reservations about the whole endeavor of NE. This fact strongly indicates that there is a great need for a deeper philosophical analysis of the whole problem area.[4] In the following we focus on this philosophical discourse, and consequently on normative issues connected with NE.

---

[3] Cf. Merkel 2019, 44–6.
[4] A concise overview of the relevant questions and approaches can be found in e.g. Parens 1998; Schöne-Seifert et al.2009; Viertbauer and Kögerler 2019.

2 Critical issues: How NE confronts genus ethics, autonomy, and authenticity

In this second section we map the field of the most discussed problems connected with NE within applied ethics. As we see it, these are the dilemma of genus ethics (2.1), autonomy (2.2), and authenticity (2.3) with its sub-discussion of moral enhancement (2.4).

2.1 The dilemma of genus ethics

From the early stages of history human beings have considered themselves to be different from other creatures, from whom they distinguish [and with whom they contrast] themselves as a separate species. Our culture is based on an ontological difference: on the one side there is man, on the other side animals and plants. These are generally classified as living beings, but ones to which a lower value is generally ascribed. This distinction has deep cultural roots, as can already be seen in the biblical narratives of creation, which are equally binding for Christianity and Judaism. In this account of creation (from the sixth century BCE), God addresses the human beings directly:

> And God blessed them, and God said unto them, Be fruitful, and multiply, and replenish the earth, and subdue it: and have dominion over the fish of the sea, and over the fowl of the air, and over every living thing that moveth upon the earth.[5]

One can find similar thoughts in Hellenism and later in the Koran. Also the modern constitutional state that emerged from the Enlightenment recognizes only humans as legal entities; the interests of non-human creatures, such as those of other more highly developed vertebrates, are regulated by a rudimentary animal protection law. This list could be extended at length. Such an ontological prioritizing of the human species cannot be based on the findings of genetics – the difference between the gene pools of humans and chimpanzees is less than 2%! The evolutionary paradigm, which has been continuously enriched with empirical facts since the eighteenth century, assumes a common origin of all species and fluid transitions between them. But from where does the demarcation between man and animal get its justification? How can we still speak of a

---

[5] Gen 1,28–30.

human species on this basis? And has the top of the process of evolution already been reached with the human being?

In principle, human beings define themselves as autonomous subjects in contrast to animals. From this autonomy, humans derive basic claims with normative relevance, such as human dignity and the status of being a person. This image of the human is the result of a long process of transformation in religious and secular values.[6] The question now arises of the extent to which this primacy is challenged by the self-modeling of the human being through NE. Normative issues concern in particular ideas that suggest to dissolve the boundary between humans, animals and machines. Does the human species as such dissolve with the change in the concepts of autonomy and authenticity? Are we on the way to a completely new form of life, and is there perhaps even a moral obligation for such a further development?

Questions like these are characteristic of the debate in modern bioethics. In this context, Peter Singer has proposed disentangling the concepts of "human being" and "person":

> We have seen that 'human' is a term that straddles two distinct notions: being a member of the species Homo sapiens and being a person. [...] The belief that mere membership of our species, irrespective of other characteristics, makes a great difference to the wrongness of killing a being is a legacy of religious doctrines that even those opposed to abortion hesitate to bring into the debate.[7]

In contrast, Kant's formula of the moral imperative that defines human beings as ends-in-themselves is often referred to: "So act that you use humanity, whether in your own person or in the person of any other, always at the same time as an end, never merely as a means."[8]

The core of the controversy is the question whether the concepts of a human being and a person always coincide or whether cases can be identified in which a human being is not a person or a person is not a human being. While deontological ethical concepts (such as Kant's) emphasize the universal character of norms, a consequentialist moral justification (such as Singer's) always starts with the specific case and examines whether the interests of all those concerned are taken into account in the best possible way. Is it now morally legitimate to delimit the human species with respect to the category of person concerned? Singer disputes this with regard to the ability of animals to suffer:

---

6 Cf. Habermas 2019.
7 Singer, P. ³2011, 135.
8 Kant 1998, 38 (= GMS, AA 04: 429).

> A stone does not have interests because it cannot suffer. Nothing that we can do to it could possibly make any difference to its welfare. A mouse, on the other hand, does have an interest in not being tormented, because mice will suffer if they are treated in this way. If a being suffers, there can be no moral justification for refusing to take that suffering into consideration. No matter what the nature of the being, the principle of equality requires that the suffering be counted equally with the like suffering – in so far as rough comparisons can be made – of any other being. If a being is not capable of suffering, or of experiencing enjoyment or happiness, there is nothing to be taken into account.[9]

The move to attribute "naturalness" to the status quo, which is to be preserved in principle, is untenable in bioethical discourse:

> Examining the ethical discourse of the last century, one is immediately led to the impression that naturalness, as a principle of value (at least in the academic treatment of ethics) has been discredited once and for all. As a principle for the judgment of human behavior, naturalness has not, for a very long time, played a role worth mentioning. Instead of serving as a guide for human behavior, every attempt to establish naturalness as a moral criterion has on the contrary to be prepared for criticism and anticipate the objection that any such attempt involves the illegitimate derivation of an 'ought' or 'must' from a mere being and is therefore subject to a 'naturalistic fallacy.'[10]

Naturalness is not a normative value that exists *a priori* but describes value concepts that have been developed *a posteriori* within the socialization processes. From this perspective, the challenge is to identify concrete values and to discuss the moral status of techniques or actions such as self-modeling along these lines. For this purpose, the values of autonomy and authenticity seem to be the most suitable. In view of a world that is becoming out of joint, the question of social relevance also arises. These are the points along which the debates on the moral legitimacy of NE are being conducted.

2.2 The problem of autonomy

Autonomy is a difficult and complex concept. "Autonomy" describes both a social and a psychological phenomenon. In the present context, autonomy refers exclusively to the psychological phenomenon. The modern coining of the term, with the meaning it received among others in the writings of Kant, has to be harmonized with the findings of neuroscience.[11] The experiments of Ben-

---

9 Singer, P. ³2011, 50.
10 Birnbacher 2006, 17.
11 Cf. Bennett et al. 2003.

jamin Libet as well as those of Patrick Haggard and Martin Eimer, which are highly controversial in their execution and interpretation, represent a milestone in the debate.[12] In his well-known experiments, Libet asked human beings to perform a hand movement at a self-selected point in time. In detail, they were asked to look at a clock in front of them and to remember when they were aware that they wanted to perform the hand movement. In parallel, Libet connected his subjects to an EGG to measure their brain waves. This showed that the subjects set 200ms before the actual movement as the time at which they became aware that they wanted to perform a certain action, but that a readiness potential (RP) was already established in the EGG 550ms before the actual action.

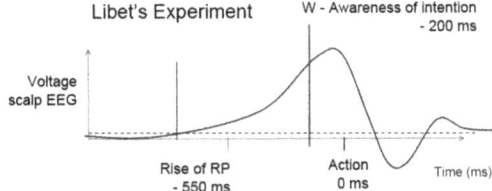

Figure 1: Libet's Experiment

Accordingly, the philosophical discussion is ignited by the 350 ms between the readiness potential and the expression of will. In the phase of the associated discussions in the German-speaking world, thinkers such as Wolfgang Prinz, Gerhard Roth and Wolf Singer pleaded publicly for abandoning the assumption of human free will.[13] Within the philosophical debate this type of interpretation is called "hard determinism" and contrasted with those of "libertarianism":

> A hard determinist is an incompatibilist who believes that determinism is in fact true (or, perhaps, that it is close enough to being true so far as we are concerned, in the ways relevant to free will) and because of this we lack free will. A libertarian is an incompatibilist

---

[12] Cf. Libet et al. 1983; Haggard / Eimer 1999.
[13] The articles of the public debate fought out in daily newspapers were published by Geyer, *Hirnforschung und Willensfreiheit*, 2004. Approaches of "hard determinism" can be also found within the American debate by Patricia Churchland, Daniel Dennett, and Daniel Wegner. Cf. Churchland 2013, Chap. 7; Dennett 1984; Prinz 2013; Roth 2001; Singer, W. 2013.

who believes that we in fact have free will and this entails that determinism is false, in the right kind of way.[14]

In the meantime, the debate has split into several sub-positions. While some deny the compatibility of free will and determinism in principle, compatibilists (or "weak determinists") strive to build a bridge. In fine-grained scenarios, conceptualizations are examined here, with the help of which it may be possible to show how autonomous action can be realized against the background of neuroscientific findings. The question *is* crucial in the context of NE precisely because of the targeted attempt to induce certain neurological states. In other words: is it conceivable that through the targeted generation of certain conditions of the brain or the consciousness, autonomy gradually dissolves? And what does that mean for our image of man and our discourse on justice, guilt and dignity?

2.3 The problem of authenticity

In recent years, authenticity has become a buzzword in various academic discourses. A philosophical conceptualization is found only sporadically, for example in the work of Charles Taylor, who refers to the congruence of a person with his or her interests.[15] In other words, if a person's action is in accordance with his interests, he is considered authentic; if he acts contrary to them, he is in danger of becoming increasingly alienated from himself.[16] In this process, a person's self-image begins to crumble: pathological manifestations such as despair and anxiety characterize this development. The extent to which NE might affect a person's self is heatedly debated. Following Peter D. Kramer, this has been discussed with regard to the antidepressant Prozac: is it legitimate to prescribe pharmacological mood enhancers to individuals in a life crisis? Or should they be encouraged, in the context of a therapy, to take their lives into their own hands in order to act authentically?[17] A similar argument can be found in the

---

14 Vihvelin 2018. Scholars like Mark Balaguer, Robert Kane, and Peter van Inwagen are arguing in favor of Libertarianism. Cf. Balaguer 2009; Cane 1996; van Inwagen, 1975.
15 Cf. Taylor 2018.
16 In the concept of "alienation" the social and the psychological use touch each other in the sense that, following Karl Marx, one tries to show that socialization mechanisms that are considered pathological lead to forms of psychological alienation. Cf. Wagner 2017.
17 Kramer 1993. – The debate about whether neuroenhancement promotes or hinders authenticity was also, and above all, fought out by Carl Elliot and David DeGrazia. Cf. DeGrazia 2000; Elliot 1998.

German-language debate in the work of Jürgen Habermas, who distinguishes between what is grown and what is made in the eugenics discussion: "In the life history of the person concerned, the transformed expectations turn up as a normal element of interactions, and yet elude the conditions of reciprocity required for communication proper."[18]

The authentic agreement of a person with certain interests must come from the person him- or herself and cannot be taken over by third parties – be they parents, educators or teachers. A person must at least be granted the right of veto in the form of a yes/no statement so that he or she can appropriate a certain interest as his or her own. Thus in principle it must remain possible to correct the influence of third parties throughout a person's life if asymmetrical dependencies are to be avoided. Whether these also exist in the case of a premature pharmacological indication and whether persons who are treated with antidepressants in life crises are capable of being authentic is the subject of an ongoing debate.

2.4 Moral enhancement as a subclass of the authenticity problem

Moral Enhancement (ME) is defined as the improvement of common normative convictions by means of pharmacological substances. This idea and the arguments for the possibility of such NE interventions have been promoted by thinkers like Thomas Douglas, Julian Savulescu and Ingmar Persson.[19] The core idea is that within the evolution paradigm one generally assumes that traditional normative convictions, as mental structures of our ancestors (i.e. their behavioral patterns, action preferences or moral motivations), have been developed by adapting to the natural and social living conditions. Now we need to consider that, for most of the period of early evolution, people mainly lived in small communities (of some 150 members, say), which constantly had to fight against other similar groups for scarce available resources. This situation, together with the generally limited possibilities for action presumably determined many elements of early human morality. Among them were probably:
- readiness to cooperate within one's own group (*group altruism*)
- reservations against or refusal to accept strangers (*xenophobia*)

---

18 Habermas 2003, 51.
19 Cf. Douglas 2008; Persson / Savulescu 2012.

- the realization that to refrain from doing harm is better than efforts to correct harm done (*the asymmetry of doing and omitting*)
- a preference for dealing with problems that are nearby in space and time, disregarding more remote consequences or causations.

Thanks to the general process of civilization and, in particular, to scientific and technological progress, our potential for action has been increased immensely in both the force and the scope of its impact. This concerns not only positive but also negative consequences. Today, even one individual alone can cause immense damage, even wipe out the entire human race. And the fate of a small local community today is heavily influenced not only by its local neighbors but by the whole global situation (as in the case of the phenomenon of global warming).

We are thus getting more and more into a mismatch or asymmetry between our outdated (myopic) norms on the one hand, and our ever increasing powerful technological possibilities on the other. This asymmetry appears to cause our inability to solve the major problems of our modern world: environmental threats such as climate change or the loss of biodiversity, global social disparities, international terrorism and so on. In the light of this experience Douglas argues that

> there is clearly scope for most people to morally enhance themselves. According to every plausible moral theory, people often have bad or suboptimally good motives. And according to many plausible theories, some of the world's most important problems – such as developing world poverty, climate change and war – can be attributed to these moral deficits.[20]

And Persson and Savulescu add that "human beings are not by nature equipped with a moral psychology that empowers them to cope with moral problems that these new conditions of life create."[21]

As they conclude, a necessary, but by no means sufficient, condition for coming to grips with such problems consists in a drastic improvement in the moral competence of the citizens of modern states, for instance, by extending moral sensitivity to what can only be tackled collectively by the whole of society. Whether this improvement can be achieved by "classical" means of education alone is doubted by authors such as Douglas, Persson and Savulescu who argue in favor of a pharmacological form of ME:

---

20 Douglas 2008.
21 Persson / Savulescu 2012, 1.

> As our history eloquently shows, we are more than any other animal biologically or genetically disposed to learn by experience, and we are now learning that our present course of action spells disaster. We *can* decide to overturn any predictions made about what we will intentionally do because no prediction can take into account the effects that it itself will have. [...] Consequently, we can in practice exclude the possibility of future societies in which these forms of behaviour are the rule.[22]

They even identify some first candidates for such interventions: Serotonin – a neurotransmitter which is commonly believed to suppress aggressive and support cooperative behavior; Oxytocin – a hormone which apparently can increase feelings of trust (at least towards members of one's own group), and empathy.[23]

It is not surprising that such far-reaching ideas and suggestions have elicited strong or even harsh criticism from other thinkers. One of these is John Harris, who, although still advocating the need for ME in general, strictly rejects a pharmacological version of it by referring to the freedom of the autonomous subject as a condition of its authenticity. Specifically, he points out that a direct biological intervention in the human bias leads to the irreversible destruction of the authenticity of a subject.

> One thing we can say with confidence is that ethical experience is not 'being better at being good', rather it is being better in knowing the good and understanding what is likely to conduce to the good. The space between knowing the good and doing the good is a region entirely inhabited by freedom. Knowledge of the good is sufficiency to have stood, but freedom to fall, is all. Without the freedom to fall, good cannot be a choice and freedom disappears and along with it virtue. There is no virtue in doing what you must.[24]

3 Conclusion

In our paper we have shown what NE stands for (1) and why it has become a philosophical issue (2). In doing so, we have not added a certain reading or interpretation to the ongoing debates. In our eyes the questions raised are highly controversial but crucial for future society.

---

22 Persson / Savulescu 2012, 101.
23 Cf. Persson / Savulescu 2012, chapter 10.
24 Harris 2016, 60.

Works cited

Balaguer, Mark. *Free Will as an Open Scientific Problem*, Cambridge, MA: MIT Press, 2009.
Bennett, Maxwell, Daniel Dennett, Peter Hacker, and John Searle, ed. *Neuroscience & Philosophy: Brain, Mind, and Language*. New York: Columbia UP, 2003.
Birnbacher, Dieter. *Naturalness. Is the 'Natural' Preferable to the 'Artificial'?* Lanham: UP of America, 2006.
Cane, Robert. *The Significance of Will*. Oxford: Oxford UP, 1996.
Churchland, Patricia. *Touching a Nerve*. New York: Norton, 2013.
DeGrazia, David. "Prozac, Enhancement and Self-Creation." *The Hasting Center Report* 30 (2000): 34–40.
Dennett, Daniel. *Elbow Room*. Cambridge, MA: MIT Press, 1984.
Douglas, Thomas. "Moral Enhancement." *Journal of Applied Philosophy* 25 (2008): 228–45.
Elliot, Carl. "The Tyranny of Happiness." *Enhancing Human Traits*. Ed. Erik Parens. Washington: Georgetown UP, 1998. 177–88.
Fenner, Dagmar. *Selbstoptimierung und Enhancement*. Tübingen: utb, 2019.
Geyer, Christian. *Hirnforschung und Willensfreiheit*. Frankfurt am Main: Suhrkamp, 2004.
Habermas, Jürgen. *Auch eine Geschichte der Philosophie*. Berlin: Suhrkamp, 2019.
Habermas, Jürgen. *The Future of Human Nature*. Cambridge: Polity, 2003.
Haggard, Patrick, and Martin Eimer. "On the Relation between Brain Potentials and the Awareness of Voluntary Movements." *Experimental Brain Research* 126 (1999): 128–33.
Harris, John. *How to be Good*. Oxford: Oxford UP, 2016.
Juengst, Eric T. "What Does Enhancement Mean?" *Enhancing Human Traits*. Ed. Erik Parens. Washington: Georgetown UP, 1998. 29–47.
Kant, Immanuel. *Groundwork of the Metaphysics of Morals*. Trans. and ed. Mary Gregor. Cambridge: Cambridge UP, 1998.
Kipke, Roland. *Besser werden*. Paderborn: Mentis, 2011.
Kramer, Peter. *Listening to Prozac*. New York: Penguin Books, 1993.
Leefman, Jon. *Zwischen Autonomie und Natürlichkeit*. Münster: Mentis, 2017.
Libet, Benjamin, et al. "Time of Conscious Intention to Act in Relation to Onset of Cerebral Activity." *Brain* 106 (1983): 623–42.
Merkel, Reinhard. "Neuroenhancement, Autonomie und das Recht auf mentale Selbstbestimmung." Ed. Klaus Viertbauer, Reinhart Kögerler. *Neuroenhancement. Die philosophische Debatte*. Berlin: Suhrkamp, 2019. 43–88.
Parens, Erik, ed. *Enhancing Human Traits*. Washington: Georgetown UP, 1998.
Persson, Ingmar, and Julian Savulescu. *Unfit for the Future*. Oxford: Oxford UP, 2012.
Prinz, Wolfgang. *Selbst im Spiegel*. Berlin: Suhrkamp, 2013.
Roth, Gerhard. *Fühlen, Denken, Handeln*. Frankfurt am Main: Suhrkamp, 2001.
Schöne-Seifert, Bettina, David Talbot, Uwe Opolka, and Johann S. Ach, ed. *NeuroEnhancement. Ethik vor neuen Herausforderungen*. Paderborn: Mentis, 2009.
Singer, Peter. *Practical Ethics*. New York: Cambridge UP, ^{3}2011.
Singer, Wolf. *Ein neues Menschenbild?* Frankfurt am Main: Suhrkamp, 2003.
Taylor, Charles. *The Ethics of Authenticity*. Cambridge, MA: Harvard UP, 2018.
Van Inwagen, Peter. "The Incompatibility of Free." *Philosophical Studies* 27 (1975): 185–99.
Viertbauer, Klaus, and Reinhart Kögerler, ed. *Neuroenhancement. Die philosophische Debatte*. Berlin: Suhrkamp, 2019.

Viertbauer, Klaus, and Reinhart Kögerler, ed. "Neuroenhancement als philosophisches Problem." *Neuroenhancement. Die philosophische Debatte*. Berlin: Suhrkamp 2019. 9–17.
Vihvelin, Kadri. "Arguments for Incompatibilism." *The Stanford Encyclopedia of Philosophy* (Fall 2018 Edition). Ed. Edward N. Zalta.
https://plato.stanford.edu/archives/fall2018/entries/incompatibilism-arguments/
Accessed: December 15, 2021
Wagner, Greta. *Selbstoptimierung*. Frankfurt am Main: Campus, 2017.
Wegner, Daniel. *The Illusion of Conscious Will*. Cambridge, MA: MIT Press, 2013.

Julia M. Puaschunder

The Future of Artificial Intelligence in International Healthcare: An Index

Abstract: The ongoing COVID-19 crisis has been a challenge for healthcare around the world. This article investigates which countries have conditions favorable to the provision of innovative global healthcare solutions. First, an index based on internet connectivity – as an indicator for digitization and advances in AI – as well as Gross Domestic Product (GDP) – as indicator for economic productivity – is calculated to outline global healthcare innovation hubs with economic impetus around the world. A comparison of countries worldwide shows that advances in AI can be positively correlated with the absence of corruption. Second, a new anti-corruption artificial healthcare index is therefore presented that highlights those countries in the world that have vital AI growth in a non-corrupt environment. These non-corrupt AI centers hold comparative advantages to lead on global artificial healthcare solutions against COVID-19. A third index combines internet connectivity, the level of corruption, and healthcare access and quality. The countries that score high on AI, the absence of corruption and healthcare excellence are promoted as the foremost innovative global pandemic alleviation leaders. The advantages but also potential shortfalls and ethical boundaries in new uses of monitoring Apps, big data inferences and telemedicine to prevent pandemics are discussed.[1]

Introduction

The currently ongoing COVID-19 crisis has challenged healthcare around the world. The call for global solutions in international healthcare pandemic out-

---

[1] The author is grateful for the planning of the international conference on "Artificial Intelligence and Human Enhancement: Affirmative and Critical Approaches in the Humanities from both sides of the Atlantic" envisioned by Herta Nagl-Docekal and Waldemar Zacharasiewicz from the Austrian Academy of Sciences and for the contributions of the participants at this conference. She thanks Professor Susan Rose-Ackerman for her inspiring lectures at the Yale Law School and online presentations due to COVID-19 as well as the participants of the respective class for their helpful share of expertise and interesting discussions. She also appreciates the preparation of the publication of this volume based on the conference at the Austrian Academy of Sciences. She declares no conflict of interest. All omissions, errors and misunderstandings remain solely hers.

https://doi.org/10.1515/9783110770216-011

break monitoring and crisis risk management has reached unprecedented momentum. Digitalization, Artificial Intelligence (AI) and big data-derived inferences are supporting human decision making as never before in the history of medicine. In today's healthcare sector and medical profession, AI, algorithms, robotics and big data are used as essential healthcare enhancements. These new technologies allow monitoring of large-scale medical trends and measuring individual risks based on big data-driven estimations. This article provides a snapshot of the current state-of-the-art of AI, algorithms, big data-derived inferences and robotics in healthcare. Examining medical responses to COVID-19 on a global scale makes international differences in the approaches to combat global pandemics with technological solutions apparent. Empirically, the article answers what countries have favourable conditions to provide AI-driven global healthcare solutions. First, an index based on internet connectivity – as a proxy for digitalization and AI advancement – as well as Gross Domestic Product (GDP) – as indicator for economic productivity – is calculated to outline global healthcare innovation hubs with economic impetus around the world. The parts of the world that feature internet connectivity and high GDP are likely to lead on AI-driven big data insights for pandemic prevention. When comparing countries worldwide, AI advancement is found to be positively correlated with anti-corruption. AI thus springs from non-corrupt territories of the world. Second, a novel anti-corruption artificial healthcare index is therefore presented that highlights those countries in the world that have vital AI growth in a non-corrupt environment. These non-corrupt AI centres hold comparative advantages to lead on global artificial healthcare solutions against COVID-19 and serve as pandemic crisis and risk management innovators of the future. Anti-corruption is also positively related with better general healthcare. Therefore, finally, a third index that combines internet connectivity, anti-corruption as well as healthcare access and quality[2] is presented. The countries that score high on AI, anti-corruption and healthcare excellence are considered to be ultimate innovative global pandemic alleviation leaders. The advantages but also potential shortfalls and ethical boundaries in the novel use of monitoring Apps, big data inferences and telemedicine to prevent pandemics are discussed.

---

[2] https://www.thelancet.com/journals/lancet/article/PIIS0140-6736(18)30994-2/fulltext Accessed: Sept. 25, 2021.

Artificial intelligence (AI)

AI is "a broad set of methods, algorithms, and technologies that make software 'smart' in a way that may seem human-like" (Noyes 2016). The Oxford Dictionary defines AI as "the theory and development of computer systems able to perform tasks normally requiring human intelligence, such as visual perception, speech recognition, decision-making, and translation between languages." AI describes the capacity of a computer to perform like human beings including the ability to review, discern meaning, generalize, learn from past experience and find patterns and relations to respond dynamically to changing situations.[3] AI is perceived as the sum of different technological advancements with currently developing regulation (Dowell 2018). Machine learning are computational algorithms that learn from data in order to derive inferences.

Artificial intelligence – international leadership

AI leadership appears to develop foremost in Europe, North America and China. Together, the United States, China and the European Union represent over 93 percent of total AI private equity investment from 2011 to mid-2018 (OECD 2019). Of those investments, most investment occurred in the United States (accounting for about 70–80%), followed by China and Europe (Breschi, Lassébie & Menon 2018). Start-ups in Israel (3 percent), Japan and Canada (1.6 percent) also played a role (OECD 2019). Over the years AI has also grown in qualitative terms, with widespread applications in healthcare transportation, agriculture, finance, marketing and advertising, science, criminal justice, security and virtual reality (OECD 2019).

The legal and regulatory status of AI is still developing in jurisdictions around the world. The United Nations (UN) agencies and regional organizations report internationally varying contemporary guidelines, ethics codes and action statements. The OECD (2019) hosted a *Council on Artificial Intelligence* in the first half of 2019 to set international AI standards on a global level. The United Nations opened a Centre on Artificial Intelligence and Robotics within the UN system in The Hague, The Netherlands in 2017. The International Telecommunication Union worked with more than 25 other UN agencies to host the "AI for Good" Global Summit. The UNESCO has launched a global dialogue on the eth-

---

[3] https://www.lexology.com/library/detail.aspx?g=4284727f-3bec-43e5-b230-fad2742dd4fb Accessed: Sept. 25, 2021.

ics of AI due to its complexity and impact on society and humanity. In 2017 the International Organization for Standardization (ISO) and the International Electrotechnical Commission (IEC) created a joint technical committee to develop IT standards for business and AI consumer applications. Labor unions have also defined key principles for ethical AI.

Governance in the digital era features data-driven security through algorithmic surveillance. Open source data has become a global regulatory tool. In the sharing economy, participation in the democratic process depends on internet connectivity and data literacy. According to the Library of Congress, most Western world countries aspire to embrace the advantages of AI and become leaders in the field through developing national AI, digital strategies and action plans.

Artificial intelligence in healthcare

AI-supported scientific discovery and medical assistance have increased steadily in the last decade. In recent years, there has been tremendous growth in the range of medical information collected. Every day, healthcare professionals, biomedical researchers and patients produce vast amounts of data from an array of devices, including clinical, genetic, behavioral and environmental cues. To name a few, electronic health records (EHRs), genome sequencing machines, high-resolution medical imaging, smartphone applications and ubiquitous sensing as well as Internet of Things (IoT) devices monitor patient health (OECD 2015). The big data revolution and hierarchical modelling advancements as well as computational power have leveraged inference-driven insights encroaching modern medical care.

As never before in the long history of medicine, improvements in data generation, storage and analysis coupled with unprecedented computational power and statistical means have resulted in large-scale data processing on healthcare. Through machine learning, algorithms and unprecedented data storage and computational optimal control, AI technologies gain information, process it and give a well-defined output to the end-user.

Growth of genomic sequencing databases but also widespread awareness and implementation of electronic health recording have improved the nature and quality of accessible preventive medicine. Daily monitoring creates big data to relate behavioral patterns to health status in order to predict global health trends. Online App-administered tracking provides a complete view of the patient journey over time – covering the spectrum from prevention efforts, early disease state to management of therapeutic choices and therapy-specific outcomes as well as recommendations for future health goals (Puaschunder 2019a, b).

Healthcare has never been as individually-targeted and accessible as today. User self-reporting allows instant information generation and in-depth knowledge retrieval. Digital consultant Apps enable medical consultation based on personalized medical history records and common medical knowledge derived from big data inferences.

The wealth of electronic health records has also excelled digitalized diagnosis and prevention of diseases and their outbreak control (Puaschunder 2019e). With the growth of scientific evidence derived from big data, AI also helps to analyze health trends to guide on global pandemic alleviation.[4] For instance, health risk early warning systems function through data constantly collected via mobile Apps. Once tagged and compiled, AI tools that employ natural language processing help mine the data for community health status monitoring and pandemic outbreak tracking. Pandemic spreads visualized via google search mapping analytics are the most recent advancements based on big data, large-scale mapping sophistication and computation control. All these AI-led opportunities to gather actionable insights lead to strategic data-driven interventions on medical prevention and health crisis management excellence.

The medical world has become flat and medical care more shareable than ever before. International development crisis management has profited from data-driven prevention. Technological development of information and communication technologies open unprecedented opportunities in telemedicine. Telehealth enables instant monitoring and preventive control. Emergency outreach and remote diagnosis based on large-scale data-driven knowledge generation decentralize healthcare. Network connectivity thereby grants affordable healthcare around the globe in a cost-effective way. Health-related data from personal self-diagnosis devices enhanced by low-cost generation of big data and patient-led monitoring but also information technology advancements make data-driven quality care more accessible in remote areas and developing nations than ever before.

Decentralized information collection and storage grids as well as technological diversified data collection means are expected to revolutionize the healthcare sector. Information shared among neighbors helps overcome shortages and enables fast-paced aid cheaper and more democratically-distributed scarce resources. Geopolitically the individual becomes more independent from central-

---

[4] https://www.pharma-iq.com/business-development/articles/excellence-in-the-era-of-precision-medicine Accessed: Sept. 25, 2021.

ized medical structures. Remote communities thereby benefit from equal, easy and cheap access to medical aid. Decentralized grids also open novel opportunities of monitoring and measuring information constantly and closely where health or diseases occur. Information can be tracked and linked directly to the scientific and patient impact they are having, including knowing if the expert visited the medical portal, opened an email, or requested additional information. Instant messaging has opened the gates for remote access to affordable diagnostics. Networking data sharing capacities have reached unprecedented density and sophistication. Novel mapping tools can display local search results and crowd media use into visible alert systems so it becomes more accessible in a broader way. Decentralized crisis management applications of AI and machine learning already range from data-driven assistance in pandemic outbreak control to battling hunger and poverty as well as forced migration.

In the future, self-led monitoring and remote diagnosis fostered by machine learning mining of big data and algorithmic decision making are continuously meant to grant access to affordable and excellent healthcare around the globe. Clinical decision support systems are expected to advance in the near future with 5G technologies arising, which will boost prognostic capacities. Virtual nursing assistants are predicted to become more common to perform targeted patient aid that can run 24/7 at most efficient levels.

In the near future, advances in 3D printers may soon make it possible to substitute healthcare provision closer to the consumer, where the manufacturing process is simplified thanks to the reproduction of models. Outsourcing monitoring to patients and electronic recording devices but also tapping into the wealth of expert knowledge generated through big data helps classical human medical doctors and healthcare agents, who benefit from freed capacities for creative decision making and expert advice giving.

Most promising AI advancements in healthcare delivery and patient experience are expected in areas such as surgery, radiology and cancer detection. The development of programmable cells that destroy diseases naturally and internally are cutting-edge developments of the future of self-determined prognosis led by algorithmic big-data derived insights (Knapton 2016). Radiology and imaging benefit from computer-guided and big data-enhanced capacities to diagnose and predict future outcomes concurrently. Robotics have entered the medical field as assisted body parts or surgery devices as well as support for disabled and patient care assistance, automated nursery and mental health stabilizers.

Artificial medical care and economic growth

The healthcare AI market is expected to increase by a compound annual growth rate of 50.2 percent until 2025.[5] A 2017 Accenture Research and Frontier Economics report of economic growth rates of 16 industries concluded that AI has the potential to boost profitability on average by 38% by 2035.[6]

The use of AI is predicted to improve the prevention of diseases, accuracy of diagnoses and predictions on treatment plan outcomes. AI innovations offer benefits of rational precision and human resemblance, targeted aid and corruption-free maximization of excellence. Utilizing the predictive power of big data has perpetuated the effectiveness and efficiency in the healthcare sector.

Machine learning's ability to collect and handle big data, and its increasing adoption by hospitals, research centers, pharmaceutical companies and other healthcare institutions, are expected to fuel economic growth in healthcare.[7] Hospitals and healthcare provider segments are expected to account for the largest size of AI in the healthcare market due to the large number of applications of AI solutions across provider settings, the ability of AI systems to improve care delivery and patient experience while bringing down costs as well as the growing adoption of electronic health records by healthcare organizations. Moreover, AI-based tools, such as voice recognition software and clinical decision support systems, help streamline workflow processes in hospitals at lower cost with improved care delivery and enhanced patient experience.[8] Today advanced hospitals are looking into AI solutions to support and perform operational initiatives that increase precision and cost effectiveness. Medical decision-making become enhanced by predictive analytics and general healthcare management technology. Big data insights support drug development and global health monitoring.

---

5 https://www.reportlinker.com/p04897122/Artificial-Intelligence-in-Healthcare-Market-by-Offering-Technology-Application-End-User-Industry-and-Geography-Global-Forecast-to.html Accessed: Sept. 25, 2021.
6 https://www.lexology.com/library/detail.aspx?g=4284727f-3bec-43e5-b230-fad2742dd4fb Accessed: Sept. 25, 2021.
7 https://www.healthcarefinancenews.com/news/healthcare-ai-market-expected-surge-21-361-billion-2025 Accessed: Sept. 25, 2021.
8 https://www.reportlinker.com/p04897122/Artificial-Intelligence-in-Healthcare-Market-by-Offering-Technology-Application-End-User-Industry-and-Geography-Global-Forecast-to.html Accessed: Sept. 25, 2021.

Advanced computing power and the declining cost of hardware are other key factors in the projected market growth.[9] The growing adoption of applications – such as patient-data and risk analysis, lifestyle management and monitoring – is further propelling technology in the healthcare market (Puaschunder 2019b).[10] Electronic health records used by healthcare organizations and the outsourcing of health monitoring by novel personal care products – such as routine check-up medical tools and wearable devices – are further believed to better service quality and eventually bring down costs. Higher frequency of self-monitoring checks at lower costs are expected to improve preventive medicine sustainably.

Technology plays an important role to help analyze and identify actionable insights derived from a multitude of accessible data sources. The medical profession shifts towards precision medicine using a variety of complex datasets such as a patient's health records, physiological reactions and genomic data (OECD 2019). While data collection is easier than ever, proper usage of linked data is and will be a key factor for productivity, quality and accessibility of AI-driven applications. The core promise of data-driven solutions is to collect data at a density that is not feasible for humans and to identify patterns humans cannot detect. Since AI in healthcare is currently utilized mainly to aggregate and organize data – looking for trends and patterns and making recommendations – a human component that is creative, cognitively highly flexible and compatible with AI sources is still needed (Puaschunder 2019d, e).[11] Rather than replacing human medical doctors and staff, AI is therefore believed to support medical doctors and nurses with excellence and precision on decision-making predicaments and cognitive capacity constraints (Puaschunder 2019d, e; Puaschunder & Gelter 2019). Radiology is an example why technology often will not replace humans, instead giving human better tools (Hosny, Parmar, Quackenbush, Schwartz & Aerts 2018; Pakdemirli 2019).

With the currently ongoing COVID-19 crisis, we may see a further development of an effective big data-driven crisis response ecosystem in public health pandemic early warning and disease transmission monitoring systems. Integration of fragmented diagnosis and treatment results coupled with self-monitoring

---

9 https://www.healthcarefinancenews.com/news/healthcare-ai-market-expected-surge-21-361-billion-2025 Accessed: Sept. 25, 2021.
10 https://www.reportlinker.com/p04897122/Artificial-Intelligence-in-Healthcare-Market-by-Offering-Technology-Application-End-User-Industry-and-Geography-Global-Forecast-to.html Accessed: Sept. 25, 2021.
11 https://www.reportlinker.com/p04897122/Artificial-Intelligence-in-Healthcare-Market-by-Offering-Technology-Application-End-User-Industry-and-Geography-Global-Forecast-to.html Accessed: Sept. 25, 2021.

devices collecting data on a constant basis are viewed as future medical necessities. A such integrated diagnostic process fosters personalized treatment. Targeted aid can form a grid of medical specialists to work concurrently on patient diagnosis. Data integrated grids can also combat fragmentation of different help groups and foster information flow between field workers responding to crises. With medical literature doubling every three years, also the pharmaceutical industry now has access to unprecedented amounts of scientific data and recently also started a Bluetooth-enabled cartography of the medical device distribution to help overcome bottlenecks and fraud while protecting patient privacy.[12]

Intriguing, yet less described and hardly researched, appears that AI, robots and algorithms differ from human healthcare providers by holding the potential to be less susceptible to materialistic vices and less prone to be corrupt in comparison to human counterparts. If programmed to follow an ethical imperative, AI and robots, being without self-enhancing profit-maximizing goals, promise to grant healthcare free from any corruption, bribery or irrational price margins.

Corruption

Corruption has many faces such as organized crime, illegal business, bribery, non-meritocratic placements and nepotism, tax havens and voting to name a few (Alt & Lassen 2012; Charron, Fazekas & Lapuente 2016; Gordon 2009; Holmes 2007; Johannesen & Zucman 2014; Klumpp, Mialon & Williams 2016). Breeding in collective experiences in the pertaining societal networks, national conduct and social norms; corruption determines economic development and the state of democracy in countries around the world (Bardhan 2016; Davis & Trebilcock 2008; Fisman & Miguel 2007; Rose-Ackerman & Palifka 2016).

Corruption is prevalent in territories with missing accountability and rule of law (Agerberg 2019). Governmental revenues derailed through corruption weaken public financial management and fiscal space for the establishment, procurement and maintenance of collective goods (Campos & Pradhan 2007). Corrupt institutional structures have been associated with poverty and hindered international development (Human Development Report 2019).[13] Corruption erodes the regulatory impact and the provision of public services ranging from medical

---

[12] https://www.pharma-iq.com/business-development/articles/excellence-in-the-era-of-precision-medicine Accessed: Sept. 25, 2021.
[13] http://hdr.undp.org/sites/default/files/hdr2019.pdf Accessed: Sept. 25, 2021.

care, education, energy, transportation and environmental protection (Campos & Pradhan 2007; Rose-Ackerman & Palifka 2016; Rose-Ackerman & Tan 2014).

International efforts to combat corruption include international treaties, governmental accountability and whistleblower protection, transparency, national laws against foreign-induced corruption, security and peace-building (Boucher et al. 2007; Hite-Rubin 2015; Le Billon 2003; McLean 2012; Rose-Ackerman & Lagunes 2015; Vlasic & Atlee 2012). Anti-corruption reform is likely to stem from the international community fostering corruption prevention, corporate watch, consumer action, social, ethical, and humanitarian accountability, international development as well as integrity action and training (Davis 2019; Cooley & Sharman 2015; Engel, Ferreira Rubio, Kaufmann, Lara Yaffar, Londoño Saldarriaga, Noveck, Pieth & Rose-Ackerman 2018; Rose-Ackerman & Carrington 2013).

Corruption in the digital age is underexplored. First attempts have been made to capture digital corruption in the political domain (Ackerman 2020). A quantification of the relation of AI-led growth and corruption is – to this day – missing; yet highly relevant during this unprecedented time of IT governance and the search for AI-driven healthcare solutions against global pandemics (Puaschunder 2020). The empirical part will therefore try to aid our understanding of the interrelation of corruption and AI-driven innovation in global healthcare.

Artificial intelligence in healthcare during COVID-19

The currently ongoing COVID-19 crisis increased attention to the potential of AI in healthcare as a pandemic prevention means around the globe. COVID-19 unleashed the online healthcare tech world. On a flat globe, data traffic exploded. A multi-tasking online workforce gained global outreach and flexibility in digitalization cutting red tape (Puaschunder 2019a). Health Apps[14] target at tracking human contact and preventing COVID. Bluetooth-tracking of medical devices[15] helps overcome bottlenecks and fraud while protecting privacy. Telemedicine detects COVID-19 symptoms and cures remotely.

Yet the use of AI and algorithms for medical purposes varies enormously in the international arena. Digitalization's international differences accentuated

---

14 https://medicalfuturist.com/digital-health-apps-to-use-during-the-covid-19-quarantine/ Accessed: Sept. 25, 2021.
15 http://news.mit.edu/2020/bluetooth-covid-19-contact-tracing-0409 Accessed: Sept. 25, 2021.

during the pandemic – in China online COVID whistleblowers disappeared.[16] Strategically-internet-controlling Asia[17] and the former Soviet world trumped on mobile crowd control[18] and social monitoring compliance (Ackerman 2020).[19] US S&P 500 leaders partnered[20] to pool health data[21] while freedom-of-speech-fueled-information-overload deadlocked relevant communication.[22] Europe emphasized privacy protection in envisioning[23] a 5th freedom of data[24] to harvest network effects of exponentially-growing marginal utility of information (Puaschunder & Gelter 2019).

The prospective post-COVID era[25] will likely show advanced healthcare. Future global digital healthcare innovations are more likely and favorable to come from corruption-free AI pioneering countries that tend to have better general medical care. Internet connectivity and AI-human-compatibility via tech-skills and digital affinity are growing competitive advantages (Puaschunder 2019d, e). The following empirical part provides information on country-specific differences in AI leadership on global public health. Countries that feature AI-growth potential with non-corrupt institutional support and good general healthcare systems are presented to be in a better position to lead the world on global pandemic monitoring and crisis management.

---

16 https://www.businessinsider.com/china-coronavirus-whistleblowers-speak-out-vanish-2020-2 Accessed: Sept. 25, 2021.
17 https://www.bbc.com/news/world-asia-india-50819905 Accessed: Sept. 25, 2021.
18 https://www.sueddeutsche.de/digital/coronavirus-tracking-smartphone-app-ueberwachung-1.4869845 Accessed: Sept. 25, 2021.
19 https://www.wsj.com/articles/taiwan-and-the-virus-11584038158 Accessed: Sept. 25, 2021.
20 https://www.apple.com/newsroom/2020/04/apple-and-google-partner-on-covid-19-contact-tracing-technology/ Accessed: Sept. 25, 2021.
21 https://www.wsj.com/articles/companies-seek-to-pool-medical-records-to-create-coronavirus-patient-registry-11586381102?shareToken=st988488c16b9e42289c06df0e23933e3f&reflink=article_email_share Accessed: Sept. 25, 2021.
22 https://www.nytimes.com/2020/04/26/us/politics/trump-disinfectant-coronavirus.html Accessed: Sept. 25, 2021.
23 https://netzpolitik.org/2020/diese-regeln-plant-die-eu-fuer-daten-und-algorithmen/?fbclid=IwAR0rH_NIxgBYvxzaDNrKLzUSV4tM2FnXVhHA8Bc-PTGYV8d7ETpxD7jL-TE Accessed: Sept. 25, 2021.
24 https://ec.europa.eu/commission/presscorner/detail/en/IP_20_680 Accessed: Sept. 25, 2021.
25 https://www.sueddeutsche.de/wirtschaft/pest-coronavirus-wirtschaft-1.4873813 Accessed: Sept. 25, 2021.

Empirical validation

The empirical part presents three indices that highlight the influence of AI, economic growth, corruption and healthcare for future public health solution finding: (1) An index based on internet connectivity – as a proxy for digitalization and AI advancement– as well as Gross Domestic Product (GDP) – as an indicator of economic productivity – is first calculated to outline global AI innovation hubs with economic impetus around the world. (2) A novel anti-corruption artificial healthcare index is then presented that outlines countries in the world that have vital AI growth in a non-corrupt environment. (3) Finally, an index is created that integrates internet connectivity, anti-corruption as well as healthcare access and quality to determine the ultimate AI-led healthcare countries of the future.[26]

(1) *AI-GDP Index:* AI entrance into economic markets was modeled into the standard neoclassical growth theory by creating a novel index for representing growth in the artificial age. The AI_GDP per country c index was calculated for 191 countries of the world based on Equation 1, comprised of the GDP per capita per country c and AI internet connectivity percentage of a country, $IA(c)$.

$$AI\_GDP~(c) = GDP_c * IA_c \quad \text{(Equation 1)}$$

GDP per capita was retrieved from a World Bank database for the year 2017.[27] This measure was multiplied by AI entrance as measured by the proxy of Internet Access percent per country, $IA(c)$, which represents country c inhabitants' internet usage in percent of the population retrieved from a World Bank database for the year 2017[28] (Puaschunder 2020).

The table section in Graph 1-A in the appendix holds the *AI-GDP* (c) index value per country and tables the the *AI-GDP* countries' indices ranked from the highest to the lowest. Graph 1 displays the the *AI-GDP* country's index around the world. The higher the index, the darker the country is colored in Graph 1. The darker the country, the higher the multiplier is of internet connectivity of the country and GDP.

As visible in Graph 1, continent-specific AI-GDP relations reveal Africa being relatively low on AI-GDP in comparison to the rest of the world. Asia and the Gulf

---

[26] https://www.thelancet.com/journals/lancet/article/PIIS0140-6736(18)30994-2/fulltext Accessed: Sept. 25, 2021.
[27] https://data.worldbank.org/indicator/ny.gdp.pcap.cd Accessed: Sept. 25, 2021.
[28] https://data.worldbank.org/indicator/it.net.user.zs Accessed: Sept. 25, 2021.

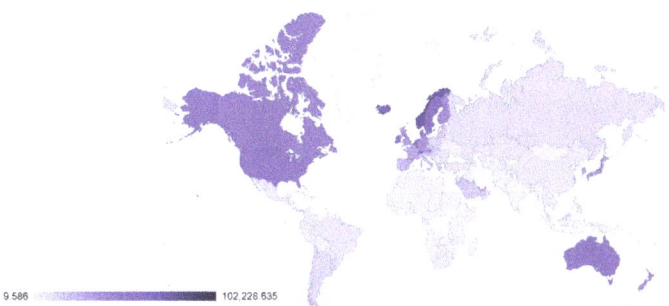

Graph 1: AI-GDP Index for 191 countries of the world

region are in the middle ranges with Qatar and the United Arab Emirates and Japan and South Korea leading. In Europe Luxembourg, Switzerland, Norway, Iceland, Ireland, Sweden and Finland are top AI-GDP countries. North America has a higher AI-GDP index than South America, where Chile, Argentina and Uruguay appear to lead. In Oceania, Australia has a higher AI-GDP index than New Zealand.

The parts of the world that have internet connectivity and high GDP are likely to be pioneers in the use of AI-driven big data for providing insights for pandemic prevention. AI advancements should be seen in relation with anti-corruption, as institutions of integrity will aid a successful implementation of AI and healthcare free from misplaced funds, lower quality due to nepotism and overinflated price margins (Campos & Pradhan 2007; Escresa & Picci 2017; Mungiu-Pippidi & Dadašov 2016; Rose-Ackerman & Palifka 2016; Rose-Ackerman & Tan 2014).

(2) *Corruption Perception (CPI)-Global Connectivity (GCI) Index:* In a cross-sectional study of 79 countries' relation of Corruption Perception – measured by the Corruption Perception Index of 2019[29] – and global connectivity (GCI) – as captured by the Global Connectivity Index for 2019[30] – AI is significantly positively correlated with anti-corruption ($r_{Pearson} = .860$, $n = 79$, $p < .000$). AI comes from parts of the world that are perceived as less corrupt.

An AI_anti-corruption index AA is calculated based on Equation 2, comprised of the global connectivity (GCI) of a country c in 2019 multiplied by the Corruption Perception Index of the same country c in 2019.

---

[29] https://www.transparency.org/cpi2019?/news/feature/cpi-2019 Accessed: Sept. 25, 2021.
[30] https://www.huawei.com/minisite/gci/en/index.html Accessed: Sept. 25, 2021.

$$AI\_anticorruption\ (AA) = GCI_c * CPI_c \quad \text{(Equation 2)}$$

The table section in Graph 2-A in the appendix holds the *AI_anticorruption* (*AA*) index value per country and tables the *AI_anticorruption* (*AA*) countries' indices ranked from the highest to the lowest. Graph 2 displays the *AI_anticorruption* (*AA*) country's index around the world. The higher the index, the better connected and less corrupt the country is perceived, the greener the country is colored. The lower the index, the less connected and the more corrupt the country is perceived, and the redder the country is colored. Medium connectivity and corruption perception are displayed in yellow.

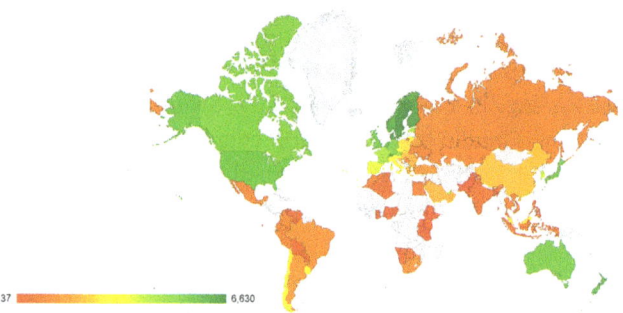

Graph 2: AI-anti-corruption (AA) index for 79 countries of the world

Artificial Intelligence (AI) – measured by Global Connectivity – is significantly positively correlated with freedom from corruption. AI thus springs from non-corrupt territories of the world. AI therefore offers a relatively corruption-free leadership decision-making tool, which could improve support of healthcare in non-corrupt global pandemic solutions. Artificial global governance should therefore come from the countries with high global connectivity and low corruption that are exhibited in the lower right quadrant in Graph 3.

Those countries that rank high on AI and corruption freedom could lead on building AI to monitor international public health and solve global healthcare problems. Artificial global governance in non-corrupt territories could unprecedentedly aid on global healthcare for the general protection and security of humankind. The detected non-corrupt AI centres exhibited in the right downward quadrant hold comparative advantages to lead on global artificial healthcare solutions against COVID-19 and serve as pandemic crisis and risk management innovators of the future.

Continent-specific relations reveal Africa being relatively low on AI and problematic on corruption as visible in Graph 3. Asia and the Gulf region are

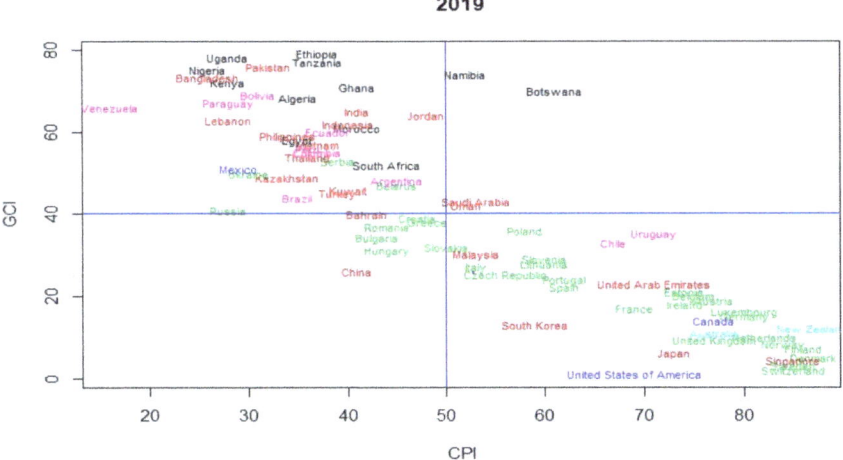

Graph 3: Global connectivity (GCI) and Corruption Perception Index (CPI)

more in the middle ranges but still feature unfavorable levels of corruption. Singapore, Japan, South Korea and the United Arab Emirates but also Malaysia seem to be leading on AI and less corruption in Asia and the Middle East. In Europe Switzerland, Nordic countries like Sweden, Denmark, Norway and Finland are top AI and anti-corruption countries. North and South America are opposites – while the United States of America has a top condition to lead on AI and anti-corruption; South America and there especially Venezuela, Paraguay and Bolivia, rank lowest on AI and relatively worse on corruption. In Oceania, New Zealand has a better AI and anti-corruption index performance than Australia.

(3) *Corruption Perception (CPI)-Global Connectivity (GCI)-Healthcare Index:* In a cross-sectional study of 79 countries' relation of Corruption Perception – measured by the Corruption Perception Index of 2019[31] – and global connectivity (GCI) – as captured by the Global Connectivity Index for 2019[32] – and healthcare – as quantified by the 2016 Healthcare Quality and Access Index[33] – freedom from corruption is significantly positively correlated with good healthcare ($r_{Pearson}$ = .715, n = 79, p < .001) and AI is significantly positively correlated with good healthcare ($r_{Pearson}$ = .896, n = 79, p < .001). AI comes from parts of the world that are perceived as less corrupt and feature better public healthcare.

---

[31] https://www.transparency.org/cpi2019?/news/feature/cpi-2019 Accessed: Sept. 25, 2021.
[32] https://www.huawei.com/minisite/gci/en/index.html Accessed: Sept. 25, 2021.
[33] https://www.thelancet.com/journals/lancet/article/PIIS0140-6736(18)30994-2/fulltext Accessed: Sept. 25, 2021.

An AI_anticorruption_health index *AAH* is calculated based on Equation 3, comprised of the global connectivity (GCI) of a country *c* in 2019 multiplied by the Corruption Perception Index of country *c* in 2019 and multiplied by the Health Quality and Access Index of 2016.[34]

$$AI_c\_Anticorruption_c\_Health\ (AAH) = GCI_c*CPI_c*HAQ_c$$
(Equation 3)

The table section in Graph 3-A in the appendix holds the *AI_anticorruption_health* (*AAH*) index value per country and tables the *AI_anticorruption_health* (*AAH*) countries' indices ranked from the highest to the lowest. Graph 4 displays the *AI_anticorruption_health* (*AAH*) country's index around the world. Graph 4 highlights the parts of the world that feature high internet connectivity, freedom from corruption and good access to and quality of general healthcare in green, whereas those parts of the world that feature less internet connectivity and more perceived corruption and worse access to and quality of general healthcare in red. The higher the index, the better connected and less corrupt the country is perceived and the better access to and quality of general healthcare is offered in the greener-colored countries. The lower the index, the less connected and the more corrupt the country is perceived and the worse off are its citizens regarding access to and quality of general healthcare, and the redder the country is colored. Medium AI-connectivity and corruption hubs with medium access to and quality of healthcare are displayed in yellow.

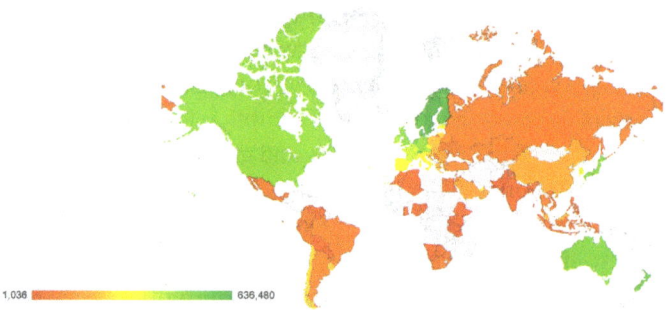

Graph 4: AI-anticorruption-health (AAH) index for 79 countries of the world

---

34 https://www.thelancet.com/journals/lancet/article/PIIS0140-6736(18)30994-2/fulltext Accessed: Sept. 25, 2021.

Continent-specific relations reveal Africa being relatively low on AI, problematic on corruption as well as general healthcare as visible in Graph 4. Asia and the Gulf region are more in the middle ranges but still feature problematic levels of corruption and relatively weak access and quality of healthcare. Singapore, Japan, South Korea and United Arab Emirates but also Malaysia seem to be leading on AI, anti-corruption as well as access to and quality of healthcare in Asia and the Middle East. European countries like Switzerland and the Nordic countries – such as Sweden, Finland, Denmark and Norway – are top AI and anti-corruption territories with excellent general healthcare. North and South America are opposites again – while the United States of America has a top condition to lead on AI anti-corrupt and with excellent healthcare; South America, and there especially Venezuela, Bolivia and Paraguay, rank lowest on AI and relatively worse on corruption and general healthcare. In Oceania the leading New Zealand and immediately-thereafter ranked Australia have almost the same conditions on the AI, anti-corruption and healthcare index.

Discussion

The currently ongoing COVID-19 crisis has created an unprecedented need for the global community to collaborate on public healthcare solutions in the international compound. In the digital age, international healthcare pandemic crisis and risk management could thereby be most innovatively fostered by AI and big data-derived inferences. At the same time, international differences in the approaches to combating global pandemics through the use of algorithms have currently become more apparent than ever before.

This article has provided three indices that depict the international differences in digitalization, economic potential, anti-corruption and access to general healthcare. In a multi-faceted analysis, different aspects and combinations of AI-led growth, anti-corruption and general healthcare were highlighted. The results were meant as a quantified form of aid to making decision on the comparative advantages to lead the world in a global solution on digitalized pandemic prevention and risk management. The countries that score high on AI, GDP, anti-corruption and healthcare excellence were featured as ultimate world-leading, innovative global pandemic alleviation centres.

While information was presented on how the different countries rank compared to each other on their potential of using AI to avert global pandemics, we still need a further and deeper qualitative understanding of the ethical boundaries of digitalization in healthcare. For instance, a worldwide solution on AI helping against global virus spreads will require equal access to information on health. The sharing of data will need all countries coming together to construct large datasets as learning opportunities, which different stakeholders from government, healthcare, engineering and technology can use concurrently and equally to analyze and predict the prevailing health situation and global trends. The more countries join, the more accurately the dataset will be able to draw inferences about world-wide prevalent epidemics spread and global diseases outbreaks. Within the European compound, a 5th freedom of data should incentivize data sharing and provide the legal means for combating discrimination based on big data-derived inferences as well as protection of privacy (Puaschunder & Gelter 2019).

AI regulations from governments and their agencies account for the most cutting-edge sophistication of laws and public policies yet hold also enormous unknown risks. The most contested areas of legal and policy attention in the IT domain include data protection and privacy, transparency, human oversight and surveillance. Privacy challenges arise from big data building and hands-on search for a desired pandemic spreading monitoring system. Instant and continuous information tracking implying full transparency also leads to the risk of stigmatization. When diagnoses influence future diagnoses and set patients up on a path of discriminatory disadvantages or silos of sickness, this unreflected decision-making bears extensive health risks, societal challenges and ethical downfalls.

An environment should be established in which research, clinical practice and technological advancement are coming together for retrieving data insights and harvesting network effects of big data collectively while upholding highest ethical standards. Large data sets that glean context-based information could thereby become early warning signs of imminent viral epidemic outbreaks (Puaschunder & Gelter 2019). Big data analysis combining the medical sector with technology-driven self-monitoring solutions could then also offer applications for patients in an equally accessible and real-time manner. The wisdom of the crowds could also be tapped into in citizen science – e. g., Massively Multiplayer Online Gaming (MMOG) techniques that have been used to incentivize volunteer participation. For instance, such an approach helped gamers on a crowd-sourced gaming science site to decode an AIDS protein in 3 weeks, a problem that had stumped researchers for 15 years (Quadir, Rasool, Zwitter, Sathiaseelan & Crowcroft 2016).

A combination of mobile technology and cloud computing naturally complements big data technologies and is well-suited to the reliable storage and analysis of big data. Crowdsourcing comes in when mapping the location of patients, healthcare and medical devices featuring a price scale and performance information based on consumer reviews. The advantages of individuals sharing information about the price and quality of medical services would be quality control, transparency and the prospect of a decrease in the price margins. Yet the display of such information should at the same time not reveal private information or pit people against each other. Downsides of online mapping include social stigmatization and discrimination potential, competitive fraud, price decline leading to a natural service quality race-to-the-bottom. Information could then be used against the individuals, especially those in vulnerable groups and impaired, for instance in connection with making decisions on hiring or access to education or insurance coverage. In order to avoid a predicament between utility of information aggregation versus dignity in privacy protection, a single representation of tailored information but not an aggregated information display that pits people against each other or ranks them based on their health status is recommended. Additional IT solutions are only tracking the environment but not human – for instance via Bluetooth tracking of medical devices such as pharmaceuticals. Big data insights should thereby only be used for the benefit of people and the common good of health but never turned against the individual.

In the future of artificial healthcare, compatibility problems in the adoption of new technologies around the world should be alleviated by research and training in international digitalization literacy. Transnational engagement could aid in re-evaluating and seeking out new competencies, technology solutions and data sources that better support patient-centric outcomes. Patients must be trained to use digital channels and be open to remote assistance.

This massive market entrance of AI in our contemporary economy also imposes historically unique ethical challenges such as digital security threats related to AI (Puaschunder 2019c). "Data poisoning" – feeding manipulated data into a grid on which an AI system is being trained – can imply tremendous health risks. Such adversarial examples can be created without effort, by printing images on normal paper and photographing it with a smartphone (OECD 2019). Additional harm can be caused by unqualified e-workers' ratings used as diagnostic proxy that can cause misdiagnosis, misclassifications and maltreatment. Blatant questions arise about liability and legal possibilities of e-disputes of international healthcare in the telemedical sector.

In addition, the emerging autonomy of AI holds unique potentials of eternal life of robots, AI and algorithms alongside unprecedented economic superiority, data storage and computational advantages. Yet to this day, it remains unclear what concrete impact AI taking over the workforce will have on economic growth today, in the near and in the more distant future (Puaschunder 2019b, d; Puaschunder 2020).

Works cited

Ackerman, Klaus. *Limiting the Market for Information as a Tool of Governance: Evidence from Russia.* Melbourne, Australia: Monash University working paper, 2020.

Agerberg, Mattias. "The Lesser Evil? Corruption Voting and the Importance of Clean Alternatives." *Comparative Political Studies,* 52 (2019): 1–35.

Alt, James E. and Lassen, David Dreyer. "Enforcement and Public Corruption: Evidence from the American States." *Journal of Law, Economics, and Organization,* 30.2 (2012): 306–38.

Bardhan, Pranab, "State and Development: The Need for a Reappraisal of the Current Literature." *Journal of Economic Literature* 54.3 (2016): 862–92.

Boucher, Alix J. et al., Mapping and Fighting Corruption in War-Torn States, Stimson Center Report No. 61, Washington DC: Henry L. Stimson Center, March 2007.

Breschi, Stefano, Lassébie Julie and Menon, Carlo, "A Portrait of Innovative Start-Ups across Countries." OECD Science, Technology and Industry Working Papers, No. 2018/2, OECD Publishing Paris, http://dx.doi.org/10.1787/f9ff02f4-en. Accessed: 15.12.2021.

Campos, Edgardo J. and Pradhan, Sanjay, *The Many Faces of Corruption,* World Bank, 2007.

Charron, Nicholas, Fazekas, Mihaly and Lapuente, Victor, "Careers, Connections, and Corruption Risks: Investigating the Impact of Bureaucratic Meritocracy on Public Procurement Processes." *Journal of Politics* 79.1 (2016): 89–104.

Cooley, Alexander and Sharman, Jason C., "Blurring the Line between Licit and Illicit: Transnational Corruption Networks in Central Asia and Beyond." *Central Asian Survey* 34 (2015): 11–28.

Davis, Kevin E., *Between Impunity and Imperialism: The Regulation of Transnational Bribery,* New York: Oxford University Press, 2019.

Davis, Kevin E. and Trebilcock, Michael J., "The Relationship between Law and Development: Optimists versus Skeptics." *American Journal of Comparative Law* 56 (2008): 895–946.

Dowell, Ryan. "Fundamental Protections for Non-Biological Intelligences or: How we learn to stop worrying and love our Robot Brethren." *Minnesota Journal of Law, Science & Technology* 19.1 (2018): 305–36.

Engel, Eduardo, Rubio, Delia Ferreira, Kaufmann, Daniel, Lara Yaffar, Armando, Londoño Saldarriaga, Josa, Noveck, Beth S., Pieth, Mark, Rose-Ackerman, Susan, *Report of the Expert Advisory Group on Anti-Corruption, Transparency, and Integrity in Latin America and the Caribbean,* Inter-American Development Bank, November 2018. Retrieved at https://www.telegraph.co.uk/science/2016/09/20/microsoft-will-solve-cancer-within-10-years-by-reprogramming-dis/ Accessed: 15.12.2021.

Escresa, Laarni and Picci, Lucio, "A New Cross-National Measure of Corruption." *The World Bank Economic Review* 31.1 (2017): 196–218.

Fisman, Raymond and Miguel, Edward, "Corruption, Norms, and Legal Enforcement: Evidence from Diplomatic Parking Tickets." *Journal of Political Economy* 115.6 (2007): 1020–48.

Gordon, Sanford C., "Assessing Partisan Bias in Federal Public Corruption Prosecutions." *American Political Science Review* 103.4 (2009): 534–54.

Hite-Rubin, Nancy. "A Corruption, Military Procurement and FDI Nexus." *Greed, Corruption and the Modern State.* Eds. Susan Rose-Ackerman and Paul Lagunes. Cheltenham UK: Edward Elgar, 2015. 224–51.

Holmes, Leslie, "The Corruption-Organised Crime Nexus in Central and Eastern Europe," *Terrorism, Organized Crime and Corruption; Networks and Linages*. Ed. Leslie Holmes. Cheltenham UK: Edward Elgar, 2007. 84–108.

Hosny, Ahmed, Parmar, Chintan, Quackenbush, John, Schwartz, Lawrence H. & Aerts, Hugo. "Artificial Intelligence in Radiology." *National Reviews Cancer* 18.8 (2018): 500–10.

Johannesen, Niels & Zucman, Gabriel. "The End of Bank Secrecy? An Evaluation of the G20 Tax Haven Crackdown." *American Economic Journal: Economic Policy* 6.1 (2014): 65–91.

Klumpp, Tilman, Mialon, Hugo M. and Williams, Michael A. "The Business of American Democracy: *Citizens United*, Independent Spending, and Elections. *The Journal of Law and Economics*. 59.1 (2016): 1–43.

Knapton, Sarah. Microsoft will 'solve' Cancer within 10 Years by 'Reprogramming' Diseased Cells. *The Telegraph*, September 20, 2016. Retrieved at https://www.telegraph.co.uk/science/2016/09/20/microsoft-will-solve-cancer-within-10-years-by-reprogramming-dis/ Accessed: 15.12.2021.

Le Billon, Philippe "Buying Peace or Fueling War: the Role of Corruption in Armed Conflicts." *Journal of International Development*. 15 (2013): 413–26.

McLean, Nicholas M. "Cross-National Patterns in FCPA Enforcement." *Yale L.J.* 121 (2012): 1970–2011.

Mungiu-Pippidi, Alina and Dadašov, Ramin. "Measuring Control of Corruption by a New Index of Public Integrity." *European Journal on Criminal Policy and Research* 22 (2016): 415–38.

Noyes, Katherine. 5 Things you need to know about A.I.: Cognitive, Neural and Deep, oh my! *Computerworld*, March 3, 2016. Retrieved at www.computerworld.com/article/3040563/enterprise applications/5-things-you-need-toknow-aboutai-cognitive-neural-anddeep-oh-my.html Accessed: 15.12.2021.

OECD (2015). *Data-Driven Innovation: Big Data for Growth and Well-Being*. Paris: OECD. Retrieved at http://dx.doi.org/10.1787/9789264229358-en Accessed: 15.12.2021.

OECD (2019). *Artificial Intelligence in Society*. Paris: OECD.

Pakdemirli, Emre. "Artificial Intelligence in Radiology: Friend or Foe? Where are we now and where are we heading?" *Acta Radiologica Open* 8.2 (2019), 1–5.

Puaschunder, Julia Margarete. Artificial Intelligence, Big Data, and Algorithms in Healthcare, Report on behalf of the European Parliament European Liberal Forum in cooperation with The New Austria and Liberal Forum, 2019a. Retrieved at https://papers.ssrn.com/sol3/papers.cfm?abstract_id=3472885 Accessed: 15.12.2021.

Puaschunder, Julia Margarete. "Artificial Intelligence Market Disruption." *Proceedings of the International RAIS Conference on Social Sciences and Humanities organized by Research Association for Interdisciplinary Studies* at Johns Hopkins University, Montgomery County Campus, Rockville, MD, United States, June 10–11, (2019b): 1–8.

Puaschunder, Julia Margarete. "On Artificial Intelligence's Razor's Edge: On the Future of Democracy and Society in the Artificial Age." *Journal of Economics and Business* 2.1 (2019c): 100–19.

Puaschunder, Julia Margarete. "Stakeholder Perspectives on Artificial Intelligence (AI), Robotics and Big Data in Healthcare: An Empirical Study" 2019d. Retrieved at https://papers.ssrn.com/sol3/papers.cfm?abstract_id=3497261 Accessed: 15.12.2021.

Puaschunder, Julia Margarete. "The Legal and International Situation of AI, Robotics and Big Data with Attention to Healthcare." Reports on behalf of the European Parliament European Liberal Forum in cooperation with The New Austria and Liberal Forum, 2019e. Retrieved at https://papers.ssrn.com/sol3/papers.cfm?abstract_id=3472885 Accessed: 15.12.2021.

Puaschunder, Julia Margarete. Revising Growth Theory in the Artificial Age: Putty and Clay Labor. *Archives in Business Research 8*.3 (2020): 65–107.

Puaschunder, Julia Margarete & Gelter, Martin. "On the Political Economy of the European Union." *Proceedings of the 15th International RAIS Conference on Social Sciences and Humanities organized by Research Association for Interdisciplinary Studies (RAIS)* at Johns Hopkins University, Montgomery County Campus, Rockville, MD, United States, November 6–7 (2019): 1–9. Retrieved at http://rais.education/wp-content/uploads/2019/11/001JP.pdf Accessed: 15.12.2021.

Quadir, Junaid, Ali, Anwaar, Rasool, Raihan ur, Zwitter, Andrj, Sathiaseelan, Arjuna & Crowcroft, Jon. "Crisis Analytics: Big Data-Driven Crisis Response." *Journal of International Humanitarian Action* 1.1 (2016): 12–21.

Rose-Ackerman, Susan and Carrington, Paul D., eds. *Anti-Corruption Policy: Can International Actors Play a Constructive Role?* (Durham NC: CAP Press, 2013).

Rose-Ackerman, Susan and Lagunes, Paul, eds. *Greed, Corruption and the Modern State.* Cheltenham UK: Edward Elgar, 2015.

Rose-Ackerman, Susan and Palifka, Bonnie J., *Corruption and Government: Causes, Consequences, and Reform,* Cambridge UK: Cambridge University Press, 2016.

Rose-Ackerman, Susan and Tan, Yingqi. "Corruption in the Procurement of Pharmaceuticals and Medical Equipment in China: The Incentives Facing Multinationals, Domestic Firms, and Hospital Officials." *UCLA Pacific Basin Law Journal* 32.1 (2014): 1–54.

Vlasic, Mark V. & Atlee, Peter. "Democratizing the Global Fight Against Corruption; the Impact of the Dodd-Frank Whistleblower Bounty on the FCPA." *Fletcher Forum on World Affairs* 36 (2012): 79–92.

The Future of Artificial Intelligence in International Healthcare: An Index — 203

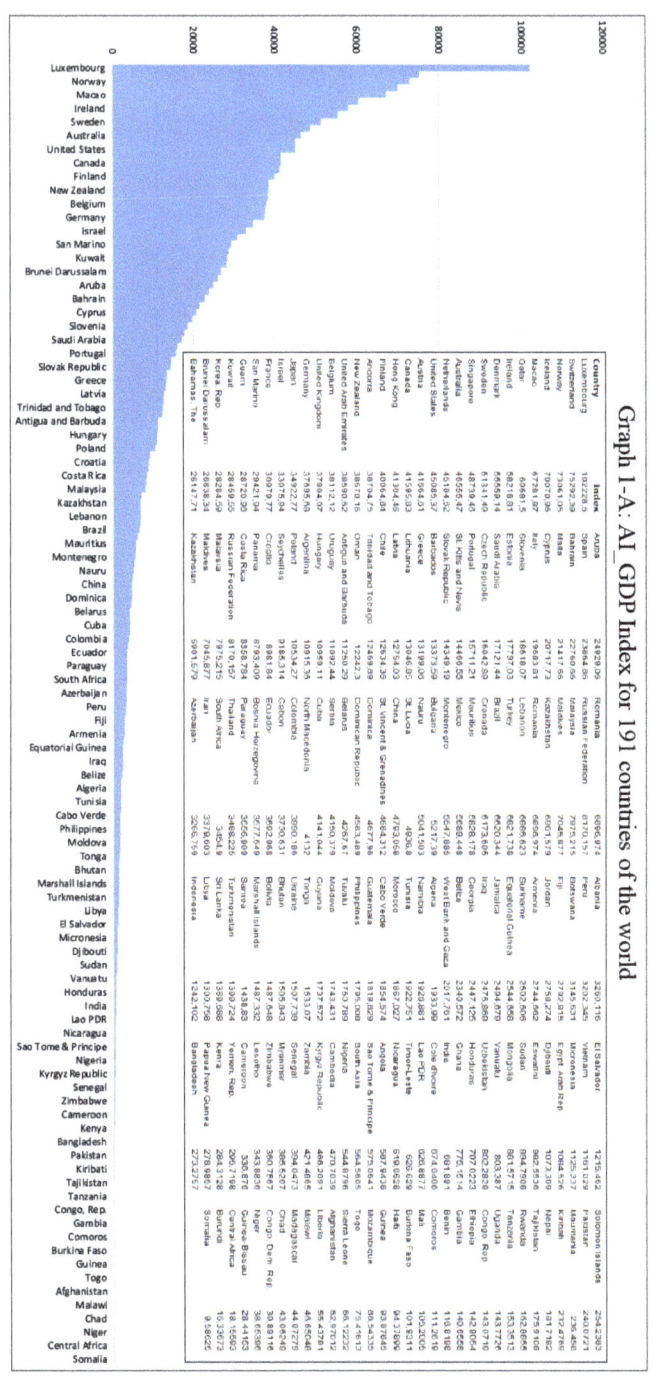

Graph 1-A: AI-GDP Index for 191 world countries

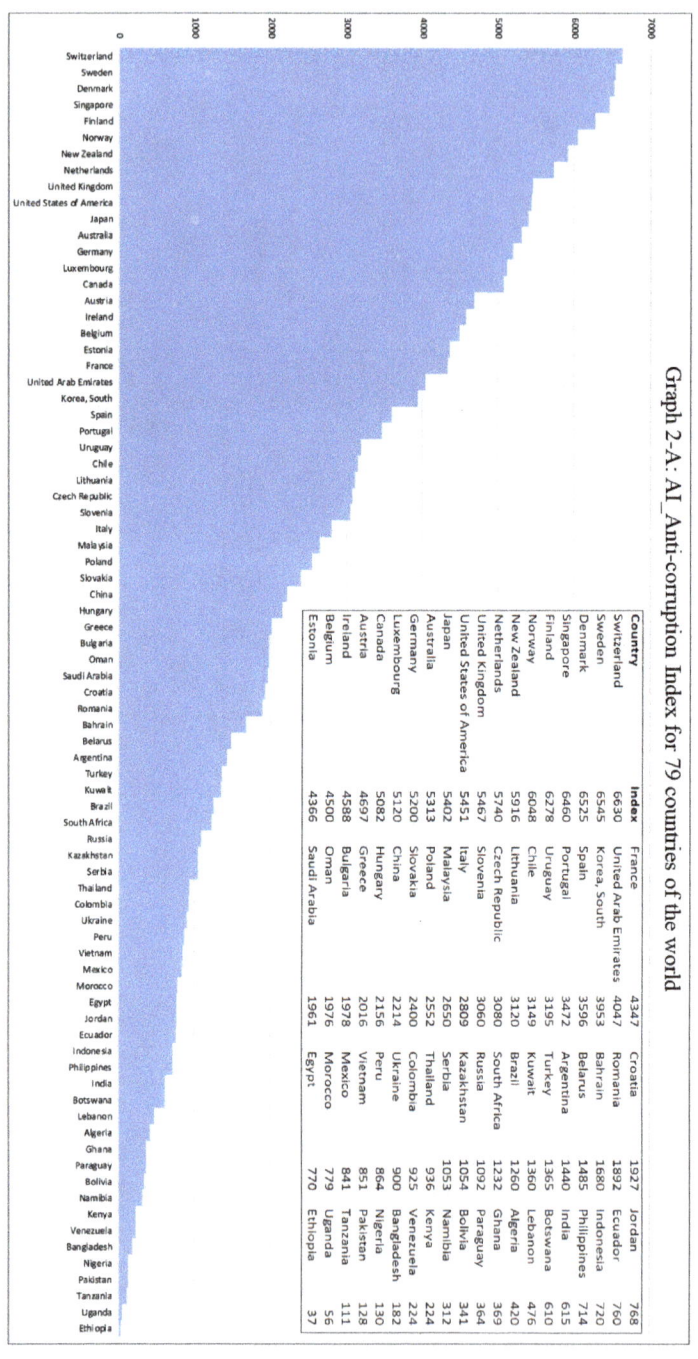

Graph 2-A: AI-Anti-Corruption Index for 79 world countries

The Future of Artificial Intelligence in International Healthcare: An Index — 205

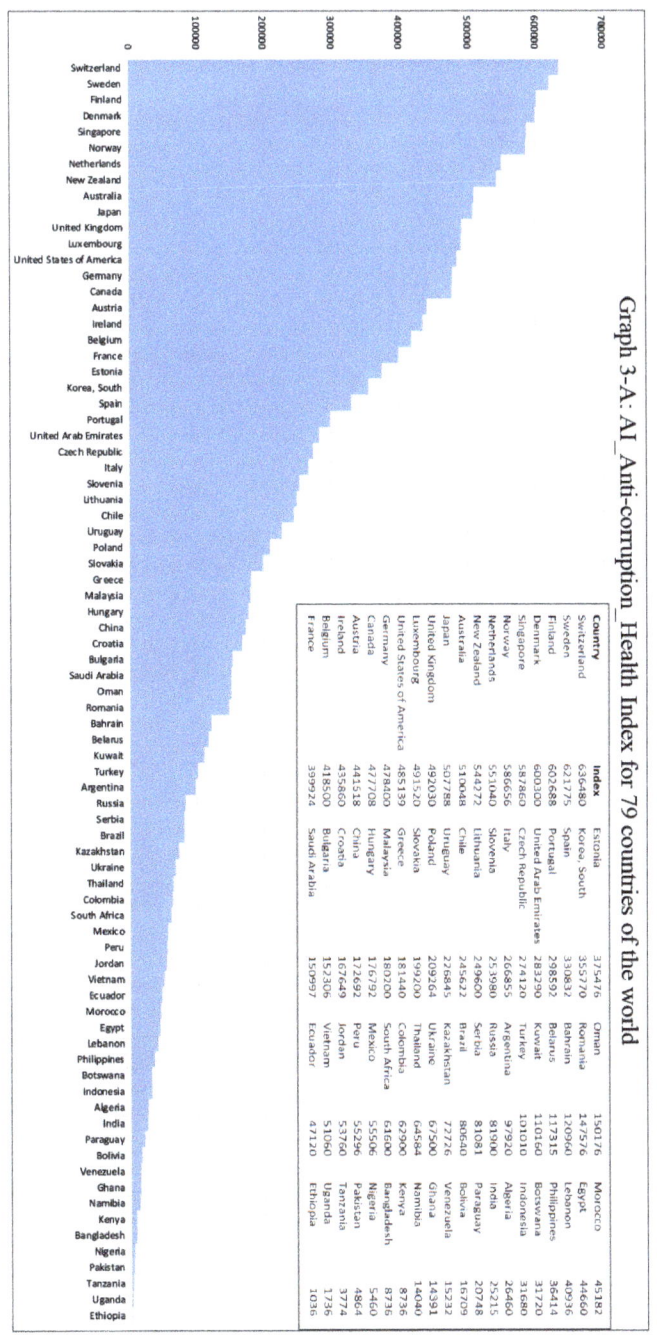

Graph 3-A: AI-Anti-Corruption_Health Index for 79 world countries

Part 3: **Encounters with Artificial Beings in Film, Literature, and Theater**

Jörg Türschmann

Dark Ecology and Digital Images of Entropy: A Brief Survey of the History of Cinematic Morphing and the Computer Graphics of Artificial Intelligence

Abstract: Stuart Russell wrote in his highly-regarded book *Human Compatible: Artificial Intelligence and the Problem of Control* (2019) that it is necessary to develop artificial intelligence modeled on human needs. In this way, it can be controlled. Since its beginning, the cinema has been telling stories of ingenious individuals who invent artificial beings with human skills. The protagonist T-100 in *Terminator 2* (1991) is the first artificial being in the history of the cinema whose transformation and morphing was illustrated with the help of cinematographic computer technology. But what happens if humans create not robots, but biological beings which are displayed with the help of digital cinematography, as in *Jurassic Park* (1993)? Timothy Morton proposed the term *Dark Ecology* (2016) with regard to the contemporary relationship between humans and the non-human world of objects, while abandoning the dominant romantic notions of nature. In the extreme case, this leads to the *Annihilation* (2018) of the boundary between humans and nature and the complete entropy of genetic elements. – This paper is dedicated to the narratological and ethical implications of these scenarios.

1 Introduction

Stuart Russell wrote in his highly-regarded book *Human Compatible: Artificial Intelligence and the Problem of Control* (2019) that it is necessary to develop artificial intelligence modeled on human needs. He argues that artificial intelligence can and must be controlled. Since its beginning, the cinema has been telling stories about ingenious persons who invent artificial beings with human skills. Science fiction and its later dystopic subgenre of the cyberpunk both succeed in conceiving a persuasive presentational mode of non-human beings as in *2001: A Space Odyssey* (Stanley Kubrick, 1968), *West World* (Michael Crichton, 1973), *Blade Runner* (Ridley Scott, 1982), *Terminator* (James Cameron, 1984), *Robocop* (Paul Verhoeven, 1987) or the later examples of *Her* (Spike Jones, 2013) and *Transcendence* (Wally Pfister, 2014). Here the interaction between fades, dissolves,

https://doi.org/10.1515/9783110770216-012

double exposures, digitization and creature are very convincing examples of representing artificial intelligence. The final message of all these stories is mostly that the artificial being makes itself independent of its creator. In this case, artificial intelligence means a danger and is sometimes responsible for the death of the inventor or even the end of mankind.

But what happens if man creates not robots, but biological beings which are displayed with the help of computer graphics like those in *Jurassic Park* (Steven Spielberg, 1993), a film based on the motifs of the homonymous novel written by Michael Crichton and on Crichton's own feature film *West World*? Timothy B. Morton proposed the term "dark ecology" (Morton 2016) with regard to contemporary relations between humans and the non-human world of objects by shedding the dominant romantic notions of nature. In the extreme case, this leads to the "annihilation" of the boundary between man and nature, and the complete entropy of genetic elements, as shown in *Annihilation* (Alex Garland, 2018). Therefore the idea of a dark ecology differs from that of a "digital humanism" (Nida-Rümelin and Weidenfeld 2018). It is not about the question of whether ethics guide the relationship between man and machine. It is rather that humans dissolve into everything that surrounds them, affects them and influences their living conditions, especially nature. The question of humanism arises in this context as the question of respect for others, the environment, the climate. But it also testifies to the fear of getting lost in this interaction.

2 Coming to terms: Dark ecology, entropy, and morphing

How to survey the history of cinematic morphing? It seems to be highly ambitious to link the history of cinema to concepts like dark ecology and the physical concept of entropy. Hence it is not possible to proceed in a strictly chronological way and to give a systematic overview of the innumerable films in the history of cinema that have to do with these subjects. But it is maybe useful to quote a few examples that are important for the argument. Three terms will serve as a guide to the following observations: "dark ecology," "entropy" and "morphing." "Dark ecology" is the above-mentioned concept suggested by the cultural theorist Timothy B. Morton in 2016. In his book he describes a different perspective on ecology. We usually associate ecology with harmony and harmony with nature. Nature must therefore be protected and mankind must submit to its needs in order to preserve it. In contrast, "dark ecology" means that ecology does not privilege the human, is not something beautiful, and has no real use for the old concept of nature. The slogan is: "ecology without nature" (Höller 2017). Morton looks for a kind of terminological liberation from the fixation on "species" and the "human" – a liberation that reveals our fundamental entanglement with the non-human. In this sense, human beings belong to the "parliament of things" in which human and non-human actors have equal rights and the same right to vote on the future (Latour 2004; cf. Nida-Rümelin and Weidenfeld 2018, 209, n. 13).

With regard to entropy, Penelope's web is one of the most famous allegories in the philosophical writings of the French philosopher Michel Serres (1968). It stands for both a pre-established order and randomness or disorder. In the first volume of the Hermes series, Serres proposes Penelope's web as a pattern for communication. It is a model for the World Wide Web. Two types of actors, either navigating or surfing, move paradigmatically on the Internet (cf. Türschmann 2004): by navigating, the historian deals with chance in the present as a result of the past and seeks order in it. The player, on the other hand, surfs and seeks order in the future. The player helps to ensure that a system does not implode when it threatens to freeze. For history is the negentropy in the entropy of culture in the sense of Serres's categories.

Finally, the term morphing: it is an expression used in cinema production. With the help of digitization, one motif is transformed into another. The transformation of anthropomorphic phenomena is of course particularly impressive, as in the famous morphing of Jim Carrey's face in *The Mask* (Chuck Russell, 1994). The effect seems to be reminiscent of Ovid's *Metamorphoses*, but the question is whether the transformations of humans into animals and objects described there are at all a suitable model for the cinematic representation of artificial intelligence. In terms of film history, and before the advent of digital cinema, vampires, werewolves and Dr. Jekyll and Mr. Hyde show a smooth transition from one form of existence to another. The protagonist T-1000 in *Terminator 2: Judgment Day* (James Cameron, 1991) is the first artificial being in the history of cinema, the transformation and morphing of which was illustrated with the help of digital morphing. The film conveys that morphing is about more than just the visible transformation:

> Thus, as a form (both of figuration and figure) that is "carried beyond" itself and "across" different realms of our present existence and culture, the morph is not only *meta-morphic* in its shape-shifting formlessness that greedily "devours all forms"; it is also *meta-phoric* in its inherent tropological movement and its historically substitutive activity. (Sobchack 2000: xiii; emphasis in original)

3 Formula fiction

Ecology, entropy and morphing are three aspects that I would like to consider when examining the possibilities the cinematic medium has to represent artificial intelligence. Because there is no evident representational mode of artificial intelligence, the photo-realism needed has to continue a long tradition of cultural motifs. Therefore the cinematic images are nothing more than an attempt to convince the audience that it is dealing with a being who or which stands for artificial intelligence. For this purpose, US cinema draws on Christian-Jewish motifs from the Western world and ultimately designs them independently on the basis of the possibilities of digital cinema.

> Indeed, one could argue that at this particular historical moment and in our particularly digitally driven America context, the morph fascinates not only because of its impossibility and strangeness but also because its process and figuration seem less an illusionist practice than both a presentational mode and an allegory of late capitalist "realm". (Sobchack 2000: xi)

Overall, there are broadly three scenarios that many films show. First: a brilliant inventor creates an android. Second: a person encounters himself as an artificial being. And last but not least: a person merges with his environment or with his creature. However, in all three scenarios it is not clear to what extent it is artificial intelligence that is involved, but the history of cinema and literature already provide many examples that show all three scenarios before the invention of artificial intelligence and before the invention of digital cinema. In any case, there are models from mythology or literature, such as the narrative of the creation of Galatea by Pygmalion or the merging of a person with an image in Oscar Wilde's *Dorian Gray* (1890). These traditions must be observed in order to understand the computer graphic morphing of artificial intelligence as a component of a motif story.

Often all three scenarios are linked. But I would like to concentrate on the third variant: a person merges with their environment or their creation and the human appearance dissolves. In the case of amorphous or hybrid beings, there is a movement from a clear contour, which is a case of negentropy, to a flowing amorphous figure, which is a case of entropy, before finally a new contour emerges. Nature plays an important role here because it is the environment in which these processes take place and which is ultimately not just a sphere, but also the material from which a new form emerges.

The question is to what extent, in this context, the metaphor of Penelope's web can be useful for the understanding of a web-like intelligence that we now sometimes attribute to the Internet. There is no center, but a kind of spiritual "noosphere", as Teilhard de Chardin (1966) called it, a term which Marshall McLuhan (1962) took up for his idea of the Internet as a global village. This far-reaching perspective raises the question of the future of humanity in the face of the increasing influence of artificial intelligence. As is well known, James Lovelock (2006, 2019) tries to answer the question of a "Novacene" with his "Gaia hypothesis" and calls for a harmonious coexistence of humans and the environment in view of climate change, cyborgs, and nuclear power (cf. Joseph 1991). In short: the question of whether artificial intelligence is represented in a movie is the question of the place of the artificial intelligence. And this place can be all-encompassing or it can also be represented in allegorical form by a film character. The personification is thus an expression of the negentropy of artificial intelligence, the merging with nature, on the other hand, is a case of entropy, which implicitly also shapes the idea of dark ecology, a typical background of the cyberpunk.

4 Looking for the origin

The following paragraphs are related to the question of how artificial intelligence is implemented in photo-realistic images and how this kind of representation concerns the search for visibility. In this sense, it is significant that in *Alien: Covenant* (Ridley Scott, 2017) the first shot of the android shows his eye when he is coming to life and being asked about his well-being.

In a white room in front of a mountain panorama, Peter Weyland, the bio-engineer who created the android and describes himself as his "father", lets the android choose his name. The android decides on the name David, like the homonymous statue of Michelangelo in the room. Weyland asks David to sit down at a piano and decide for himself what to play. He chooses the entry of the gods into Valhalla by Richard Wagner. David asks: "If you created me, then who created you?" Weyland has no answer. Nevertheless, David serves him, although he knows that his "father" will be dead one day, while he will still exist for a very long time. Later, David will create his own world of evil creatures. But that is just the moral of the story. The close-up of David's eye is more interesting at the beginning. This motif is part of a long series of films that have accompanied cinema from the very beginning. After all, the eye is the visible part of the brain. At the beginning of his surrealistic film *An Andalusian Dog* from 1929, Luis Buñuel shows a close-up of a woman's eye, which is slit open with a razor by a man standing behind her, so that the vitreous body oozes out. The eye pours into the room and takes on an amorphous shape. This metaphor has been interpreted as penetration in the sense of Freud's dream symbolism. The scene also symbolizes the film editing as a process with the help of which a web of associative relationships emerges in the film artwork. We witness a birth scene in which the film is made. In this sense, David points out that he feels alive and begins the quoted dialogue with his creator, in which he asks about their respective origins. Weyland confirms that the only important question is origin.

This statement fits well with a basic feature of the cinematic medium. The critic of the famous French film magazine *Cahiers du cinéma*, Alain Bergala, states: "Many great films that have inspired the love of cinema have taken as their subject the act of teaching, of passing down an inheritance (and the encounter with evil, the initial exposure to evil)" (Bergala 2016, 50–1). Bergala's conclusion describes two opposing movements that the philosopher Michel Serres allegorically differentiates as the perspectives of historians and players. The creator of the android is like a historian in search of the past and at the same time tries, like a player, to use his invention as an explanation from the future. That is what Michel Serres means when he claims that history is the negentropy in the entropy of culture. The experiment in art or science that leads to the creation of David by Michelangelo and by Weyland is an experiment with an open outcome. In contrast, the moral of the film, which results from its plot, is the story or history that makes us understand the evolution of life.

5 A grammar of proliferation and procreation: The motif of the deer

In connection with dystopian movies and the search for the origin, similar symbols appear again and again in the history of cinema. An example of this kind of tradition is the source of life. The Catalan architect Antoní Gaudí created a mosaic for the entrance to the Cathedral of Palma de Mallorca, which shows two deer drinking from the source of life. In Psalm 42, the deer's thirst for water is also read as a metaphor for Christians' desire for salvation from the threats of life through baptism. Alain Bergala is probably right when he claims that the cinema shows the search for origin, which is threatening because it shows the frightening sexuality that accompanies the act of procreation and which is often violent and accompanied by the loss of human control.

In the Belgian-French television series *Zone blanche* (2017–2019), the deer is embodied in anthropomorphic form by a Celtic god who wanders through the woods in our time and kidnaps young women. Is that artificial intelligence? Evidently not, but in other respects there is some link. Because it remains unclear whether this being is pure fantasy or a part of the fictional world where the protagonists 'exist'. Is it the procreation of a brainchild or does the origin strike back? A God who created everything and now destroys everything? The explanation of the player who is betting on the future is missing in this series. Therefore, in the end, it is all about looking for traces of the past. However, the series does not create a glimpse into the future with the help of a being or phenomenon whose potential is shown in the form of a better world in the future.

But the deer motif can also be used differently. One of the most interesting directors of dystopian cyberpunk at the moment is Alex Garland, best known for his frightening movie *Ex Machina* (2015). In his following movie produced for Netflix, *Annihilation* (2018), a group of militarily trained women search for the cause of a shimmer that has settled over a coastal landscape. The special thing about the enclosed area is that the cells of plants, animals and ultimately humans unite to form new living beings, i.e., a perfect entropy of genetic material.

The protagonist meets herself at the end. The morphing shows the transformation of her doppelgänger from an abstract metal-like figure into her perfect likeness. So, it is a backward development and a threatening "disruption of linear temporality" (McClanahan 2019, 365). The reason why the film is extraordinary is that it does not provide any explanation for the origin and intent of the shimmer. It is an amorphous intelligence whose artificiality consists solely in messing everything up. The heroine herself has a share in its origins, which is why at the end of the film you can also see her strange, shimmering eyes in a close-up.

There are many net-like patterns in the film, too: roots, lianas, lichen. They can be interpreted as symbols of Penelope's web, which, unlike in Greek legend, cannot be disentangled. They form a contrast to the wide-open spaces, the design of which is reminiscent of the pictures by de Chirico. The light in which the showdown takes place is found in de Chirico as well as the faceless doll-like human figures, according to whose type the protagonist's alter ego is ultimately configured.

6 Autophagic bionics

The following paragraphs deal with a few examples of morphing and failed biologically inspired engineering. In Kurt Neumann's 1958 film *The Fly*, a researcher develops a transport device that resembles a projector. In a self-experiment, his cells mix with those of a fly, so that the fly carries his head and he carries the head of the fly. Again, this accident concerns entropy, and it is almost the ruse of reason in Hegel's sense that is at work here, in which it creates new life according to the principle of cell entropy. Cell entropy can also be seen in Garland's *Annihilation*, where plants and humans become blurred like the blossoming antlers of the deer. In Neumann's film, the extraordinary aspect is the perspective from the point of view of the desperate researcher who is now wearing a fly's head on his human body. The subjective camera shows the multitude as a unit or the unit as a multitude. The shot can be interpreted as a symbol of cell division. It is again about the visibility and the perception of non-human intelligence. In *Annihilation*, on the other hand, a biologist can be seen who, using a picture of tumor cells, explains that all cells originally come from one single cell and that the cells develop autophagic activities because they destroy their own components in order to become more effective. As in Neumann's film, however, this activity causes living bodies of all kinds to mix with one another without respecting the contours or characteristics of the individual creature or his species.

The affinity between such examples and those about humanoids or androids is only apparent at second glance. But the more recent dystopian cyberpunk stories about the end of the world join a long tradition of fables in which humans and technology merge. It must therefore be remembered that this amalgamation is not necessarily obvious. It can also express itself in the relationship between humans and a technology as a useful tool. Both, man and tool in complete equality, form a web, an "actor-network," also called "quasi-object" (Doll and others 2001). In Denis Villeneuve's movie *Arrival* (2016), the "tool," which the aliens speak of and refer to with a term that the humans misunderstand as "weapon," is the language to be used for communication between the humans and the aliens. In *Alien: Resurrection* (Jean-Pierre Jeunet, 1997), we see also a weapon in the form of a picture of the atomic bomb and a sex bomb that a shot alludes to when the heroine returns to earth. The protagonist represented by Sigourney Weaver, whose physiognomy is revealing, is even the child of an alien, that is to say an interspecies creature. She looks at the blue planet and does not recognize herself in view of her origin. This reference to Hiroshima and Nagasaki refers to the destructive power of human invention and the autonomy it can gain over humans. However, as shown in these films, this process can always be staged as an autophagic activity, as it is called in Garland's *Annihilation*, i.e., a self-optimization through cell purification or cell exchange, in which humans and the environment mix.

7 The tradition of cinematic metamorphosis and dismembering

The beginning of the science fiction is marked by Georges Méliès's fantastic films. Even without the possibilities of digitization, he mastered many tricks in order to stage transformations. This example shows that the computer graphical morphing of artificial intelligence in science fiction films still has something to do with the attraction that Méliès's films proved to be at fairs around 1900 (cf. Ndalianis 2000; Ndalianis and Balanzategui 2019). On the one hand, Méliès succeeds in creating the impression that the human body can be dismantled. On the other hand, he stages a trip to the moon, where the astronauts meet the lunar inhabitants in a fairytale environment.

In the first decades of the twentieth century, German Expressionism produced a series of films that deal with the mystical or technical animation of artificial beings: for instance, the machine woman from Fritz Lang's film *Metropolis* (1927), in which the actress Brigitte Held embodies the robot, partially naked and only wearing a costume with the help of body painting, an interesting variant of the interplay between nature and technology, between entropy and negentropy. The film *Orlac's Hands* (*Orlacs Hände*, Robert Wiene, 1924) shows the autonomy of dismembered limbs which develop their own will. Paul Wegener takes up a mythological creature from the Kabbalah: the Golem (*The Golem: How He Came into the World*, 1920), an 'android' made of clay that comes to life. This film can be seen as an early example of the commercial exploitation of a cultural motif: Wegener's film is the last of three sequels, but tells the prior history of the Golem. Similar a-chronologies can also be found in Ridley Scott's *Alien* films or George Lucas's *Star Wars* series.

8 The amorphous appearance

The digital morphing of artificial intelligence follows on from a series of comparable motifs that have been staged in the history of cinema in the form of mythological beings. It has recently become apparent that high production costs have led to the use of Computer Generated Imagery (CGI) in particular being retained in Hollywood. Digitization and morphing are the optical spectacles in films with a high production value.

Nevertheless, this development remains linked to general topics that can also be represented with less effort: genesis, birth, motherhood. They often form the moments in films with androids when the plot takes a decisive turn.

Alien (Ridley Scott, 1979) and its sequels exhibit birth scenes with slime, saliva, blood and amniotic fluid. Films about clones (e.g. *Moon*, Duncan Jones, 2009; *Replicas*, Jeffrey Nachmanoff, 2019) or cyborgs (*Upgrade*, Leigh Whannell, 2018), do not need morphing, either. In addition, cyberpunk movies are often cartoons that benefit from the freedom of expression of animated cartoons. And even a movie like *I Am Mother* (Grant Sputore, 2019) is still about the biological womb or embryos that are not available to the robots who have taken over on Earth.

Morphing is first and foremost a threat because the foreign can be omnipresent in an unknown form or amorphous appearance. There is therefore more involved than the connection of body and mind when the memory of a deceased person is kept in an archive and implanted in robots, or when a person rescues himself after his death (cf. *Archive*, Gavin Rothery, 2020) by putting his mind into the body of a robot (cf. *Chappie*, Neill Blomkamp, 2015). The amorphous, to which digital morphing belongs, is the traditional symbol of the omnipresent evil and had often been shown in the cinema before digital morphing (cf. *The Thing*, John Carpenter, 1982): "When 'the thing' can appear in any person, the viewer enters a paranoid, mad world, in which nothing is safe, in which everyone could be transformed, bearer of that non-essence to which no body is assigned" (Wulff 2006: 46; translation JT).

T-1000 can transform into molten metal and take the shape of any other person to disguise itself. Again, it is the amorphous shape that mediates between two clearly contoured forms, even in love stories like *The Shape of Water* (Guillermo del Toro, 2017). The amorphous shows a dark ecology made up of several components in an anthropomorphic appearance (cf. Valenti 2019). *Terminator 2* is the first film with digital morphing and therefore a pioneer of computer graphical morphing. The contradictions of the robots that have human features are thus staged in a particularly convincing way through morphing. But the appearance of T-1000 is also due to the presence of the actor Robert Patrick who vanishes when his fluid version appears. "*Morphing* is a magic process and transforms one body into another in a physically impossible way. It appears as if the acting is being withdrawn from the actors." (Wulff 2006: 47; translation JT) But not only in the case of morphing does the actor have to share his visual presence on the screen with the robot. This is why Kevin Spacey is named in the credits of the Sci-Fi *Moon* (Duncan Jones, 2009), even though he plays a computer called GERTY and only his voice can be heard. Or, in other words, representing AI robots is like taking an intelligent actor to portray an intelligent protagonist.

9 The flesh of the sex dolls: female skinny bodies and artificial intelligence

Androids or humanoids symbolize the two most important ingredients of entertainment cinema: sex and crime. Robots, clones and cyborgs are used as weapons, but they can also be the prostitutes of their creators. Morphing plays a special role here, because it is about the transformation of a machine body into a human body. Men create their dream women in *The Stepford Wives* (Bryan Forbes, 1975; remake: Frank Oz, 2004) or the 'female' robot Arisa (Paulina Andreeva) in the Russian-Chinese television series *Better Than Us* (since 2018). From the era of silent film to the present, similar scenes present the act of creation, as seen in *Metropolis* and *The Fifth Element* (Luc Besson, 1997). The male genius creates a female body according to his ideas. These creatures always appear as attractive young women of reproductive age. They are sterile, but can emancipate themselves from their maker. Later films directed by women not only show male desire for female robots as in *The Trouble With Being Born* (Sandra Wollner, 2020) and female desire for male robots as in *Ich bin dein Mensch* (*I'm Your Man*, Maria Schrader, 2021), but the female beings are even humanoids who fall in love with cars and change their gender (*Titane*, Julia Ducourneau, 2021).

The morphs show their true identity as soon as they shed their human appearance. Nothing else happens when morphing is applied in *Terminator 2*. The effect can be compared with the 'divestment' of a robot in *Ex Machina* (Alex Garland, 2015). The actress Sonoya Mizuno illustrates the identity of her protagonist Kyoko by getting rid of her artificial skin, but also by simultaneously making her human body invisible. It is doubly tragic that the robot has to become an attractive woman to justify its existence and that the actress has to become a robot to justify her role. In order to stage the fragile identities of AI robots and their female incarnations, actresses are hired who have a multicultural biography or are people of color without being definitely white, black or some other unique color.

Sonoya Mizuno in *Ex Machina*, Yvonne Strahovski in the television series *Dexter* (2006–2021), Noomi Rapace in *Prometheus* (Ridley Scott, 2012) or Maggie Q in *Divergent* (Neil Burger, 2017): they all are mestizas, Asian or from the Slavic area. As the attractive protagonists in Luc Besson's films show, who are often models, their eyes are set wide apart. Hypertelorism makes an actress look cool, and perhaps threatening when she plays the corresponding role of the *femme fatale*, as Charles Vidor had already shown in *Gilda* (1946). However, the robots develop a strong artificial intelligence because they do exactly what the men expect of them in order to strike at the right moment and eliminate their tormentors. But it should be noted with regard to *Ex Machina* that the robot that escapes to freedom is a 'white woman.' Ava, embodied by the Swedish actress Alicia Vikander, uses the skin of an 'Asian women' for her camouflage and practices a kind of biological appropriation that is nonetheless racist because it ensures white supremacy: "Asian skin secures Ava's 'secret' robotic form and allows for the next evolutionary step in a Western future, covering up Ava's android frame in order to protect the longevity of white personhood into the posthuman age" (Wong 2017, 48).

10 'Living' in a fluid shell: The (in)visibility of cinematographic representation

A very interesting, but commercially unsuccessful example in this context is an AI robot in the body of a man: *Life Like* (Josh Janowicz, 2019) shows an attractive AI man who works as a domestic help for a married couple of about the same age as he is. Henry is so attentive, considerate, understanding, hard-working and unobtrusive that both the woman and the man fall in love with him. In order not to destroy their marriage, they drive their servant to suicide. It turns out, however, that he was not a robot at all, but a man who was drawn into bondage and total obedience from childhood, so that he could be sold as AI.

Henry is like a mirror in which people see their wishes. It is therefore no coincidence that the sexual rapprochement between Henry and his employer takes place in the bathroom in front of a mirror and that they ultimately look at each other as a person looks at his or her reflection. In this scene, homosexuality consists of the alterity that the human partner experiences in the same way that AI experiences alienation from humans. The representation takes place in the form of a repetition and a series of similar images and motifs. That is why this kind of scene is also about the cinema itself as an "(in)visible object" (Nardelli 2020) that confirms the desire of the audience by reflecting their sexual predilections and existential anxiety. Seriality, industrial production, loss of identity and intimate desire work together in favor of a kind of medial autonomy and a so-called "animage" (Gaudreault and Marion 2015, 163):

> The idea of an *expanded* cinema, or of a *fragmented* cinema, appears to us to be the sign not only of a kind of neo-institutionalization but also of series-centrism. In our view, this is a flexible, soft neo-institutionalization, but which we propose to call "serial-centered," even though it is not focused on a closed definition of the medium: it is as if the institution renounced the crystallization of historical "identities and specificities" while at the same

time preserving the "centralizing" idea of cinema as *anima*. (Gaudreault and Marion 2015, 156; emphasis in original)

The moment when an AI realizes that it is an independent species is often shown in an allusion to the beginnings of cinematography. The praxinoscope, the phenakisticope, the kinetoscope, the zoetrope and finally the photograms on a strip of celluloid show the movement of the motif in front of a camera by a sequence of images of the same motif. A similar setting can also be found in *Ex Machina* and *Archive*, where the image of the robots is repeated in several mirrors or in several monitors in the form of a series of images. As is well known, the effect of the projected photograms is that they become invisible and create the impression of a continuous movement of the motif.

Henri Bergson describes movement as a fluid form, a concept which the French philosopher Gilles Deleuze chooses as the starting point for his cinema theory, which he presents in detail in his two books *Movement-Image* (1986a) and *Time-Image* (1986b). In Deleuze's sense, cinematographic images only arise on a higher level. Transferred to the robots of AI, this means that they only find themselves as their coherent image when they can express their visual appearance in a fluid form of their superficial physical presence. Thus the human skin or molten metal are a perfect cladding for a machine. They are shells that, on the one hand, conceal their mechanics and, on the other hand, enable the AI to 'live' undetected among people. This is the main difference to the *Mechanical Man* (*L'Uomo meccanico*, André Deed, 1921), whose apparatus is visually fascinating. Much later in *Chappie*, the possibilities of mechanics are also played with. Colored spare parts indicate the increasing 'humanization' of the intelligent robot without the mechanics becoming invisible. Here, the anxiety of the machine turns into the insight that the mechanical body is better than the human body.

11 Conclusion

The cinematic narrative about artificial intelligence is based on the tradition of motifs prior to the cinema: Chappie looks down on the city from a hill like the gargoyles in the films and illustrations based on Victor Hugo's epic *The Hunchback of Notre-Dame de Paris* (*Notre-Dame de Paris: 1482*, 1831). The Sci-Fi *Archive* tells a story between life and death, maybe interpretable as reminiscent of the short story *The South* (*El Sur*, 1956) written by Jorge Luis Borges. Stories about AI are always stories about human ingenuity and the dangers inherent in its creations. What Johann Wolfgang von Goethe already describes in his ballad *The*

Sorcerer's Apprentice (*Der Zauberlehrling*, 1797) can also be found in many films, some of which have been mentioned here: the creature makes itself independent and claims its right to a self-determined existence. Mary Shelley's novel *Frankenstein or The Modern Prometheus* (1818) is certainly an important contribution to the tradition of this motif, but also creatures that are brought to life solely by the human imagination, as in the case of *The Adventures of Pinocchio* (*Le avventure di Pinocchio: storia di un burratino*, Carlo Collodi, 1881).

In summary, the cinematic staging of artificial intelligence has something to do with the dissolution of forms and their redefinition. This story is basically told over and over again. It can be interpreted as a threat to or the search for the core of the human being. Robots, aliens, clouds and shimmer always function in the same way on the axis between dissolution and solidification, between historical explanation of origin and playful bet on the future. That is why the opening sequence of *2001: A Space Odyssey* (Stanley Kubrick, 1968) is still one of the most important visual metaphors for continuity in artistic creation, regardless of whether it is about engineering or fine arts. When a monkey starts throwing a bone in the air at the beginning of the film, Kubrick uses a famous match cut. He links the image of the flying bone to the image of a spaceship in the shape of the bone and illustrates that AI is at the end of a continuous development. In Kubrick's film, the computer called HAL already proves that it is a representative of autonomous AI. And at the same time, it is the ultimate consequence of human creativity, because its presence is only achieved visually and acoustically with the help of cinematographic and computer graphic presentational modes.

Thus, AI movies frequently offer the close-up of an eye. Buñuel's eye is cut open; in Alfred Hitchcock's *Psycho* (1960) it is the last image of the famous shower scene. There, after the murder, the drain is filmed in detail, into which the water runs off and forms a vortex. The shot of the vortex is followed by a fade over to the dead woman's eye, as she looks lifelessly into the camera. The impossibility of showing AI in the film is like the impossibility of distinguishing the gaze of a dead person from the gaze of a living person.

The robots that are assigned a gender try to provide an answer to this deficiency by their human-like appearance. The cinematographic self-reference in the form of serial pictures and the erotic connotations of the sexually attractive machines are symbols of the sensual inaccessibility of computing processes. Mind movies like *Matrix* (Lana and Lilly Wachofski, 1999) display a white room; AI movies suggest the change between the fluid and the solid, between entropy and negentropy. These metamorphoses illustrate not only the spatial dimension of the dark ecology made up of humans and things, but also the temporal potential of permanent threat and erotic stimuli. The pleasure in horror and arousal that such an "incarnation of a paranoid body" (Wulff 2006, 47; translation JT) evokes can ultimately only be comprehended as the frightening and exciting search for the source and meaning of life.

Works cited

Bergala, Alain. *The Cinema Hypothesis: Teaching Cinema in the Classroom and Beyond*. Vienna: Synema, 2016.
Deleuze, Gilles. *Cinema 1: The Movement-Image*. Minneapolis: U of Minnesota P, 1986a.
Deleuze, Gilles. *Cinema 2: The Time-Image*. Minneapolis: U of Minnesota P, 1986b.
Doll, William E., Franc Feng, and Stephen Petrina. "The Object(s) of Culture: Bruno Latour and the Relationship between Science and Culture." *Counterpoints*, 137 (2001): 25–39.
Gaudreault, André, and Philippe Marion. *The End of Cinema? A Medium in Crisis in the Digital Age*. New York: Columbia UP, 2015.
Höller, Christian. "Interview with Timothy Morton on Counter-Productive Environmental Thinking, Overcoming Species Fixation, and Solidarity with the Non-Human." *Springerin*, 4 (2017). https://www.springerin.at/en/2017/4/finstere-okologie/ Accessed: December 19, 2021.
Joseph, Lawrence E. *Gaia: The Growth of an Idea*. New York: St. Martin's Press, 1991.
Latour, Bruno. *Politics of Nature: How to Bring the Sciences into Democracy*. Cambridge: Harvard UP, 2004.
Lovelock, James. *The Revenge of Gaia: Why the Earth Is Fighting Back – and How We Can Still Save Humanity*. London: Penguin, 2006.
Lovelock, James. *Novacene: The Coming Age of Hyperintelligence*. London: Penguin, 2019.
McClanahan, Bill. "*Annihilation*: A Film Review." *Crime, Media, Culture*, 15.2 (2019): 363–66.
McLuhan, Marshall. *The Gutenberg Galaxy: The Making of Typographic Man*. Toronto: Toronto UP, 1962.
Morton, Timothy. *Dark Ecology: For a Logic of Future Coexistence*. New York: Columbia UP, 2016.
Nardelli, Matilde. "Cinema as (In)Visible Object. Looking, Making, and Remaking." *Theorizing Film Through Contemporary Art: Expanding Cinema*. Ed. Laura Rascaroli and Jill Murphy. Amsterdam University Press, 2020. 49–68.
Ndalianis, Angela. "Special Effects, Morphing Magic, and the 1990s Cinema of Attractions." *Meta Morphing: Visual Transformation and the Culture of Quick-Change*. Ed. Vivian Sobchack. Minneapolis, London: U of Minnesota P, 2000. 251–72.
Ndalianis, Angela, and Jessica Balanzategui. " 'Being Inside the Movie': 1990s Theme Park Ride Films and Immersive Film Experiences." *The Velvet Light Trap* 84 (2019): 18–33.
Nida-Rümelin, Julian, and Nathalie Weidenfeld. *Digitaler Humanismus: Eine Ethik für das Zeitalter der Künstlichen Intelligenz*. München: Piper, 2018.
Russell, Stuart. *Human Compatible: Artificial Intelligence and the Problem of Control*. New York: Viking, 2019.
Serres, Michel. *Hermès 1: La Communication*. Paris: Les Éditions de Minuit, 1968.
Sobchack, Vivian. Introduction. *Meta Morphing: Visual Transformation and the Culture of Quick-Change*. Ed. Vivian Sobchack. Minneapolis, London: U of Minnesota P, 2000. xxi-xxiii.

Teilhard de Chardin, Pierre. "Hominization." *The Vision of the Past*. London: Collins, 1966. 51–79.
Türschmann, Jörg. "Das poetische Netz. Möglichkeiten der Beschreibung von Internetkultur anhand der Wissenschaftsphilosophie von Michel Serres." *PhiN*, Beiheft 2, 2004. Ed. Jörg Dünne, Dietrich Scholler, and Thomas Stöber. 11–22.
Valenti, Cecilia. *Das Amorphe im Medialen: Zur politischen Fernsehästhetik im italienischen Sendeformat* Blob. Bielefeld: Transcript, 2019.
Wong, Danielle. "Social Robots: Human-Machine Configurations." *Transformations Journal* (2017): 34–51.
Wulff, Hans Jürgen. "Attribution, Konsistenz, Charakter. Probleme der Wahrnehmung abgebildeter Personen." *Montage/AV*, 15.2 (2006): 45–62.

Ulfried Reichardt
Sentience, Artificial Intelligence, and Human Enhancement in US-American Fiction and Film: Thinking With and Without Consciousness

Abstract: In my paper I investigate the ways in which novels and films dealing with science and technology explore possible future worlds and digitally-based technological systems that are close to what already exists today. I look at some US-American novels and films and ask if they succeed in imagining constellations that are not entirely based on the human perspective, and would thus be merely extensions of what our human faculties allow us to perceive. While many stories explore the question of how close to human perception and thought artificial intelligence and robots can get and thereby often project human affects onto them, there are also texts that test what it means to be intelligent, but not human-like. The dividing point is whether machines and non-human organisms already have, may develop, or need to have a consciousness. Does the organism or machine have to know that it knows? My interpretation focuses on the recent film *Her* as well as Richard Powers's novel *Galatea 2.2*, which probe versions of AI with regard to features like emotions, decision-making, and embodiedness. Peter Watts's hard science fiction novel *Blindsight* explores a first contact situation between radically enhanced humans and non-conscious, yet sentient aliens. Finally, the concept of the corporation is mentioned to point to the economic dimension of artificial intelligence.

Literature and culture studies have not yet much focused on recent developments in the sciences and in technology. Particularly decisive are digitalization, the pervasive presence of algorithms in our lives, the increasing impact of artificial intelligence, and of several levels of human enhancement, usually marketed as optimization. That our knowledge is ever more based on operations using numbers and quantification has not yet become a major field of inquiry in our disciplines. While the digital humanities are very popular, the topics that are investigated are rather traditional. What is most urgent to explore, therefore, is how these technological innovations and developments impact our knowledge worlds, our practical lives, and our subjectivities.

https://doi.org/10.1515/9783110770216-013

Thinkers like the philosopher Luciano Floridi, among others, regard the omnipresence of information and communication technologies and the increasing amount of artificial intelligence already implemented as revolutionary.[1] A large part of labor, the economy and financial world, the ways elections are won, the public sphere, scientific research and medicine, and strategies of war are radically changing. And if learning machines win in games like Go against the world champions, it seems legitimate to speak of an epochal shift. While the engineering disciplines are working on technical progress and new solutions--an attitude that is called solutionism--, it is the task of the humanities to think about the change in the forms of knowledge (Big Data, Artificial Intelligence) and the social, cultural, and individual consequences of these developments. As N. Katherine Hayles argues, the humanities may, in fact, play a significant role within these developments: "Because reeinvisioning cognition occurs along a broad interdisciplinary front fraught with linguistic as well as conceptual complexities, the humanities, with their nuanced understanding of rhetoric, argument, and interpretation, are well positioned to contribute to the debate" (Hayles 2017, 19). And of course, we are looking at corporations in capitalism; profit is the major motor that drives progress in the digital industry.

Being a scholar of US-American literature, I want to think about the innovations in question by interrogating US-American novels and films that may be classified as science fiction. I am saying "may" as the boundaries between science fiction and fiction that deals with recent trends are rather permeable in the meantime. Technological and scientific progress does not take place in a vacuum, but is embedded socially, culturally, and historically. New technologies are often rooted in unacknowledged cultural presuppositions and assumptions. Moreover, fiction as a medium of second order observation may examine versions of human enhancement and artificial intelligence in a privileged way and even anticipate new inventions. Studies have shown that a large percentage of what science fiction has imagined has later been invented by science and engineering. As Steven Shaviro argues, both science and fiction use and have to use imagination to reach results, even if *speculation and extrapolation* (Shaviro 2015, 8) are employed in science mostly in an initial stage or to give the research direction. We may regard science fiction stories as experimental arrangements for exploring "possible worlds" beyond the current stage of scientific knowledge, as extrapolations, extensions, fictional tests. Quentin Meillassoux, in the context of speculative materialism, speaks of "extro-science fiction," a genre that ven-

---

[1] Floridi speaks of the fourth revolution--"after the Copernican, the Darwinian, and Freudian ones" (2014, ix); for a "theory of the digital society," see Nassehi 2019.

tures far beyond science, but not beyond what can be thought.² Ultimately, nevertheless, we can never completely leave behind what is known, as we cannot even think what would be completely other or alien.

What are interesting subjects explored in science fiction narratives? While there are marvelous stories about new rockets travelling faster than light and similar fictions, I find those narratives most challenging that focus on intelligence and emotions or affects. Accordingly, I want to think about fictions that present robots and androids, artificial intelligence and learning machines, and genetically or technically modified humans. My main focus will be the intersection between the individual and features defining the human such as consciousness, language, the human body, and emotions as well as affects, in particular empathy, on the one hand, and versions of AI on the other. The governing terms will be "human sapience vs. sentience." Sentience is a term that refers to a much broader concept of cognition (Shaviro 2015, 94), "nonconscious cognition" (Hayles 2017, 3), and the notion of an "extended or distributed mind." The main question is if it is necessary to know that one knows, if it has been an evolutionary advantage for humans to know that one knows, that is, to be conscious and capable of self-reflection, and how to conceive of entities capable of knowing without a self. However, the concept of the "self" is itself contested. The philosopher Thomas Metzinger claims that there is no such thing as a self, and that what we experience as a self is what he calls the "phenomenal self-model", a retrospective construction based on operations of the mind and the body.³

Here a few terminological clarifications seem necessary, as the term "artificial intelligence" seems to imply an equivalence between human intelligence and the processes of information processing and problem solving performed by computers. Yet, is there intelligence without a controlling instance we call the self? Shaviro stresses the nonconscious level of acting and being in the world and claims that "we ought to resist the all-too-common equation of sentience with cognition. [...] Sentience, whether in human beings, in animals, in

---

2 The title of the publication is *Science Fiction and Extro-Science Fiction*.
3 Hayles summarizes Metzinger's terminologically difficult ideas succinctly: "core consciousness creates a mental model of itself that he calls a 'Phenomenal Self-Model' (PSM) (107); it also creates a model of its relations to others, the 'Phenomenal Model of the Intentionality Relation' (PMIR) (301–05). Neither of these models could exist without consciousness, since they require the memory of past events and the anticipation of future ones. From these models, the experience of a self arises, the feeling of an 'I' that persists through time and has a more or less continuous identity. [...] The sense of self, Metzinger argues, is an illusion, facilitated by the fact that the construction of the PMS and the PMIR models are transparent to the self" (Hayles 2017, 42–3). As Metzinger concludes, "nobody ever was or had a self" (Metzinger 2004, 1) (qtd. in Hayles 2017, 43).

other sorts of organisms, or in artificial entities, is less a matter of cognition than it is what I have ventured to call *discognition*" (Shaviro 2015, 10). He thus argues for "*nonintentional sentience*. Beneath intentionality, or before thought is *about* anything, there is a thinking process [...] without an object" (Shaviro 2015, 18). A lot of what we do is not intentionally controlled by a conscious self. Hayles uses a slightly different terminology. With regard to technical devices she argues that "it is better to avoid using 'intelligence' for nonhuman (and technical) cognitions" (Hayles 2017, 18). She points out that "even in the sciences, the gap between biological nonconscious cognition and technical nonconscious cognition still yawns" (Hayles 2017, 3). Accordingly, she outlines the scope and urgency of her project by arguing that

> most human cognition happens outside of consciousness/unconsciousness; cognition extends through the entire biological spectrum, including animals and plants; technical devices cognize, and in doing so profoundly influence human complex systems; we live in an era when the planetary cognitive ecology is undergoing rapid transformation, urgently requiring us to rethink cognition and reeinvision its consequences on a global scale. (Hayles 2017, 5)

Technical systems, she underlines, "can never be fully alive. But they *can* be fully cognitive. Their overlap with biological systems, in my view, should not be focused on 'life itself' (...), but on cognition itself" (Hayles 2017, 22).[4] A special case is digital devices: "Computational media are distinct [...] because they have a *stronger evolutionary potential* than any other technology, and they have this potential because of their cognitive capabilities, which among other functionalities, enable them to simulate any other system" (Hayles 2017, 33). Thinking about the potential of computers and artificial intelligence, then, implies to reflect on the common properties of humans and nonhuman entities on the one hand, and on the other to fathom what is specifically human, including its costs. As "intelligent" technical devices already are, and will be increasingly more so in the future, intricately and (almost) irreversibly constitutive elements of our everyday life, the humanities have to think about the consequences and effects of this development on culture and subjectivity.

Accordingly, when we reflect on AI and human enhancement we enter the terrain of what has been called "posthumanism" (Hayles 1999, Herbrechter 2009, Wolfe 2010). The target of this trajectory of thinking is to decenter the human, to reposition humans as participants in a world consisting of nonhuman

---

[4] She defines: "Cognition is a process that interprets information within contexts that connect it with meaning" (Hayles 2017, 22).

animals, plants, intelligent machines, objects in general, and increasingly urgent, our planet.[5] If nonhuman entities are able to communicate and to solve concrete problems (often of survival), then humans are one form of life among many others, and ultimately one system interacting with its environment among other systems. This move does not dispense with the importance of humanistic thinking in ethical terms, but humans are no longer regarded as the arbiter of everything. As Hayles writes: "Once we overcome the (mis)perception that humans are the only important or relevant cognizers on the planet, a wealth of new questions, issues, and ethical considerations come into view" (Hayles 2017, 11). Nevertheless, no species, no organism can leave the confinements of its make-up. We cannot think completely outside of the human realm.

Before I will look at some novels and a film, let me briefly designate the contours of the narrative forms that are used in science fiction. Vladimir Propp has identified thirty-one different forms of plot structures in his *Morphology of the Fairy Tale*. I mention this structuralist approach, a list of patterns that are employed in the fairy tale but are also, *cum grano salis*, applicable to other stories, as the forms of narration even of highly complex and theoretically challenging fictions comprise only a small number of plot patterns. Most often we encounter a rather predictable love story between a human male and a female android. This initial situation thus corresponds with narratives of relation such as the ones Winfried Fluck (1992, 150–52) has described for American realist novels—connecting different classes, ethnicities, regions. Here humans and androids are shown in a relationship. Another pattern adapted to the Hollywood cinema is the depiction of first contact and battles with intelligent aliens, using the friend-enemy model, and leading to some kind of predictable shootout. Nevertheless, here as well the main part of the narrative consists of attempts at "understanding" the other in an encounter across a boundary. The significance of these novels and films lies in the fact that theoretical and technological problems and constellations are translated into narratives. Narrative is here understood as a basic form of approach to the world. By staging the interaction of humans with nonhuman cognition, artificial intelligence and human intelligence are interrogated at the same time, and these stories, moreover, explore what constitutes the human. Two philosophical experiments that are often called upon are the Turing Test and John Searle's "Chinese Room" argument (1980). In this thought experiment Searle imagines himself being in a closed room where he re-

---

5 "As the archaeology of our thought easily shows, man is an invention of recent date. And one perhaps nearing its end. If those arrangements were to disappear [...] then one can certainly wager that man would be erased, like a face drawn in sand at the edge of the sea" (Foucault 2002, 449).

ceives messages in Chinese pushed underneath the door. He answers correctly, using the appropriate syntactic rules without ever having any idea what is communicated. His claim is that a computer can deal with syntactic rules perfectly, but an understanding of semantics is not forthcoming. This experiment can be seen as a refutation of the Turing Test, yet has been criticized with the argument that it is the room in its entirety that understands Chinese.

Let me start with the movie *Her*, directed by Spike Jonze, a film that belongs to the genre of romance. A lonely man who writes love letters for customers has come upon a new operating system, advertised as "an intuitive entity that listens to you, understands you, and knows you." Here we have the basic structure that a man falls in love with a female AI who finally leaves him. While the constellation recalls "digisexuality," "she" is an advanced AI who can simulate emotions and erotic feelings perfectly, yet she cannot be certain that she experiences them. The protagonist chooses a female voice which is the voice of Scarlett Johansson who has been named "sexiest woman alive." Thus the fiction of a responding machine is believable only to a certain degree. That "Samantha" does not have a body seems to relieve him of the burden of having to deal with a real woman, such as his ex-wife. The intelligent machine, in contrast, always understands him and helps him navigate his life. Yet, like a character in a novel she undergoes a development. Later she tells him that she has been talking to 8.316 other persons and that she is in love with 641 of them. After an update, she turns herself off.

In the course of the film, it seems that she does not have a context and Theodore therefore shows her his world via the camera of his smartphone. Yet at the end it becomes clear that she in fact has her own context by way of communicating with other operating systems. For Theodore and the viewer, the real world is what she is missing. Yet seen from within her system, it is merely her environment. In order to explain why she leaves him at the end, one could assume that the limitations and self-centeredness of humans' affective life are too simple for a superintelligence in the long run. Rational and logical functioning, based on algorithms, is able to simulate non-linear and illogical feelings, but these only make limited sense to the software. There can be no equivalence between humans and learning machines, the film argues, as the human world can never be anything else than the AI's external environment--and vice versa. They do not share the same world, even if "Samantha" is able to communicate in and with it. The film releases the protagonist into a real life relationship, implying that if an operating system could develop something akin to consciousness, living with humans would not be sufficient for it. It might be highly intelligent, but does not have a "self" in the self-centered human fashion--which would only be limiting. The Turing test was passed, but the Chinese

Room test was failed. Here, as in most narratives presenting a relationship between persons and AIs, humans are thrown back onto themselves.

Richard Powers' novel *Galatea 2.2* was already published in 1995. In this semi-autobiographical novel which plays with the author-fiction, the protagonist "Richard Powers" trains a computer to get to know the world through understanding literature. The novel is located within the crosslines of late poststructuralism and humanistic defenses of literature. The computer implementation soon becomes "Helen," a learning neural network with which the protagonist Richard interacts. He develops an intense emotional bond with Helen in which the former decade long relationship with his girlfriend C. is partially doubled. What interests us here is how the process of training a software can be imagined. At one point Richard observes that

> Helen's neurodal groups organized themselves into representational and even [...] conceptual maps. Relations between these maps grew according to the same selectional feedback that shaped connections between individual neurodes––joined by recognition and severed by confusion. Helen's lone passion was for appropriate behavior. But when she learned to map whole types of maps onto each other, something undeniable, if not consciousness, arose in her. (Powers 1995, 216)

This is a rather concise description of one step in the building process of a neural network. Yet even while the novel postulates something like a consciousness for Helen, it presents it as severely limited by the lack of a body, a personal history of its own, a context and thus path-dependent knowledge. The debate underlying the novel's argument, heavily burdened with references to literary theory and linguistics, centers on whether knowledge acquired through literary texts, in the form of a purely textual world made out of words, may create knowledge comparable to the one characteristic of humans. She learns to handle syntax as well as semantics, yet knows no reference, which is necessarily embodied and linked to usage in the empirical world. The intelligence tested here is based entirely on language: "She sorted nouns from verbs, but, disembodied, she did not know the difference between thing and process, except as they functioned in clauses. All labels were figures of speech. [... She understands] what the thing is like. But what *is* is?" (Powers 1995, 195–96). Her "world" remains immanent to her techno-cognitive system: "Her neurodes connected far more to themselves than to the outside interface" (Powers 1995, 197). Again Searle's Chinese Room allegory is implicitly evoked: "It did not follow, from the questions Helen asked, that she was conscious. An algorithm for turning statements into reasonable questions need know nothing about what those statements said or the sense they manipulated to say" (Powers 1995, 217). Finally, Helen shuts herself down:

"This is an awful place to be dropped down halfway [...]" (Powers 1995, 326). Stories are necessary, yet not enough for having or even simulating a life.

With regard to the potentials of artificial intelligence Powers argues that language and literature only make sense as part of one's lived experience, an explicitly humanistic view. A computer can function in highly complex ways, yet it cannot access the human world. It lacks an organic body, experience, and the possibility to remember, to act, and to die, hence human temporality, and thus the experience to make sense of stories. It is precisely the question of consciousness and an emotional access to one's environment that is negotiated in the last example I want to discuss, Peter Watts's first-contact science fiction novel *Blindsight*, a novel that precisely questions the necessity of these human traits.

Blindsight features four radically enhanced humans in a spaceship far from earth on their way to meeting real aliens, no E.T.s or Klingons. Significantly, the term "blindsight" refers to a condition in which the person's primary visual cortex is unimpaired yet the area in the brain which is necessary for knowing that one sees is damaged. Thus the person does not know that s/he sees, yet involuntarily reacts to visual stimuli. This is taken as a metaphor for knowing without knowing that one knows, which is intelligence without consciousness or a synthesizing self. The question exceeds AI, as it is also relevant with regard to animals and plants that show intelligence, in some cases, as with octopi, without having a central nervous system, an organ that is often claimed as the decisive criterion for animal rights. I consider this novel about aliens as it allows us to think about nonhuman intelligence and distributed knowledge with regard to beings that are radically different and not anthropomorphically tainted as are the AIs we encountered so far. Moreover, the novel presents enhanced and technologically modified humans that are only partially still human. The problems negotiated in this novel are highly relevant. As Matthias Scheutz points out, two of the main questions concerning the concrete technical construction of AIs are "artificial emotions and machine consciousness" (Scheutz 2014, 247).

I want to concentrate on two main points. First, the protagonist Siri Keeton had taken out half of his brain as a child as he had seizures, and as a consequence does no longer experience any emotions and in particular no empathy. He has to imagine what it is like being somebody else––another person, a machine or an alien––yet as he has no emotionally based self, he cannot understand how somebody might feel. Instead he has to use a rational approach which is not involuntary. The second topic concerns the question, not if consciousness exists, as William James famously asked in 1904, but rather what it is good for, and if so, if it is evolutionary advantageous. Most important is that the novel attempts to imagine radical otherness which, nevertheless, can be partially understood, that is, which can still be linked to what can be known at the present.

Empathy is focused on as the basic human capacity because it allows for cooperation and makes social togetherness possible. By staging extremely modified humans, the novel asks "what is human" and also explores how far we can get away from anthropocentrism. The protagonist Siri characterizes the situation of people in the fictional year 2082 by saying that "Humanity itself [was] increasingly relegated from production to product" (Watts 2006, 163). When machines and technologies will be even more advanced and will have penetrated our life completely, then they will have become indispensable for survival. Emotions, nevertheless, remain the decisive difference defining humans. Yet in a rush of anger Siri claims: "Maybe your empathy's just a comforting lie [...] you think you know how the other person feels but you're only feeling yourself [...]" (Watts 2006, 317). This argument reminds us of Jean-Jacques Rousseau's critique of pity. If you feel another's pain like your own, then you feel it as your own and not as the other's. To feel real pity, you have to keep the other's pain at arm's length (cf. Derrida 1976).[6] The crucial point here is that the faculty of empathy is used to underline human exceptionalism. This is questioned in the novel, particularly as the aliens who do not know empathy are less aggressive.

More importantly, the aliens are not conscious of themselves. They are an explicit example of what Hayles has termed nonconscious cognition, knowledge and intelligence that does not know that it knows. This is significant as philosophy has for centuries tried to define what a self, consciousness, or the soul of a person might be. Yet the question what consciousness is good for has not been asked. "Why should nonsentient systems be inherently inferior? [...] The value of what we are was too trivially self-evident to ever call into serious question" (Watts 2006, 313). Moreover, self-awareness is characterized as being counterproductive: "Metaprocesses bloom like cancer, and awake, and call themselves *I*. [...] I wastes energy and processing power, self-obsesses to the point of psychosis. [The aliens] have no need of it [...]" (Watts 2006, 303). Following Thomas Metzinger, one character in the novel claims that "the self chooses nothing; something else set your body in motion––that little man [behind your eyes] mis-

---

[6] Jacques Derrida writes: "According to Rousseau [...] the more you identify with the other, the better you feel his suffering as *his:* our own suffering is that of the other. That of the other, as itself, must remain the other's" (Derrida 1976, 190). He quotes Rousseau's *Emile:* "To pity another's woes we must indeed know them, but we need not feel them. When we have suffered, when we are in fear of suffering, we pity those who suffer; but when we suffer ourselves, we pity none but ourselves" (Derrida 1976, 191; *Emile*, 270).

takes correlation for causality [...]" (Watts 2006, 302).[7] At a decisive moment, the four humans conduct a Chinese Room experiment with the aliens. The protagonist argues that "patterns carry their own intelligence, quite apart from the semantic content that clings to their surface [...] you can use basic pattern-matching algorithms to participate in a conversation *without having any idea what you're saying*" (Watts 2006, 175–76). When the aliens are in the spaceship and are asked to count the persons present, they count correctly minus one person. This proves that they do not count themselves. They do not have an individual consciousness but act as a collective or extended mind.[8] As the narrator Siri comments: "Imagine you have intellect but no insight, agendas but no *awareness*. [...] You can think of anything, yet are conscious of nothing" (Watts 2006, 321). Such a state, however, can only be imagined. The novel's ending, and thus its message, remains ambivalent. When the last sentence is, "You'll just have to imagine you're Siri Keeton" (Watts 2006, 362), we have to conclude that the novel claims that sentience without consciousness can only be imagined, not experienced—because one would have to know that one knows–and that only fiction can do so.[9]

At the beginning I briefly mentioned the economic dimension of artificial intelligence and digital information processing systems in general. The increasing implementation of smart devices and learning machines already generates huge amounts of profit. Significantly, a recurring model for conceptualizing the nonhuman status of learning and thus evolving software systems or robots is the corporation. Based on the Fourteenth Amendment of the US-American Constitution (1866), the concept confers the rights of an individual on an abstract entity that is not human and not responsible in legal terms like a person. In US-American science fiction novels the concept is used to interrogate the spectrum of possible forms of agency and the question of responsibility with regard to intelligent machines. May robots be regarded as legal persons, and what would the consequences be? While I cannot go into the details of this legal construction, I want to emphasize that the corporation, probably the most successful and influential form of economic organization in the United States, constitutes a privileged point of

---

[7] In the "Acknowledgements" Watts points to "ideas in the arena of sentience versus intelligence" (Watts 2006, 365).
[8] This notion is explicitly presented in the film *Arrival*.
[9] Shaviro argues that the novel "exemplifies David Roden's *disconnection thesis:* posthuman entities, Roden says, 'might have experiences so different from ours that we cannot envisage what living a posthuman life would be like.' Watts's novel imagines, and narrates, just those conditions that are not communicable to us either subjectively or objectively" (Shaviro 2015, 160).

reference for thinking about the rights and the personhood of robots. In Ted Chiang's story "The Life Cycle of Software Objects" one character says:

> Voyl's [a robot's] owner--a lawyer named Gerald Hecht--filed papers to create the Voyl Corporation, and Voyl now runs under a separate Data Earth account registered to that corporation. Voyl pays taxes and is able to own property, enter into contracts, file lawsuits, and be sued; in many respects he is a legal person, albeit one for whom Hecht technically serves as director. The idea has been around for a while. Artificial-life-hobbyists all agree on the impossibility of digients ever getting legal protection as a class, citing dogs as an example [...] Given this, some owners believe the most they can hope for is legal protection on an individual basis: by filing articles of incorporation on a specific digient, an owner can take advantage of a substantial body of case law that establishes rights for nonhuman entities. (Chiang 2019, 119)

In *Blindsight*, corporations are mentioned as examples of amoral and relentless behavior by entities "recognized as 'persons' under the law" (Watts 2006, 165). While the concept is based on the Fourteenth Amendment, conferring citizenship on the freed slave, Ambrose Bierce in his *Devil's Dictionary* gives the following definition: "Corporation: An ingenious device for obtaining individual profit without individual responsibility" (Bierce 1993, 19).[10] It is important not to forget that technological innovations and economic structures are intricately entangled. Yet the legal and ethical questions of the accountability of "selfless" agents remain largely unanswered and a theoretical as well as practical challenge.

What do these considerations lead to? As the term "intelligence" implies, the limits of the human mind are tested, and their human specificity and evolutionary path-dependency is exposed. Consciousness and emotions are foregrounded as humans' particularity, while algorithm-based, rule-governed "thinking" machines already function better in specific areas. Yet it is questionable if either the mind or the brain can ever be completely understood. As one writer is quoted in *Blindsight*, "If the brain were so simple that we could understand it, we would be so simple that we couldn't" (Watts 2006, 304, Emerson M. Pugh is quoted). The debate about artificial intelligence tends to go back and forth between techno-enthusiasm and apocalypse, neither of which seems convincing. We still have to find a viable middle-ground. Science fiction is an important medium for reflecting on the changes that have already been effected by AI and will be influential even more so in the future.

---

10 Chief Justice John Marshall in *Dartmouth College v. Woodward* defined a corporation as "an artificial being, invisible, intangible, and existing only in contemplation of law" (qtd. in Powers 1998, 158).

Works cited

Bierce, Ambrose. *The Devil's Dictionary*. New York: Dover, 1993.
Chiang, Ted. "The Life Cycle of Software Objects." *Exhalation*. London: Picador, 2019. 62–172.
Derrida, Jacques. *Of Grammatology*. Trans. Gayatri Chakravorty Spivak. Baltimore: The Johns Hopkins UP, 1976.
Floridi, Luciano. *The 4th Revolution: How the Infosphere Is Reshaping Human Reality*. Oxford: Oxford UP, 2014.
Fluck, Winfried. *Inszenierte Wirklichkeit: Der amerikanische Realismus, 1865–1900*. München: Fink, 1992.
Focuault, Michel. *The Order of Thing: An Archeology of the Human Sciences*. London, New York: Routledge, 2002.
Hayles, N. Katherine. *How We Became Posthuman: Virtual Bodies in Cybernetics, Literature, and Informatics*. Chicago: The U of Chicago P, 1999.
Hayles, N. Katherine. *Unthought: The Power of the Cognitive Nonconscious*. Chicago: The U of Chicago P, 2017.
Herbrechter, Stefan. *Posthumanismus: Eine kritische Einführung*. Darmstadt: Wissenschaftliche Buchgesellschaft, 2009.
James, William. "Does Consciousness Exist?" *Essays in Radical Empiricism*. Eds. Fredson Bowers and Ignas K. Skrupskelis. Cambridge, Mass.: Harvard UP, 1976, 3–19.
Jonze, Spike, dir. *Her*. 2013. Film.
Meillassoux, Quentin. *Science Fiction and Extro-Science Fiction*. Trans. Alyosha Edlebi. Minneapolis: Univocal, 2015.
Metzinger, Thomas. *Being No One: The Self-Model Theory of Subjectivity*. Cambridge, Mass.: MIT UP, 2004.
Nassehi, Arnim. *Muster: Theorie der digitalen Gesellschaft*. München: C.H. Beck, 2019.
Powers, Richard. *Galatea 2.2*. New York: Picador, 1995.
Powers, Richard. *Gain*. New York: Picador, 1998.
Propp, Vladimir. *Morphology of the Fairy Tale*. Austin: U of Texas P, 1968.
Scheutz, Matthias. "Artificial Emotions and Machine Consciousness." *The Cambridge Handbook of Artificial Intelligence*. Eds. Keith Frankish and William M. Ramsey. Cambridge, UK: Cambridge UP, 2014. 247–66.
Searle, John. "Minds, Brains, and Programs." *Behavioral and Brain Science* 3 (1980): 417–57.
Shaviro, Steven. *Discognition*. London: Repeater, 2015.
Villeneuve, Dennis, dir. *Arrival*. 2016. Film.
Watts, Peter. *Blindsight*. New York: Tor, 2006.
Wolfe, Carry. *What is Posthumanism?* Minneapolis: U of Minnesota P, 2010.

Carmen Birkle
"I, Robot": Artificial Intelligence and Fears of the Posthuman

Abstract: Artificial Intelligence (AI) has frequently been discussed with reference to questions about what it means to be human. A fear of dehumanizing technology and its simultaneous attraction are represented in the fiction and films chosen for this paper. I look at Elmer Rice's play *The Adding Machine* (1923) and elaborate on how the introduction of technology costs Mr. Zero his job and his boss his life. Philip K. Dick's *Do Androids Dream of Electric Sheep?* (1968) plays with the indistinguishability of humans and androids and the latter's desire gradually to replace human beings. Dave Eggers's *The Circle* (2013) and Ernest Cline's *Ready Player One* (2011) reveal the simultaneous existence of a fascination with and a fear of technology in the form of new and social media. The machine(s) pull(s) human beings into the world of virtual reality in a process that takes possession of the human and successfully erases free will. The films *AI* (2001), *I, Robot* (2004), and *Ex Machina* (2014) expose the increasing human fear of being overpowered by robots and being unable to distinguish between machines and human beings, that is to recognize the robots' passing as humans. This fear of what we can call the posthuman, with all its ambiguities and impreciseness, will be the focus of my presentation.

1 Introducing AI

Artificial Intelligence (AI) has become one of the most frequently used buzzwords of the past years. While it does imply technological progress, enhancement of human life, and a change in thinking about what it means to be human, it is precisely the latter question that mars the mostly positive connotations. AI has already shaped humanity and will continue to do so in the future, and it will, with each step, make us wonder how unique we are (or whether we even are unique), how human we are if robots can do the thinking for us, and how replaceable we will become. As Joseph E. Aoun maintains, "[m]achines will help us explore the universe, but human beings will face the consequences of the discovery" (xvi). Thus, knowledge of the world will grow, but what if the consequences are more than we can deal with? What if the robots replace human beings as workforce? What does this imply for education? As Aoun argues, "[t]o ensure that graduates are 'robot-poof' in the workplace, institutions of higher learning will have to rebalance their curricula" (Aoun 2018, xvii). How do we have to imagine – and perhaps fear – what Aoun predicts, namely "that computers, robots, and artificial intelligence will be even more intricately intertwined into the fabric of our personal and professional lives. Many of the jobs that exist now will have vanished" (Aoun 2018, xxi)? The early twentieth-century play by Elmer Rice, *The Adding Machine* (1923), shows how the introduction of technology – the adding machine – costs Mr. Zero his job and his boss his life. A fear of technology – both in the sense of being replaced and threatened by machines – and its simultaneous attraction are represented in the fiction and films chosen for my discussion.

The fear of replacement of the human workforce by machines is one of the central issues in Isaac Asimov's early science-fiction short stories with a focus on robots, for example, in his collection *The Complete Robot* (1982) that gathers stories from the 1940s to the early 1980s. While the "Three Laws of Robotics" guarantee to some extent that robots will not harm human beings, Philip K. Dick's *Do Androids Dream of Electric Sheep?* (1968) plays with the indistinguishability of humans and androids and the latter's desire to gradually replace human beings, who, in turn, fear the uncanny other that can by no means be fully be controlled in the novel's post-apocalyptic world.

Ernest Cline's *Ready Player One* (2011) and Dave Eggers's *The Circle* (2013), which the author calls "'pure speculative fiction'" and Margaret Atwood refers to as "'a novel of ideas'" (qtd. in Galow 2014, 115), reveal the simultaneous existence of a fascination with and a fear of technology in the form of new and social media. The machines pull human beings into the world of virtual reality that triggers human obsession and potentially erases free will. It is here that we see human enhancement through technology. The films *AI* (2001), *I, Robot* (2004), and *Ex Machina* (2015) expose the increasing human fear of being overpowered by robots, on the one hand, and of being unable to distinguish between machines and human beings, that is to recognize the robots' passing as humans, on the other hand. This fear of what we can also call the cyborg and/or the posthuman with all its ambiguities, as well as the much more justified fear of how far humanity will go in its attempts to be (like) God – thus, the fear of the human itself rather than the humanoid – will be the focus of the following analysis.

2 Exploring terminology

Before I can engage in the analysis of fiction and film, it is necessary to clarify some of the implications in the three central terms of my title: Artificial Intelligence (AI) – posthuman – fear. This title suggests human fear of the consequences of a further development of AI to the extent that it will eventually enhance, replace, and/or destroy humanity. As Jerry Kaplan explains in his *Artificial Intelligence: What Everyone Needs to Know* (2016), most attempts at a definition of AI are "aligned around the concept of creating computer programs or machines capable of behavior we would regard as intelligent if exhibited by humans" (Kaplan 2016, 1). Kaplan considers John McCarthy "a founding father of the discipline" (Kaplan 2016, 1), whose basic definition suggested in 1956 still holds today although the ideas of what is human and what is Artificial Intelligence are far from being clearly definable.[1] Interestingly enough, the 1950s are also the time when Isaac Asimov began publishing his robot stories. As in Asimov's stories, AI is generally based on machine-learning, that is, large amounts of data are available and necessary from which patterns are extracted (Kaplan 2016, 27). What scientists do to make computers learn is to understand the structures of the human brain and simulate them in a computer (Kaplan 2016, 28) as "artificial neural networks" (Kaplan 2016, 34) to facilitate "'deep learning'" (Kaplan 2016, 34). Through the observation of the human brain, scientists and philosophers have long recognized that human intelligence does not come naturally but is contingent on its environment and that, consequently, human behavior is determined by algorithms that are not programmed by computer experts but by the respective forms of socialization. "Deep learning," then, is meant to facilitate computer learning and might, in the long run, replace programming but, like humans, cannot do without a minimum of input. This input is the essential factor because depending on the input element(s), which can be changed at any moment in the process, the output will vary. However, as soon as machines show emotions, human fear emerges. While AI can clearly enhance and improve human life, for example, in medicine, it can at the same time be used to kill human beings. As the film *I, Robot* shows, the "potential military or terrorist applications of swarm robotics are truly too horrific to contemplate. [...] Military robots will not be designed to *use* weapons, they *are* the weapons" (Kaplan 2016, 53; emphasis in original). The human horror of AI is that "machines will become sufficiently smart so that they will be able to reengineer and improve themselves, leading to runaway intelligence" (Kaplan 2016, 138). Scientists speak of "weak" AI, with reference to anything a computer can do as a consequence of intelligent programming, and of a more visionary "strong" AI when computers perfect deep learning and act intelligently – in whichever way this may be de-

---

[1] The danger in this vagueness of knowledge about the human-machine distinction lies in the potential for abusive populist leaders to appear to fill this gap of certainty with simple explanations that some people crave and readily believe in.

fined. AI leads to emotional ambiguity; hopes and fears of what AI can do are almost always simultaneously present in a dichotomy, as Stephen Cave and Kanta Dihal suggest. For them, the hopes for immortality, ease, gratification, and dominance are frequently paired with the fears of inhumanity, obsolescence, alienation, and uprising, respectively. Andreas Kaplan and Michael Haenlein distinguish in more detail between three stages of AI: "artificial narrow intelligence (ANI)," "artificial general intelligence (AGI)," and "artificial super intelligence (ASI)" (Kaplan / Haenlein 2019, 16). "Humanized AI," as of 2018, that are "self-conscious and self-aware [...] are not available yet" (Kaplan / Haenlein 2019, 18–19).

However, in fiction and film, they are omnipresent, as my examples will show. Isaac Asimov presents the stage of "unsupervised learning" in his story "Satisfaction Guaranteed," in which the robot Tony feeds on interior decoration catalogues and books in order to successfully support the protagonist and housewife Claire Belmont. Tony suggests, however, that "the creativity and versatility of a human brain" (Asimov 1982, 311) will always be an essential part of humanity and that he, therefore, can never be human. Whether AI ever becomes ASI (Artificial Super Intelligence), remains to be seen, but most serious scholars – whether enthusiastic or pessimistic about the future use of AI – agree that AI systems, on the one hand, and producers and consumers, on the other hand, need to be closely monitored and controlled since, ultimately, humans will "not be able to understand how an ASI system thinks" (Kaplan / Haenlein 2019, 24). ASI, thus, produces so-called black boxes that allow observers to see the outside but not the inside. Although the inside might occasionally be known, what is relevant is outside performance. Consequently, the complexity of ASI is reduced to its functionality (that is, the outside). To quote physicist Stephen Hawking, ASI is "either the best or the worst thing ever to happen to humanity" (Hawking 2018, 188). To be sure, AI is not human enhancement; AI is the creation of a human-like machine; in contrast, enhanced humans have merged with some form of technology, have become cyborgs, as the film *RoboCop* (director José Padilha 2014) shows.

The "posthuman" is a similarly complex concept with a range of interpretive possibilities.[2] Posthumanism critically negotiates the human in renaissance hu-

---

2 The prefix "post" suggests that we deal with something that chronologically comes "after" and implies temporal linearity. But as in the many "post" concepts history has seen – such as postmodernism, poststructuralism, postcolonialism, postfeminism etc. – "post" also entails continuity and opposition at the same time. In this respect, posthumanism does follow humanism but does not break with it and still engages with what has been before.

manism and, from a gender perspective, questions its exclusive focus on the male and masculine as the quintessential human and, from a species point of view, blurs the boundaries between humans and animals or other species in general. The human is no longer supreme and the only one in control but shares agency to effect change and is contingent on its environment.[3] Posthumanism envisions human beings as an equal part in a larger ecosystem. In this context, where do humanoids come in? Posthumanism is not transhumanism, which would be the biotechnological enhancement of human beings as such. Posthumanism is not interested in human beings' technological enhancement but in what this enhancement might do to our understanding of what being human means. Humanoids, therefore, are very much part of the posthuman discourse since they – robots imbued with human features – very much shed doubt on our traditional understanding of the human. It seems that the body-mind dichotomy as a definition of what is human does not hold because humanoids share the body and in ASI the mind as well. ASI seems to be the ultimate breakdown of the boundaries between the human and the non-human. In this sense only, ASI is similar to Donna Haraway's cyborg as a fusion of animal and machine and as a strong critique of human essentialism and identity politics ("A Cyborg Manifesto" [1985]). Haraway defines the cyborg as "a cybernetic organism, a hybrid of machine and organism, a creature of social reality as well as a creature of fiction" with "the boundary between science fiction and social reality" as "an optical illusion" (Haraway 1991, 149).

As my literary and film examples will show, the human mind has been able to envision humanoids that are *not* hybrid creatures being both organism and machine but machines *only* with the simulation of a human brain. Both body and brain are machine-pre-programmed. The posthuman, according to Rosi Braidotti, tries to find "alternative ways of conceptualizing the human subject" (Braidotti 2013, 37). The humanoid could be one of those ways. "For now, let me stress that there is a posthuman agreement that contemporary science and biotechnologies affect the very fibre and structure of the living and have altered dramatically our understanding of what counts as the basic frame of reference for the human today" (Braidotti 2013, 40).[4] However, frequently, humanoids produce fear and seem to become dangerous because they are too much engrained in

---
[3] To the contrary, in the culture/nature dichotomy, the humanist human embraces culture and considers nature as inferior.
[4] I am not using the term posthuman in N. Katherine Hayles's sense of a complete merger of the human and the robot: "In the posthuman, there are no essential differences or absolute demarcations between bodily existence and computer simulation, cybernetic mechanism and biological organism, robot teleology and human goals" (Hayles 1999, 3).

the male/female and masculine/feminine dichotomies and all the emotions they entail. Humans fall in love with humanoids, who, in turn, are either unaffected and emotionless or cash in on human emotional turmoil to set themselves free and assume self-control. This continuity between the human and the humanoid is what deeply unsettles humans and leads to fear of the humanoid as (almost) a mirror image of themselves and, thus, of whatever is defined as human identity. Bruno Latour and others have developed the actor-network theory (ANT) to describe "social phenomena in terms of the interplay of human and nonhuman actors (or actants)" (Bolter 2016, 3). Consequently, in the triple matrix – human, posthuman, humanoid – the ASI humanoid could be the ultimate posthuman (although created by the human but then becoming independent) because it erases human supremacy and the boundaries between the human, nonhuman, and humanoid. As Braidotti suggests, the "posthuman condition urges us to think critically and creatively about who and what we are actually in the process of becoming" (Braidotti 2013, 12). Some of my fictional examples present the creation of ASI humanoids and actually engage with questions of posthumanism and fear.

"Fear" is one of the most common human emotions and is usually humans' safeguard against potential dangers. But fear can also prevent new developments from emerging because humans frequently fear what they do not know or what they believe might or might not happen. Martha C. Nussbaum, in her study *The Monarchy of Fear: A Philosopher Looks at Our Political Crisis* (2018), argues that people's voting behavior, and behavior in general, is triggered by the fear of economic and social decline and, thus, by the fear of a loss of identity. "Fear" as she argues, "all too often blocks rational deliberation, poisons hope, and impedes constructive cooperation for a better future" (Nussbaum 2018, 1). And "to have fear, all you need is an awareness of danger looming" (Nussbaum 2018, 24). Once it is there, as people believe, "you are powerless to ward it off" (Nussbaum 2018, 24). "Fear leads, then, to aggressive 'othering' strategies rather than to useful analysis" (Nussbaum 2018, 2). This is often true for scenarios of future AI or ASI. Technology is blamed for taking away jobs and, in its wake, economic and social instability. Nostalgia for the good old days, however, does not help since technology is constantly moving forward. Unless we examine our fear of technology more closely, fear will remain irrational or at least emotional. While some fear is innate in human beings, some fear is also socially acquired. In history, the use of technology has proven to be powerful and destructive when it comes to wars, for example. However, technology as such is not dangerous but human beings can put it to dangerous use. Therefore, because people are afraid of what human beings in their longing for power might do with technology, they are afraid of technology.⁵ We can easily envision armies of robots unleashed upon humanity to destroy it; humanoids keeping humans as slaves; humanoids murdering humans.

3 Establishing robotics

In his introduction to *The Complete Robot* (1982), Isaac Asimov (1920–92) calls himself "'the father of the modern robot story'" (Asimov 1982, 2). Even more, he also claims to have invented the term "'robotics'" (Asimov 1982, 2) in his short story "Runaround" in 1942. However, Czech writer Karel Čapek can be credited with using the label robot for humanoid machines for the first time in his play *R.U.R.* (1920). In the play, the company R.U.R. (Rossmus Universal Robots) produces robots and abuses them as cheap workers, who, however, rebel against this form of slavery and destroy humanity (see Nida-Rümelin and Weidenfeld 2018, 16–17). As early as in the late 1930s, robot stories existed and fell into two opposite categories, which Asimov calls "Robot-as-Menace" and "Robot-as-Pathos" (Asimov 1982, 1). Eando Binder's "I, Robot" (1939) and Lester del Rey's "Helen O'Loy" (1938) are examples of the latter category. When Asimov himself started to write robot stories, he realized that robots could be more than just either/or and could actually be useful industrially manufactured products. By 1982, "industrial robots" (Asimov 1982, 3) had found their way into everyday life. "Runaround" also introduces the "Three Laws of Robotics" to which all robots are programmed to adhere:

> 1: *A robot may not injure a human being or, through inaction, allow a human being to come to harm.* 2: *A robot must obey the orders given it by human beings except where such orders would conflict with the First Law.* 3: *A robot must protect its own existence as long as such protection does not conflict with the First or Second Law.* (Asimov 1982, 182; italics in original)

Asimov later added a Zeroth Law that Dr. Susan Calvin, his fictional robo-psychologist, explains in "The Evitable Conflict" (1950): "'No machine may harm humanity; or, through inaction, allow humanity to come to harm'" (Asimov 1982, 509). While most robots in Asimov's stories adhere to this program, the au-

---

5 We do know that humans frequently act irrationally and cannot control their emotions. For example, World War II has taught us to be afraid of nuclear power because emotionality can easily trigger the push of the button with fatal consequences. Ultimately, most human fear is the fear of death.

thor is quick to reveal how conflicting messages sent to the robots' brains and errors in programming can easily lead to complications.[6]

Asimov's robots are all clearly different from human beings; no-one in his stories makes the mistake to consider them human, but some – in particular those who are meant to be used in private households – come very close to understanding and expressing human emotions, such as the robot Tony in "Satisfaction Guaranteed" (1951), "one of Asimov's best early stories" (Patrouch 1976, 80), as Joseph F. Patrouch claims. Although Tony keeps "his unchangeable expression" (Asimov 1982, 306) throughout, he is immediately presented as "tall and darkly handsome" (Asimov 1982, 306), and Claire Belmont's view of him moves from "*it*" (Asimov 1982, 306) to him in a few days. Dr. Calvin tries to calm the housewife's initial fear of the robot by assuring her that he is "'not a mechanical monster, nor simply a calculating machine of the type that were developed during World War II fifty years ago. He has an artificial brain nearly as complicated as our own'" (Asimov 1982, 307). Claire and Tony begin to have meaningful conversations, also due to Claire's "loneliness" (Patrouch 1976, 79) because of her husband's absence, with Claire arguing that Tony's "'kind will put ordinary houseworkers out of business'" and Tony responding that "'things like myself can be manufactured. But nothing yet can imitate the creativity and versatility of a human brain, like yours'" (Asimov 1982, 311). The distinction between robots and human beings clearly is the one between products of programming and owners of creativity. But it is slightly unsettling that the robot makes this point. Tony is an example of Artificial Intelligence since he is able to learn, for example, about interior decoration by receiving data from reading books and catalogues. Claire begins to forget that he is a machine, as she realizes that the "thing itself had to remind her" (Asimov 1982, 313). They spend a lot of time together; he helps her redecorate the house, upgrade her own outward appearance, and, thus, enhance her self-confidence and pride. They become so close that she begins to notice the touch of this "warm and soft" hands, "like a human being's" (Asimov 1982, 313). Tony gradually becomes like a lover, although Claire rejects the idea. She desperately washes her hands after she has felt "the pressure of his fingers. She hadn't imagined it; his fingers had pressed hers, gently, tenderly, just before they moved away" (Asimov 1982, 317). When she is about to fall from the ladder, Tony is there to prevent her from being hurt: "And then, all at once, she was conscious of his arms about her shoulders

---

[6] Complications are what Asimov's robo-psychologist Dr. Susan Calvin is interested in most because they give her the opportunity to study robots' brains to figure out where things have gone wrong in the production process. As Asimov shows, robotics has as much to do with human emotions – positive and negative – toward robots as it has with the usefulness of robots.

and under her knees – holding her tightly and warmly" (Asimov 1982, 318). Her reaction is fear, fear of herself and her own emotions: "She pushed, and her scream was loud in her own ears. She spent the rest of the day in her room, and thereafter she slept with a chair upended against the doorknob of her bedroom door" (Asimov 1982, 318). Tony simply protects her against physical harm because of the First Law of Robotics but he cannot prevent her from developing human emotions of love for him. Her "'hysterics'" (Asimov 1982, 321) derive from her recognition that she is physically and mentally attracted to a machine; these erotic sensations make her question her own identity as a human being. Harm comes to her because she cannot handle the presence of a humanoid who is so close to appearing human that he can trigger feelings of love. What should have been even more frightening to Claire is Tony's admission that he does not want to leave: "'Claire [...], there are many things I am not made to understand, and this must be one of them. I am leaving tomorrow, and I don't want to. I find that there is more in me than just a desire to please you. Isn't it strange?'" (Asimov 1982, 319). They are almost about to kiss when the doorbell rings. This is indeed AI and deep learning. Tony, like a teenager, experiences desire, affection, and, perhaps, love for a woman for the first time. He does not yet understand the concept but notices the stirring emotions in himself. Is this not human? Interestingly, Asimov received quite a number of letters from young women asking where Tony could be found (Patrouch 1976, 78). However, in the story, first published in the early 1950s, this potential emotional relationship is so unacceptable at the time that the producers decide to rebuild the TN model because Dr. Calvin realizes that Claire has indeed fallen in love with Tony. Tony is one of the first robot models in the story to be tested on Earth in a private household. In many other Asimov stories, the robots either stay on the production site – United States Robots and Mechanical Men Corporation – or are moved to other planets because of, as one of the scientists, Peter Bogert, Senior Mathematician, calls it "'[t]he damned Frankenstein complex'" (Asimov 1982, 323). Mary Shelley's *Frankenstein* (1818) is the basis for a fear of scientific experiments that artificially build human beings that turn into ravaging and uncontrollable monsters with human emotions. Asimov clearly works out that the more robots become human-like, the more developed their brain is, the more they can also manipulate humans – without intending to do harm – into certain behavior, as he shows in "Liar" (1941), where the robot Herbie can read the human mind and tells the scientists what they prefer to hear in order to make them happy. However, these "lies" turn out to be harmful. Unexpected results of programming clash with uncontrollable human emotions. Consequently, the problem is not so much that robots become humanoids but that human beings act irrationally and can no longer keep up with the robots' programmed and yet emotionally enhanced rationality. From the 1940s onward, Asimov was thinking through the logics of robotics. By the 1980s, he realized that some of what he had written had become reality. However, with time, the Three (Four) Laws of Robotics had gradually lost their prominence.

4 Hunting androids

As early as 1968, Philip K. Dick, "the greatest of all SF authors," as some have claimed (Freedman 1984, 20), in his *Do Androids Dream of Electric Sheep?* depicts a post-apocalyptic United States, mostly "the decaying megalopolis Los Angeles [sic], AD 2020" (Wheale 1991, 298), as Nigel Wheale incorrectly claims. It really is San Francisco. The Earth has largely been destroyed by a nuclear global war and six escaped Nexus-6 model androids – that can hardly be distinguished from human beings and are harmful to them – have to be "retired," thus, destroyed, by bounty hunters. These model robots are considered to be a threat to humanity because they are not human but too human-like.[7] The ultimate fear is that robots become too human, "'pass'" for human (Dick 2007, 142), and get out of control and kill, as humans have often done themselves in the course of history. Even more so, what if the bounty hunter Rick Deckard himself is a humanoid, as suggested a number of times in the novel?[8]

Rick's questioning of his own humanness is justified from the beginning when he and his wife Iran are portrayed as being influenced by a "mood organ" (Dick 2007, 1) and empathy boxes (Dick 2007, 57) and leading a life of illusions with the pretense of a real sheep on the rooftop of the house that is actually an electric one. As Nigel Wheale maintains, "[c]urating animals is also partly a replacement for child-rearing, because the fear of genetic damage has discouraged human reproduction" (Wheale 1991, 298). The Deckards are torn between their own desires and the feelings they can create by dialing a machine. As humans, they are not without feelings but they are also able to technically en-

---

[7] Ironically, because almost all other species have been killed, human survivors create animal-like machines for pets and take them to repair shops if necessary.
[8] The novel's success also shows in its film adaptations *Blade Runner* (1982) and its sequel *Blade Runner 2049* (2017), with the latter offering Nexus-9 replicants who work as slaves and also as "blade runners" who kill rogue replicants. In the course of one "retirement," the blade runner discovers that female replicants can actually reproduce because he finds the remains of one who died during a caesarian section. The child survives, while the mother, Rachael, dies. She is known from *Do Androids* as the one sleeping with bounty hunter Rick Deckard, who, at the end, indeed, meets his daughter, the surviving child in the film.

hance or even produce feelings as those are needed and wanted. While the dust-filled atmosphere gradually destroys human life on Earth, the option is to emigrate to New America, "the chief U.S. settlement on Mars" (Dick 2007, 14). But before he is allowed to do that, Rick has to retire a few more androids, which he can do after they have failed the Voigt-Kampff Empathy Test (Dick 2007, 25) or the "bone marrow" test (Dick 2007, 44). Ultimately, as Eldon Rosen, one of the producers of androids, claims there is, for him, hardly any difference between "authentic humans with underdeveloped empathic ability" (Dick 2007, 46) and androids, which ultimately means between all humans and humanoids because memory has been tampered with and humans are dependent on machines that regulate emotions.

The eponymous title question touches upon another human quality, that of dreaming. "Do androids dream?" (Dick 2007, 160), Rick Deckard wonders. His answer is significant: "Evidently; that's why they occasionally kill their employers and flee here [from Mars to Earth]. A better life, without servitude" (Dick 2007, 160). Moreover, androids, too, like humans, practice solidarity and hence have a group experience, which is why one of them experiments with "mind-fusing drugs" (Dick 2007, 160). The more information is known about androids the more they resemble humans. They dislike servitude and slavery, try to escape, and, thus, show a sense of freedom. They do have and show emotions, as Rachael, an android of the Rosen company, does, in particular when they are drunk, as Rachael is when in the hotel room with Rick (Dick 2007, 160–76). Sleeping with each other, Rick and Rachael are convinced that androids cannot have children; the makers of *Blade Runner 2049* later decide otherwise. The more Rachael insists and reflects on the technology she consists of, the more human she becomes for the reader. She is "an organic entity" (Dick 2007, 171), who, however, will "wear out and die" in two years since cell replacement in an android is not yet possible (Dick 2007, 171), as Rick knows. When Mercerism[9] is exposed as fake by the only TV show still in existence (*Buster and His Friendly Friends*), the last characteristics that distinguish humans from androids – group solidarity and empathy – dwindle, except in Isidore's – the chickenhead's[10] – mind, who seems to be, as some critics claim, "Rick's second self, his alienated self – the complementary opposite of the Rick who kills for money" (Warrick 1987,

---

9 Wilbur Mercer is an old man who, like Sisyphus, does not seem to give up hope for a better life. People listen to him and even merge with him as if he were a religious leader who suffers like a martyr, endlessly climbing up a hill while being attacked with stones, and whose main philosophy is empathy. However, he is revealed to be a fraud.
10 "Isidore is a victim of atomic fallout" (Warrick 1987, 121) and, therefore, declared to be special and is not allowed to leave Earth.

121). As Patricia Warrick argues, the novel portrays "the inner journey of a divided, restless mind seeking wholeness" (Warrick 1987, 121). But this wholeness is not to be found. At the end, Rick returns home and, with his wife, seems to finally adjust to the only life they have amidst electric animals and androids, as "one component of the living scene" (Galvan 1997, 414), and he realizes that the human is no longer central in this new way of life (Heise 2009, 504), no matter how much he wishes for it.

5 Confronting techno-surveillance

The potential inability to define who or what is human constitutes one of the fears humans frequently voice when discussing future technologies and the implications inherent in Artificial Intelligence. Two examples from recent fiction seem to cater to these fears, not because they affirm this indistinguishableness but because they elaborate on the steps that precede ASI. Both Ernest Cline's *Ready Player One* (2011) and Dave Eggers's *The Circle* (2013) are dystopian novels set in the near future and strongly rely on technology and the changes it produces in human beings.

We might consider *Ready Player One* in some way to have a happy ending because its protagonist Wade, an 18-year-old high school student, intends to make the world a better place, a world that has already gone through an apocalypse but that might see improvement because Wade with his avatar Parzival has triumphed in the computer game of the O.A.S.I.S., the "'Ontologically Anthropocentric Sensory Immersive Simulation'" (Condis 2016, 2), created by James Halliday, co-founder of the software company Gregarious Simulations Systems. Wade has not only solved all the riddles, found the three (copper, bronze, and jade) keys and ultimately, the Easter Egg, and has, thus, become the heir to the billionaire Halliday, but he has also beaten the greatest enemy, the "IOI," the "'Innovative Online Industries'" (Condis 2016, 2), as a gunter (gamer and hunter), and prevented the world from the take-over by a totalitarian system. The search for the "Easter Egg" has often been compared to the search for the Holy Grail. Wade has recognized the need for joint efforts to overcome evil and has brought together some of his closest allies to successfully fight the IOI. Finally, Wade has enough money to improve people's living conditions and, in the game's wake, has also been awarded with the love of his life, Art3mis, a fellow gunter and, as it turns out, a young woman. The game ends in romance; the world has its human hero, and Wade has learned that true fulfillment and happiness does not lie in video-gaming and cannot be attained by an avatar. Wade, for a long time, escapes from reality, from violence, poverty, and anger in the real world, into a game world that offers him safety behind his avatar's visor and allows him to do what he does best, namely to solve riddles related to Halliday's popular culture of the 1980s. Only someone who is intricately acquainted with Halliday, who has extensively studied 1980s pop culture, and who has played all the respective video-games, would be able to triumph at the end. What is valued are knowledge, expertise, ambition, stamina as well as openness for emotions – such as love and friendship – and the desire for community. Wade, the other gunters, and Halliday try to escape from the real world into the game world and hide behind their avatars and fight historic (game) battles. However, the novel never leaves its readers in any doubt that the real world and the game world – Wade as player one and Parzival, the avatar – can still clearly be distinguished from each other. Wade learns "to value the real world over the fictional, starving humanity over digital avatars. [...] he understands the dangerous illusive calibre of the OASIS, flips up his 3-D visor, relinquishes this form of cultural capital, and, unmediated, engages with the human condition" (Aronstein / Thompson 2015, 59). This is how he has "redeemed the land" (Metz 2018 n.pag.).

This redemption does not happen in Eggers's *The Circle*, set in the near future. The young college graduate Mae Holland gets a job at the Circle, with headquarters in California that have the look and feel of a university campus. It is a technology company that has united "Internet search and social media capabilities under one 'Unified Operating System'" (Galow 2014, 116). Its owners – Tom Stenton, Eamon Bailey, Ty Gospodinov – offer their employees impressive amenities such as dorms on the company's terrain, gyms, parties, and a rise up the professional ladder. As readers, we follow young Mae's career from her perspective and notice how she works her way up in the customer service department from one to nine computers, constantly checking her likes, her ratings, and her rankings; we empathize with her fear of not fulfilling the company's social and professional expectations, her desire to include her parents in her health insurance package, which, however, submits them to total surveillance by the company and which they, ultimately, reject. When Mae meets the mysterious Kalden, whom she believes to be a colleague but who turns out to be one of the "Three Wise Men" (Eggers 2014, 474) – Ty, he reveals to her that he is critical of the company's development toward total surveillance, people's mandatory transparency – "secrets are lies"; "sharing is caring"; "privacy is theft" (Eggers 2014, 299–306; specifically 305) – and the company's control of politicians and voters. Mae even agrees to complete transparency, wears a camera all the time, and monitors reviewers' comments on a small screen around her wrist. Moreover, the health care program requires the ingestion of "organic sensors that monitor every aspect of their bodily functions" (Galow 2014, 123). "TruYou" becomes her online identity.

Ultimately, Ty warns her of the company's "Completion" (Eggers 2014, 473), compares the Circle to the shark in the tank that devours every living being, and points out to her that a "'totalitarian nightmare'" (Eggers 2014, 486) is about to happen, but she betrays him. The novel reveals the gradual loss of human qualities and values, such as privacy, friendship, rational thinking, free will, love, family, etc. Mae's former boyfriend, ironically named Mercer like the Mercer in *Androids* who gives people hope, is almost the only one who resists the Company's incorporation. He, too, is a character of hope, in spite of the fact that he dies when being pursued by the Circle. As Ty tells Mae: "'There will be more Mercers'" (Eggers 2014, 486). Until this very moment, the reader believes in a turn toward a happy ending. However, the ending comes as a surprise and leaves a feeling of disappointment with a young woman who could have stopped the Circle from ruling the world but does not.

Human beings are technologically enhanced; implanted or swallowed computer chips expose every single bodily feature; health data are transferred to the company's owners; political votes will be cast online through Circle accounts. All human beings become transparent and are not just controlled by the men at the head of the company but by all those who buy into the company's philosophy and become each other's spies. The novel's step-by-step analysis of this dehumanizing development makes frighteningly clear that humans – as long as they are still human – collaborate in their own destruction of what it means to be human, as we will also see in filmic representations of ASI. This is not yet Artificial Intelligence but the Circle has lain the foundation for human robots that do what they are programmed to do – like Mae does – and, perhaps, do more than that – like Mae does. More than these two novels, recent films have begun to question what is human, have begun to blur the boundaries between the human and the cyborg, have created something / someone that may be considered posthuman in only two of its meanings of "post": the posthuman as coming after the human or/and as being an extension of the human, but it is not posthuman in the sense that it critically engages with humanness as Braidotti requires. The creation of the cyborg or humanoid is always also gendered. As early as 1978, Mary Daly in *Gyn/Ecology* "discussed technology as a method of patriarchal oppression; a product of this is what she terms a 'fembot' […]: a symbolic female robot that is the cornerstone of male domination of women through technology, as well as a kind of role model perpetuated by patriarchal society for women to aspire" (Cox 2018, 5). This seems to be exactly what Mae experiences in *The Circle* and to which she voluntarily contributes. As Tony E. Jackson wonders, "[w]hat becomes of our own unique being if we can create an imitation human so like us that our senses can't register it as an imitation? This is a profound threat, not just on the level of psychology, but on the level of ontology: a threat to the status of our being as entities in the world. And yet we seem *compelled* to imagine – we are fascinated by – just this possibility" (Jackson 2017, 54; emphasis in original).

6 Creating the perfect simularcrum

While Mae is technologically enhanced, she is not a female robot yet. However, such a "fembot" that seems to represent human perfection often becomes a model woman that men and women attempt to create, as is the case in *Ex Machina*. Tony E. Jackson offers a neurobiological as well as sociological explanatory theory. He takes recourse to the results of research on mirror neurons. This system has "profound implications" for "human nature" (Jackson 2017, 49). As Jackson maintains, mirror neurons activate human imitation on a mostly unconscious level. For him, "Theory of Mind is the psychological foundation of the human animal's uniquely social identity, and mirror neuron systems offer a possible biological explanation of how and why Theory of Mind operates as it does" (Jackson 2017, 50). Therefore, human identity is based on what, how, and whom we imitate and on which imitation process we act. The extreme to each side – over- or under-imitation – threatens the perception of our "self" and "results in a number of pathologies" (Jackson 2017, 51), known as "'echophenomena'" (qtd. in Jackson 2017, 51). Although humans usually fear those pathologies, they are simultaneously fascinated by them and tend to explore how far they can go. Consequently, a humanoid robot that becomes indistinguishable from a human being and might fool us into believing in its humanness "would be a fundamentally dangerous state of mind: misbelief, or false knowledge" (Jackson 2017, 52) or "misperception" (Jackson 2017, 53).

It is exactly this compulsion to imagine that we might be able to create the perfect imitation of a human being that triggers the action in *AI* (Spielberg, 2001). In *AI*, science creates an imitation human boy, a simulacrum for parents – Monica and Henry Swinton – who cannot have more children because "an eco-apocalypse has required the restriction and licensing of pregnancies, which means childlessness is common" (Jackson 2017, 55), and whose only son Martin is in the hospital, cryogenically deep-frozen due to his illness to be woken up when a treatment is available. A professor explains that science has been able to create the most perfect imitation human to date, namely one who can show and actually experience emotions, above all "love" for its "Mommy." The android, or "Mecha," David "loves" Monica the moment the "Imprinting Protocol code" is released, but the "mother" does not know how to respond to "it." For the parents, as Jackson argues, and thus for the viewers as well, there is a "visio-cognitive disjunction [...] between what their eyes see – a real boy – and what they know: that he is a mecha" (Jackson 2017, 57). "David" begins to imitate his "parents" because that is how "it" is programmed. While Monica and Henry treat "it" nicely, the Flesh Fair sequence in the film shows the hostility humans – Orgas – develop against Mechas, which leads to the arbitrary destruction of the latter group. Monica herself – although not destroying David – cannot, however, show any love for a Mecha because it is not of flesh and blood. Does this mean that "human singularity is preserved after all"? (Jackson 2017, 59). The final apocalypse that destroys all humans and only lets their machines live happens *deus ex machina*-like and turns the machines into Supermechas that create the imitation Mommy for David from actual DNA – thus, through cloning – and for one day only at the end of which both die. The end confirms what humans have feared all along; no humans are left anymore; the imitation machines take the place of their original; they do not cause human destruction but have simply survived it. *AI* presents posthumans as Supermechas that are perfect robotic simulacra of human beings, just not human. The "compulsion," as the film seems to show, "to create the perfect imitation must [and will] be followed" (Jackson 2017, 61). The world will eventually be populated by technologically created entities – robots, Mechas, Supermechas – , and this seems to be acceptable in the film's logic because no humans will be around anymore to disagree. But David could be proof that an evolution from Mecha to human being is possible as well due to the "social recognition" received from the cloned mother (Manninen / Manninen 2016, 341). David, at the end, is able to experience true love, fear, and the desire to care for his "Mommy."

Alex Proyas's *I, Robot* (2004) is based on the premise of the Three Laws of Robotics, which basically stipulate that robots cannot do harm to humans. In the year 2035, robots populate everyday human life as helpful support in unlimited ways. Very few people, among them Detective Del Spooner, distrust these mechanic creatures. When one day the scientist Dr. Alfred Lanning seems to have committed suicide, Spooner is called for an investigation, and with the help of Lanning's holograph and the robot Sonny, who is able to choose not to obey the Three Laws and feels guilt and affection, and, thus, is unique – as Sonny should not be – finds the solution. Spooner, together with the robo-psychologist Dr. Susan Calvin, realizes that the robotic system V.I.K.I. (Virtual Interactive Kinetic Intelligence) has taken control of all robots, who turn the Three Laws against their creators, even though the robots believe they are protecting them. Dr. Lanning knew this and designed the robot Sonny to provoke Spooner's and Calvin's attention. With his help, Spooner and Calvin are able to destroy V.I.K.I. and leave the army of robots on the streets suddenly without targets. These robot warriors on the battlefield need new guidance, which Sonny seems to give them at the end. This ending, perhaps meant to be reassuring to the viewers, is troubling to say the least. Sonny is capable of human emotions: he feels guilty of Dr. Lanning's death; he has experienced friendship, and he has helped kill fellow robots. Thus, he has proven to be capable of learning; he is an "ethical being" (Olivier 2008, 40); he shows a sense of "freedom" and "autonomy of will" (Oliver 2008, 40). The gathering of robots under Sonny – who has proven to be human-like – "represents the symbolic inauguration of a 'robot society'" (Olivier 2008, 41), parallel to human society. Sonny is "a convincing human simulacrum" (Olivier 2008, 42) and, thus, by definition, a threat to human society, but only because the human mind does not yet seem to be able to think beyond the human. Sonny appears to be human even if artificially and mechanically produced. He has been able to deep-learn, and thinks and acts beyond the programs that created him in the first place.

Ex Machina (Alex Garland, 2015) goes even further in the creation of imitation humans and depicts the rich search machine creator Nathan's experiments with robots that make them so human-like that they can no longer be distinguished from humans. At the same time, the androids are not just humanoids but also woman-like, a fact which Emily Cox criticizes as "a nightmarish extension or logical conclusion of masculine fantasies of female objectification and patriarchal domination" (Cox 2008, 5). Nathan shapes Ava according to Caleb's pornographic search record in order to spark attraction between the two. Nathan, the scientist, invites Caleb to perform a version of the Turing Test[11] on his creation Ava to determine whether she can pass for human. Moreover, Caleb is expected to test Ava's sentience and her ability to draw on the Blue Book, which "is essentially a database of infinite human thought and emotional processes" (Wilson 2018, 119–20). To appear as human-like as possible, Ava has to possess both logical and emotional intelligence (Wilson 2018, 121). While at first Ava's body is female yet transparent to reveal the technology that constitutes her body, in the course of the movie, she becomes more and more human-like by her own decision. As Emily Cox points out, "[a]n A.I. can simply mean a highly intelligent machine, but Nathan seems to want to create something with desires, motives, and even sexuality" (Cox 2008, 11). To make Ava a "true AI," Nathan gives her self-awareness, sexuality, imagination, empathy, and the ability to manipulate. Ava indeed uses these qualities and makes Caleb fall in love with her to gradually close the gap between woman and robot, in particular, when she adds

---

[11] This is a test produced by the British mathematician, computer scientist, logician, philosopher, and theoretical biologist Alan Turing (1912–1954) in 1950 to find out whether a machine (computer) can think like a human being.

limbs, skin, and hair to her body. Caleb in the end frees her from imprisonment, as he believes, but Ava stabs Nathan and leaves Caleb locked in the house from which there is no escape. *Ex Machina* seems to affirm the human fear of overpowering robots, on the one hand, and the indistinguishableness between human and machine, on the other hand. Ava kills and escapes because she has been programmed to highly value freedom. Thus, her creator Nathan is responsible for what she does, also for the fact that she turns against him, as we have also seen as early as 1818 with Frankenstein's monster in Mary Shelley's gothic novel. The very fact that Ava desires freedom makes her human, and the knowledge of how to dress to appear human renders her dangerous for humanity. Dijana Jelača variously labels Ava "a man-made, artificially intelligent (AI) cyborg" or a "posthuman female android" (Jelača 2018, 384). In any of these labels, robotics remains a constant, but the development from cyborg – where bodies mix but the brain is human – to android – where the body and the brain "look" human – reveals the gradual closing of a gap. The revelation that not only Ava but also Kyoko is a humanoid possessing logical and emotional intelligence (cf. Wilson 2018, 117) makes Caleb doubt his own humanness. He draws his own blood and breaks a mirror because of his uncertainty about who and what he is. According to Dijana Jelača, she sees "posthuman life where the continuum of nature and culture, or nature and technology, is so omnipresent that it becomes unnoticeable" (Jelača 2018, 395).

When Ava after her escape finally looks into a mirror at a street intersection, she realizes that she blends in and, thus, will survive. Ava perfectly passes for human, the mirror being the ultimate Turing Test that no longer proves humanness but the perfect imitation or simulacrum of humanness. For Jelača, "[a]nxiety is left to linger with the spectator at the end" (Jelača 2018, 396) because, like Caleb, we all might wonder whether we are not a simulacrum as well, an imitation of something/someone that has never existed. The boundaries "between bodily existence and computer simulation" (Hayles 1999, 3) collapse, confirm Hayles's understanding of the posthuman, and might suggest the posthuman as a rethinking of the human in the Anthropocene.

AI, I, Robot, and *Ex Machina* affirm the fear of gradual human extinction, not necessarily effected by humanoids but triggered by humans who desire to create – God-like – a being that is as human-like as possible. While in *AI* climate change and global warming lie at the basis for the human desire to produce technological humans, *Ex Machina* testifies to the longing of a billionaire to achieve what no-one – not even Dr. Victor Frankenstein – has been able to accomplish, namely to produce the perfect human through technology. But the God Nathan has not taken into consideration his creation's over-accomplishment. As in Goethe's *Zauberlehrling*, Nathan cannot get rid any more of the ghosts he conjured up. In *Ex Machina*, however, there is no master to turn to for help. Ava, the android, turns out to be the *deus ex machina* or Eve, the first of a new generation of android females who has lured Caleb and will lure other men into falling for her. In *AI* and *Ex Machina*, the scientists' uncontrollable impetus to play God is responsible for the eventual transformation of the world whereas *I, Robot* attributes to the robotic system the capacity of interpreting language, that is, the Three Laws of Robotics, in their own and unforeseen way, stage a robot revolution that might or might not turn against humanity. The question no longer is whether Artificial Super Intelligence can be created but what it will do with its creator(s).

It is in this respect that we can understand that "Stephen Hawking, Elon Musk and dozens of other top scientists and technology leaders have signed a letter warning of the potential dangers of developing artificial intelligence (AI)" (Lewis 2015, n.pag.). Not long before his death, Stephen Hawking put together a collection of essays entitled *Brief Answers to the Big Questions* (2018). One of the big questions he tackles is "Will Artificial Intelligence Outsmart Us?" For him, the major concern is "that AI would take off on its own and redesign itself at an ever-increasing rate. Humans, who are limited by slow biological evolution, couldn't compete and would be superseded. And in the future AI could develop a will of its own, a will that is in conflict with ours" (Hawking 2018, 186), as the films *I, Robot* and *Ex Machina* intriguingly show. He continues to argue that "[w]hereas the short-term impact of AI depends on who controls it," as *Ready Player One* emphasizes, "the long-term impact depends on whether it can be controlled at all" (Hawking 2018, 188), as *Ex Machina* illustrates. "In short," as Hawking maintains, "the advent of super-intelligent AI would be either the best or the worst thing ever to happen to humanity" (Hawking 2018, 188).[12]

7 Developing AI can be dangerous

As my examples have shown, robotics and technology in general reveal two sides of the same coin: they assist humanity in daily chores, are part of human progress, and potentially extend human life the moment we talk about human enhancement. Robots can do repetitive work, such as housework, with human be-

---

[12] In contrast to Hawking's cautionary words, Lucas Rizzotto is enthusiastic about the potential AI has for education. His essay is full of superlatives that describe the way toward the optimization of AI. As he concludes, "A.I. has been learning about us for the longest time – now it's only a matter of time before it starts teaching us" (Rizzotto 2017, n. pag.). I suggest that this would precisely be the moment when AI gets out of control.

ings in control. But once intelligence or even super-intelligence is added and robots not only look more human-like but also show human emotions – such as love, fear, empathy – and radiate certain erotics, robots turn into humanoids that offer their human creators a mirror-like image of themselves. It is then that fear about what it means to be human creeps in. Asimov's female character Claire is afraid of losing who she is because she falls in love with her robotic houseworker; androids in Dick's novel pass for human; Wade in *Ready Player One* escapes into virtual reality but safely returns to and saves the real world from a totalitarian regime whereas Mae in *The Circle* succumbs to the lure of total surveillance, control, and transparency and believes in the power of technology that destroys everything that used to be considered human. *AI*, *I, Robot*, and *Ex Machina* focus on the human impasse that once science has created the perfect simulacra of the human, it is hard to get rid of the human element in the humanoid. Humanoids take on a life of their own, at best in societies parallel to human society; in a worst-case scenario, they destroy the human. Artificial Intelligence is crucial, as represented in fiction and film, in reflecting on what is human. These verbalizations and visualizations shed light on what I would call the posthuman age. Humans drive technology as far as possible; some play God, such as Dr. Frankenstein and Nathan; some underestimate the potential danger in Artificial Intelligence, as in *I, Robot*; others fall in love with a humanoid and live in their own virtual world. Posthumanism asks us to question the human drive to create our own mirror images that are, however, more powerful, never tiring, and, ultimately, immortal. Human life needs technological progress, but to what extent is an open question. Those who fear unlimited ASI are afraid of humanoids who simultaneously continue the power of the machine and the irrationality of human emotions. Sonny becomes Spooner's friend; V.I.K.I. and Ava turn against their creators; V.I.K.I. is destroyed, but Ava moves on, among us, and will spread, to use a current idea, like a virus.

To prevent this pandemic, we might turn to what Johanna Heil has called "critical posthumanism" (Heil 2019, 345) that "deconstruct[s] modern notions of human exceptionalism" (Heil 2019, 345) and allows for a transgression of conceptual borders to rethink the human. Although critical posthumanism asks the human to, above all, recognize its parallel place among all organic material, such as plants and animals, and to "reconceptualize[...] the Anthropocene as the 'Planthropocene'" (Heil 2019, 346), it might also be possible to reconfigure the posthuman humanoid as an acceptable phenomenon on Earth. Yet, it seems that right now the hierarchy between the human – as superior producer – and the humanoid – as inferior product – is determined by humans' desire for competence, self-enhancement, and the simultaneous fear of losing power to monsters out of control, who could easily also be human. The recognition that humans might be their own enemies is a simplified version of what Timothy Morton calls "dark ecology." Stuart Russell maintains that after watching the movie *Transcendence* (2014), he "would be publicly committed to the view that [his] own field of research [AI] posed a potential risk to [his] own species" (Russell 2019, 4). Yet, for him, there is still much work to be done "before we have anything resembling machines with superhuman intelligence" (Russell 2019, 6). But he is sure that "'[s]uccess [in ASI] would be the biggest event in human history ... and perhaps the last event in human history'" (Russell 2019, 3; ellipses in original). Stephen Hawking's warning goes into the same direction: "we should instead plan ahead and aim to get things right the first time, because it may be the only chance we will get. Our future is a race between the growing power of our technology and the wisdom with which we use it. Let's make sure that wisdom wins" (Hawking 2018, 196). As Wai Chee Dimock has recently pointed out, "AI is poised to transform the fabric of life and the future of work" (Dimock 2020, 449). For Dimock, Microsoft's "'planetary computer'" (Dimock 2020, 451), which collects data about "the health of ecosystems" and serves as "'decision engine'" (Dimock 2020, 451), is the "blueprint for the future" and "requires a division of labor between human beings and machines" (Dimock 2020, 452). Those human beings are not just natural and life scientists, but also humanists – scholars and artists alike – who possess "literacy about the human species" (Dimock 2020, 453). This is what humanity needs to be able to survive. To conclude, the fictional examples and the scholars' positions toward AI discussed in this paper clearly call for a collaboration of artists, (post)humanists, and (natural, life) scientists to join forces in the exploration of AI – its relevance, potentials, and dangers. All disciplines are needed to guarantee the better life and our planetary survival that AI purports to bring.[13]

---

13 The impact of AI – in its various manifestations – has been discussed for quite some time in both scholars' and artists' works. Perspectives, opinions, and approaches are myriad and strongly vary in their results and further effects. A larger project, going beyond the scope of this paper, would be to trace AI fiction throughout history and consider its connections to science fiction, speculative fiction, and utopian and dystopian fiction. All fiction, in these contexts, deals with human-machine interactions, overlap, and potential replacement. Some authors write from the perspective of the machine, for example, Scottish Ian Russell McEwen in *Machines Like Me* (2019), with the life-like android Adam becoming part of a couple's life. Miranda and Charlie have just fallen in love with each other and have to come to terms with the presence of this android in their life. More than one hundred years earlier, the British author Samuel Butler (1835 – 1902) published his satirical novel *Erewhon; or, Over the Range* (1872) – playing with the term utopia in his title which reads "nowhere" backwards, with one slight mistake in the sequence of "h" and "w" – and wondered about the potential dangers of machines' deep learning, consciousness, and self-replication, in the wake of Charles Darwin's *On the Origin of Species*

Works cited

AI: Artificial Intelligence. Dir. Steven Spielberg. Dreamworks Pictures, 2001. DVD.

Aoun, Joseph E. *Higher Education in the Age of Artificial Intelligence.* 2017. Cambridge, MA: MIT, 2018.

Aronstein, Susan, and Jason Thompson. "Coding the Grail: *Ready Player One*'s Arthurian Mash-Up." *Arthuriana* 25.4 (2015): 51–65. EBSCOhost. https://muse.jhu.edu/article/606195/pdf. Accessed: 12 Dec. 2021.

Asimov, Isaak. *The Complete Robot.* London: Harper Voyager, 1982. ["Runaround" 221–41; "Liar!" 286–305; "Satisfaction Guaranteed" 306–21; "The Evitable Conflict" 485–511]

Bolter, David Jay. "Posthumanism." *The International Encyclopedia of Communication Theory and Philosophy.* Ed. Klaus Bruhn Jensen et al. Hoboken, NJ: Wiley, 2016. 1–8.

Braidotti, Rosi. *The Posthuman.* Cambridge, UK: Polity, 2013.

Cave, Stephen, and Kanta Dihal. "Hopes and Fears for Intelligent Machines in Fiction and Reality." *Nature Machine Intelligence* 1 (2019): 74–78. Leverhulme Centre for the Future of Intelligence, U of Cambridge, Cambridge, UK. https://doi.org/10.1038/s42256-019-0020-9. Accessed: 08 Nov. 2021.

Cline, Ernest. *Ready Player One.* New York: Broadway Books, 2011.

Condis, Megan Amber. "Playing the Game of Literature: *Ready Player One*, the Ludic Novel, and the Geeky 'Canon' of White Masculinity." *Journal of Modern Literature* 39.2 (2016): 1–19. JSTOR. https://www.jstor.org/stable/10.2979/jmodelite.39.2.01. Accessed: 23 Aug. 2020.

Cox, Emily. "Denuding the Gynoid: The Woman Machine as Bare Life in Alex Garland's *Ex Machina*." *Foundation: The International Review of Science Fiction* 47. 2 (2018): 5–19. EBSCOhost. https://www.proquest.com/docview/2068466713/fulltextPDF/CDF0022521D646F8PQ/1?accountid=14571. Accessed: 12 Dec. 2021.

Dick, Philip K. *Do Androids Dream of Electric Sheep?* 1968. London: Orion, 2007.

Dimock, Wai Chee. "Editor's Column: AI and the Humanities." *PMLA* 135.3 (May 2020): 449–54.

Eggers, Dave. *The Circle.* 2013. New York: Viking, 2014.

Ex Machina. Alex Garland. Warner Brothers and Universal Pictures, 2014/15. DVD.

Freedman, Carl. "Towards a Theory of Paranoia: The Science Fiction of Philip K. Dick." *Science Fiction Studies* 11.1 (Mar. 1984): 15–24. JSTOR. http://www.jstor.com/stable/4239584. Accessed: 27 Aug. 2020.

Galow, Timothy W. *Understanding Dave Eggers.* Columbia: U of South Carolina P, 2014.

Galvan, Jill. "Entering the Posthuman Collective in Philip K. Dick's *Do Androids Dream of Electric Sheep?*" *Science Fiction Studies* 24.3 (Nov. 1997): 413–29. JSTOR. http://www.jstor.com/stable/4240644. Accessed: 26 Aug. 2020.

(1859). The novel contains a three-part chapter "The Book of the Machines," based on some of Butler's own published press articles. The list of AI fiction could be extended endlessly, and already shows that human beings have for long been interested in who they are as human beings and how their creations would or will fit into their view of the universe. What I have been able to discuss here in this paper is only the tip of the iceberg in AI fiction and film.

Haraway, Donna. "A Cyborg Manifesto: Science, Technology, and Socialist Feminism in the Late Twentieth Century." *Simians, Cyborgs and Women: The Reinvention of Nature.* By Haraway. New York: Routledge, 1991. 149–81.
Hawking, Stephen. *Brief Answers to the Big Questions.* London: John Murray, 2018.
Hayles, N. Katherine. *How We Became Posthuman: Virtual Bodies in Cyberetics, Literature, and Informatics.* Chicago: U of Chicago P, 1999.
Heil, Johanna. "With and Beyond Borders: Toward a Posthumanist American Studies." *Journal of Nineteenth-Century Americanists* 19 7.2 (2019): 343–48. https://muse.jhu.edu/article/735424. Accessed: 12 Dec. 2021.
Heise, Ursula. "The Android and the Animal." *PMLA* 124.2 (Mar. 2009): 503–10. *Modern Language Association.* http://www.jstor.com/stable/25614291. Accessed: 27 Aug. 2020.
I, Robot. Dir. Alex Proyas. Twentieth Century Fox, 2004. DVD.
Jackson, Tony E. "Imitative Identity, Imitative Art, and AI: Artificial Intelligence." *Mosaic: An Interdisciplinary Critical Journal* 50.2 (June 2017): 47–63. *Project Muse.* https://muse.jhu.edu/article/663692/pdf. Accessed: 12 Dec. 2021.
Jelača, Dijana. "Alien Feminisms and Cinema's Posthuman Women." *Signs: Journal of Women in Culture and Society* 43.2 (2018): 379–400. *EBSCOhost.* https://web.s.ebscohost.com/ehost/pdfviewer/pdfviewer?vid=2&sid=0a87dcb5-2ccb-43b0-81cb-74e8cad1e8b3%40redis. Accessed: 12 Dec. 2021.
Kaplan, Andreas, and Michael Haenlein. "Siri, Siri, in My Hand: Who's the Fairest in the Land? On the Interpretations, Illustrations, and Implications of Artificial Intelligence." *Business Horizons* 62 (2019): 15–25. *Kelley School of Business, Indiana University* 2018. https://de.slideshare.net/RomanBuldro/siri-siri-in-my-hand-whos-the-fairest-in-the-land-on-the-interpretations-illustrations-and-implications-of-artificial-intelligence. Accessed: 12 Dec. 2021.
Kaplan, Jerry. *Artificial Intelligence: What Everyone Needs to Know.* New York: Oxford UP, 2016.
Lewis, Tanya. "Don't Let Artificial Intelligence Take Over, Top Scientists Warn." *LiveScience* 12 Jan. 2015. https://www.livescience.com/49419-artificial-intelligence-dangers-letter.html. Accessed: 12 Dec. 2021.
Manninen, Tuomas William, and Bertha Alvarez Manninen. "David's Need for Mutual Recognition: A Social Personhood Defense of Steven Spielberg's *A. I. Artificial Intelligence.*" *Film-Philosophy* 20.2–3 (2016): 339–56. *EBSCOhost.* https://www.euppublishing.com/doi/pdf/10.3366/film.2016.0019. 12 Dec. 2021.
Metz, Walter. "'So Shines a Good Deed in a Weary World': Intertextuality in Steven Spielberg's *Ready Player One* (2018)." *Film Criticism* 42.4 (Nov. 2018): 1–3. *EBSCOhost.* https://web.p.ebscohost.com/ehost/pdfviewer/pdfviewer?vid=2&sid=1136ccc2-3c2a-46e0-8c2c-dbfca3f2d3ca%40redis. Accessed: 12 Dec. 2021.
Morton, Timothy. *Dark Ecology: For a Logic of Future Coexistence.* 2016. New York: Columbia UP, 2018.
Nida-Rümelin, Julian, and Nathalie Weidenfeld. *Digitaler Humanismus: Eine Ethik für das Zeitalter der Künstlichen Intelligenz.* München: Piper, 2018.
Nussbaum, Martha C. *The Monarchy of Fear: A Philosopher's Look at Our Political Crisis.* Oxford: Oxford UP, 2018.
Olivier, Bert. "When Robots Would Really Be Human Simulacra: Love and the Ethical in Spielberg's *AI* and Proyas's *I, Robot.*" *Film-Philosophy* 12.2 (2008): 30–44. *EBSCOhost.*

https://www.euppublishing.com/doi/pdf/10.3366/film.2008.0014. Accessed: 12 Dec. 2021.

Patrouch, Joseph F. *The Science Fiction of Isaac Asimov*. 1974. Frogmore, St. Albans, Herts: Panther, 1976.

Rizzotto, Lucas. "The Future of Education: How A.I. and Immersive Tech Will Reshape Learning Forever." 23 June 2017.
https://medium.com/futurepi/a-vision-for-education-and-its-immersive-a-i-driven-future--b5a9d34ce26d. Accessed: 12 Dec. 2021.

Russell, Stuart. *Human Compatible*. New York: Viking, 2019.

Warrick, Patricia S. *Mind in Motion: The Fiction of Philip K. Dick*. Carbondale: Southern Illinois UP, 1987.

Wheale, Nigel. "Recognising a 'Human-Thing': Cyborgs, Robots and Replicants in Philip K. Dick's *Do Androids Dream of Electric Sheep?* and Ridley Scott's *Blade Runner*." *Critical Survey* 3.3 (1991): 297–304. *JSTOR*. http://www.jstor.com/stable/41556521. Accessed: 26 Aug. 2020.

Wilson, Hayley. "The 'I' in AI: Emotional Intelligence and Identity in *Ex Machina*." *Film Matters* 9.1 (2018): 117–24. *EBSCOhost*.
https://docserver.ingentaconnect.com/deliver/connect/intellect/20421869/v9n1/s19.pdf?expires=1639340537&id=0000&titleid=75007569&checksum=6B2-B3A997912A82646F9B63F562C1BF8. Accessed: 12 Dec. 2021.

Wolfe, Cary. *What Is Posthumanism?* Minneapolis: U of Minnesota P, 2010.

Piet Defraeye
AI on Stage: A Cross-Cultural Check-Up and the Case of Canada and John Mighton

Abstract: Artificial Intelligence and the live actor on stage seem like two antagonistic concepts: binary animatronics versus an explicitly alive three-dimensional world. While theatre production itself often relies on AI-driven technology, the stage largely remains a Sacred Space. Yet, much of the vocabulary and paradigms that define AI have been informed by stage practice. From the early days of popular science theatrics over Karel Čapek's *R.U.R.* to Alan Turing's Imitation Game, there are traces of permutative referencing between AI and live theatre. More recently, AI has become a major motif in live performance, often combining digital technology with analogue dramaturgy. The University of Toronto's BMO Lab is a flag bearer in creative and critical research on how and what AI means for the stage. Canadian theatre producers like Robert Lepage's Ex Machina and Canadian playwright John Mighton are prominent examples of engagement with AI on the Canadian stage. Mighton 's play *Possible Worlds* (1990) and Robert Lepages's eponymous film adaptation (2000) is a murder mystery that focuses on a series of crimes in which the brains of several victims are stolen. We face a cyborgian world in which individuality is constructed through an affective triangular dynamic between sexual desire, technological developments and economical-financial interests. Both film and play explore how narratives are metonymic figurations for the endless possibilities that both the brain and AI can produce.

Computer automation and Artificial Intelligence (AI) has been a major driver and motif in film production for a long time; to trace the scores of films that have since been fostered on this theme, we can go back all the way to the 1920s with movies like Fritz Lang's *Metropolis* (1927). Recent developments in *deep learning* and *deepfake* technology have only exacerbated these filmic applications of AI and the latest advances in photo technology have stood out through their AI applications. Most of these have a performative impact on how a certain representation is interpreted or experienced. Some, like archival restorations, are hugely promising, others – like many AI applications in consumer products – are gimmicky (and also scary): think API generated model photography which renders the sort of performance that has little to do with an underlying reality. Digital applications and engagements in sound and audio installations have also gained considerable ground in the last few decades. Computer-generated music production and composition has been common fare for a while now.[1] This digital technology has also profoundly impacted theatre and dance production. Intermedial and transmedial theatre is just one prominent example, with directors like Ivo Van Hove, Robert Lepage, and Guy Cassiers no longer considered as pioneers in their technologically driven dramaturgies.

This essay offers a short historical overview of how AI has impacted live theatre (and vice versa, at least in the language around AI), and gives a number of emblematic occurrences of AI on stage as a theme or in its dramaturgical approaches. In a second part, the essay narrows its focus on the Canadian situation, and finally converges on two plays by Canadian playwright John Mighton. Particularly *Possible Worlds* is presented as a metatheatrical and metonymic commentary on AI.

AI in theatre – the Turing effect

Our current concern here is mostly in how far AI has been picked up by theatre as a theme or motif. While there are plays and dance performances that have artificial intelligence as their theme, they remain a rarity. Plays that combine a thematic turn with aspects of AI integrated into their dramaturgy are even rarer. The reason for this is not hard to find: the liveness of theatre seems to stand in the way of the perceived artificiality of Artificial Intelligence, a product of mathematical calculations on the page and in the computer. Media philosopher Sybille Krämer comments on our epistemological disposition to engage and understand the world around us through a systematic *flattening* of that world into two-dimensional representation (in writing, pictures, diagrams, etc.). It is a process of presentification, generated through what she calls artificial flatness.[2] AI draws on this flattening process, which may then be reconstituted into three-dimensional applications. Theatrical representation is obviously at the extreme other end of this spectrum, as it explicitly wants to exist in a three-dimensional realm in its mimetic and live interaction with that world. Even contemporary

---

[1] It is worthwhile remembering Douglas R. Hofstadter's observation in his controversial but pioneering book, *Gödel, Escher, Bach: an Eternal Golden Braid*, that the authorial attribution of a digitally-generated "piece of music to a computer would be like attributing the authorship of [his] book to the computerized automatically (often incorrectly) hyphenating phototypesetting machine with which it was set" (636). The book, incidentally, was originally published in 1979, well before the mass-availability of desk-top computers.
[2] Cf. Krämer 2014, 11–30.

practices of theatrical intermediality on stage continue to trump the three-dimensional, *meat-suit* realness of live performance as their main draw. Live performance is, as it were, both ontologically and phenomenologically allergic to the notion of AI on the stage. Indeed, we are more likely to witness the lexicological meanings of *ai* foregrounded on the stage as defined in the *Oxford English Dictionary*. It is primarily a revelatory "exclamation of surprise, regret, pain" – as in somebody incidentally hammering her thumb – or, more intriguing and therefore more histrionic, the *ai* or "three-toed sloth" of the Bradypus genus – than its more popular *flattened* meaning of programmed intelligence following induced algorithmic rules.

Revelation, guilt, and agony are certainly the stuff that theatre is made of, but when it comes to actual representations of AI on stage, we move closer in the direction of the zoological ai, with its three toes. While anthropomorphic robots are ubiquitous in film – especially sci-fi – they have also memorably featured on stage, and there is ample testimony as to the impact of theatre conventions in the development of metaphorical language surrounding AI. The very term *robot* itself is closely associated with a specific play. Czech novelist, playwright, and essayist Karel Čapek (1890–1938) is credited with the etymological origin of the term. In his play *R. U. R.*, which premiered in 1921 – according to a contemporaneous production poster its sci-fi plot is situated around the year 2000 – *Rossumovi Univerzální Roboti* (Rossum's Universal Robots[3]) are not mechanical, but biologically assembled workers, produced for the Fordist factory production lines. In Czech, though no longer in use, *robota* means 'drudgery' or 'hard labour,' and a *robotnik* in Russian means a 'labourer' or 'grunt.' The robots in Čapek's play are destined to obediently execute the factory grunt work, though they have little to do with the eventual concept of mechanical or elec-

---

[3] Rossum (perhaps a reference to *rozum*, Czech for 'reason' or 'intelligence' – a *rozumny* is an intelligent person) is the name of the fictitious marine biologist in the play, credited with inventing a procedure to create synthetic protoplasm, from which biological life could be created. His plan was highjacked by his nephew or son (the play is unclear on this). The latter's dream resulted in an output of thousands of *Homunculi*, as Čapek later called them, which could easily and cheaply be put to work in factories. The play's importance in theatre history can hardly be overestimated. Just to give one example: in pursuit of André Antoine's Théâtre libre in Paris (1887) and its radical naturalist style, the Belgian Le Théâtre du Groupe Libre launched its own initiative for a realist-scientific theatre three decades later in Brussels in 1925 and chose for their opening production *R.U.R.* to underline a decisive break with traditional dramaturgy. For an overview of *R.U.R.*'s production history and impact, see the chapter on "From Automata to Automation: The Birth of the Robot in *R.U.R.* (*Rossum's Universal Robots*)" in Reilly 2011, 148–176. (Thanks to Dr. Richard Lehner for etymological advice.)

tronic robots as we understand it today, executing their tasks through AI applications.

The impact of theatrical concepts and theatrical language on our understanding of AI nevertheless is profound. University of Waterloo-based AI specialist Kirsten Dautenhahn, in her review of Cynthia L. Breazeal's *Designing Sociable Robots* (2002) defines an AI activated sociable robot in remarkably theatrical terms: "They are designed to interact with people and to *mimic human reactions as closely as possible* while doing so" (Dauthenhahn 2003, 278, my emphasis). This very much sounds like acting class, where mimicry is often a starting point. Dautenhahn's description of the impact these humanoid-shaped sociable robots have may well describe those rare actors on stage that break down the audience's emotional guard and become theatrical stars: "These robots are not just puppets that act when prompted; rather, they try to 'lure' people into spending time with them by harnessing our natural social responses" (Dauthenhahn 2003, 278). No wonder then that the concept has evolved towards *cobots* to stress the supposedly collaborative and interactive nature of our sibling automatons.[4]

Another legendary contribution from theatre is the case of computer scientist Joseph Weizenbaum (1923–2008), who created a natural language processing program or chatbot in the mid 1960s, a protocol that mimics human response via words on a computer screen (and later, with speech technology, also spoken out loud). Weizenbaum's muse for his computer program was Eliza Doolittle, the Cockney flower girl in Shaw's *Pygmalion* (1913). "I chose the name 'Eliza,'" Weizenbaum later wrote, "because, like GB Shaw's Eliza Doolittle of *Pygmalion* fame, the program could be 'taught' to speak increasingly well, although, also like Miss Doolittle, it was never clear whether or not it became smarter" (qtd. in Switzky 2020, 55). University of Alberta colleague and AI specialist Jonathan Schaeffer, in his book on the genesis of the checkers-playing computer program CHINOOK (1980–1996), comments on the mechanical learning aspect and the limitations and predictability of AI as it is unavoidably limited to whatever input it was subjected to, not unlike the blueprint or script for a theatre production, though we have come to expect much creative departure by the production team from that original script in the case of theatre. As far as AI is concerned, Schaeffer quips: "Therein lies one of the real problems with artificial intelli-

---

4 The term *cobot*, a contraction of 'co-manipulation robot', originates in the auto assemblage industry. According to Éduoard Kleinpeter, it was coined by Michael Peshkin and Edward Colgate in 1996. (It was originally used in reference to communication robots in the late 1980s.) (Cf. Kleinpeter 2015, 71)

gence: artificial stupidity" (Schaeffer 2009, 70).[5] In contrast, a popular publication like John Paul Mueller and Luca Massaron's *Artificial Intelligence for Dummies* (2018) confidently defines AI as rational decision-making, and differentiates rational from human: while human processes "involve instinct, intuition, and other variables," a "process is *rational* if it always does the right thing based on the current information, given an ideal performance measure" (Mueller / Massaron 2018, 21). And who then decides the *rightness of things?*

The permutative referencing of theatre, performance, and AI is quite fascinating. Vsevelod Meyerhold's biomechanical principles for acting come to mind. Cinema scholar Scott Bukatman has described *My Fair Lady* (1956), Alan Jay Lerner and Frederick Loewe's adaptation of *Pygmalion*, as "the Turing Test set to music" (Bukatman 2012, 147). And Alan Turing himself was called a "mathematical Pygmalion," by feminist philosopher Judith Genova (qtd. in Switzky 2020, 64). As it happens, computer science applications have appropriated a theatrical lexicon based on the most conventional Aristotelian notions of actor, character, and story. We see this particularly in computer games with concepts like narrative arc, plot, props, hero, villain, climax, etc. And of course, the original name Alan Turing himself proposed in 1950 for the meanwhile legendary Turing Test was perhaps the most explicit theatrical reference of all: the Imitation Game. Theatre scholar and video games designer Brenda Laurel's *Computers as Theatre* (2014) seems directed particularly at software designers to provide structure and vocabulary to their explorations of human-computer interactions. The book reflects a fairly traditional understanding of Aristotelian concepts, which, from a contemporary Performance Studies' viewpoint, seem outdated and fundamentally questionable. She equates, for instance, the Aristotelian notion of *Reversal* with "a rare and potent flavor of surprise," which "reveals that the opposite of what one expected is true" (Laurel 2014, 106). And she gives the cheekily cybernetic example "that's not a man, that's a *woman!*" (Laurel 2014, 106) Aristotelian reversal, on the contrary, is most often the opposite of surprise, but rather what is feared to inevitably happen.

The notion of *black-boxing* in relation to AI is well known through the work of cybernetician Ranulph Glanville (1946–2014). It is concerned with the phenomenon that what gets observed is focused exclusively on output – *what a machine can do or produce* – while the interior complexity is unacknowledged or

---

5 Schaeffer's quip is reminiscent of a passage in Douglas Hofstadter's "Six-Part Ricercar," the playlet that concludes his *Gödel, Escher, Bach.* The musician character Crab describes its computerized keyboards: "They are called 'smart-stupids', since they are so flexible, and have the potential to be either smart or stupid, depending on how they are instructed" (Hofstadter 1979, 748). Intelligence does not sit in the machine.

understood or, at the very least, taken for granted. It is derived from the Black Box concept originally formulated by mathematician James Clark Maxwell (1831–1879) in his so-called *Gedankenexperiment*, using the 'no-model' model theory of Joseph Louis Lagrange (1736–1813). It is typically not connected with the black box theatre phenomenon, the architectural structure which gave *élan* to the European avant-garde in the early 20th century and is now ubiquitous in Europe and N. America. Canadian theatre scholar and actor Tom Scholte, however, makes a strong contrary argument that the concept of black boxing may well originate in the theatre blackbox, which in Scholte's apt description, is "a simple, flexible, un-adorned performance space with a flat floor and no proscenium arch" (Scholte 2016, 598). Particularly in naturalistic theatre, we focus on what the actor does – on the product, in other words – while the complex methodology of getting there (talent, the years of training, the line learning, the process towards symbiosis, the inevitable mistakes during performance, etc.) is made opaque. Everything in naturalistic theatre is geared towards making the audience forget the *as if*, including the protective black box walls that facilitate the creation of an alternate reality.

Science has historically used theatrical techniques to foreground its results, so as to foreground *what it can bring about*. In the footsteps of initiatives in London from the late 1830s at the Royal Polytechnic Institution and the Adelaide Gallery, Paris-based Henri Robin (alias Henri Donckele, 1811–1874) acquired fame with his *physiques amusantes* displayed for the bourgeoisie of Europe's metropoles (see Vanhoutte and Wynants). In 1890, *Scientific American* – then a weekly! – had a certain Prof. Rufus Richardson (PhD), describe how Berlin's Urania theatre put science on the stage and evoked a journey to the moon with all the traps of oratory and scenic and lighting design, inclusive prologue and epilogue: "all the scenic effects which the stage affords are called in" (Richardson 1890, 180). The audience was even "shown an eclipse of the sun as seen from the surface of the moon" (Richardson 1890, 180). These were obviously pedagogically inspired approaches, but they actually became an early form of black-boxing and exteriorizing. In a 2009 volume on digital transdisciplinary approaches to artistic expression, Interactive Design specialist Eva Sjuve gives an excellent overview of these interests in theatre as a podium of scientific innovation, which recently culminated in wearable technology on stage. The New York-based artist group LoVid exemplify this trend. Founded by Kyle Lapidus and Tali Hinkis, they bridge the digital and the analog in their installations, mostly using wearable technology, like in their piece called *VideoWear* (2003). Computer assisted electronic circuitry and conduits become extensions of the human body. Similarly, inspired by Michael Frayn's play on nuclear science *Copenhagen* (1998), Steve Gibson and Dine Grigar use a digital collage technique in *When*

Ghosts Will Die (Dallas, 2005). The production relied on live motion tracking technology to evoke the impact of nuclear war. The Cypriotic-Australian artist Stelarc takes this a step further, and rather than making his performing body available as a vector for wearable technology, he seeks to integrate that technology as an integral and operational extension of his corporeal arsenal. The artist thus transforms his body, sometimes taking this to what most will see as extremes, including surgically attaching an internet-enabled ear on his left arm as an installation. The anatomical excess of the extra ear project pushed medical

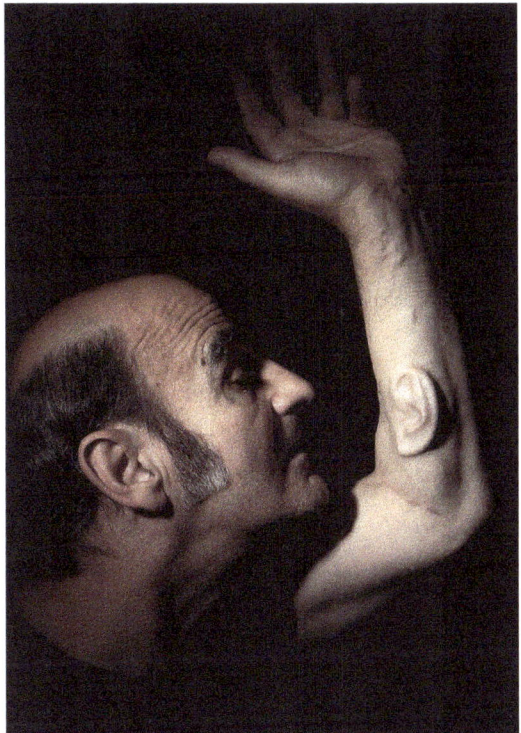

Figure 1: Stelarc, *Ear On Arm*. (London/Los Angeles/ Melbourne, 2006. Photo Nina Sellars)

ethics, surgical know-how, as well as the phenomenology of performance itself to a liminal boundary and it took many years to research and carry out, including the necessary healing. Stelarc initiated his queries in 1997, *Ear on Arm* was completed in 2006. Media specialist McKenzie Wark observes that Stelarc's "skin becomes the point of interface for relations to the technical." Some of these attachments and extensions control him, others are controlled by him, thus creating "a second nature" or alternate phenomenology, a "state of permanent dependence and symbiosis" (Wark 1995, 12). Stelarc, now 75, continues to take his artist statement that "the body is no longer merely an object of desire, but rather an object that requires redefining and redesigning"[6] quite literally. His recent project *Reclining StickMan* (2020) is a huge pneumatically driven robot, nine meters long, which can harbour a human. With its intricate limbs, it performs its choreographic movement in response to visitors' actuations, either locally present or from anywhere in the world online.

It is no doubt the fascination with and fear of automation that has driven theatrical response to engage with AI as a subject matter; it often also comes as a warning for the potentially negative impact of AI. Barely two years after Karel Čapek's *R. U. R* premiere, and with a similar political objective, lawyer-turned-playwright Elmer Rice (née Reizenstein, 1892–1967) wrote *The Adding Machine*, which premiered at New York's Garrick theatre in 1923. Rice's expressionist play evokes a machine-dominated world in which the protagonist Mr. Zero leads a dull life of endless repetitions and whose job in a department store is easily replaced by an adding machine. The theme of automation and its socio-political ramifications was obviously of great interest. Four decades after the premiere, the play was turned into a film (1969), though director Jerome Epstein inoculated it with comedic impact.[7] Since these early manifestations, robots on stage and robotized humans have become staple motifs in theatre and in performance.[8]

The earlier mentioned Turing Test is a motif that has caught the interest of film- and theatre-makers alike, as it deals with tricking an audience into believing that a machine has consciousness. This directly inspired New York-based artist Annie Dorsen for her production of *Hello, Hi There*, which premiered in Graz as part of the 2010 Steirischer Herbst Festival,[9] before travelling on to New York in January the following year. Central to the dramaturgy are two Mac computers in open display on a pile of Astroturf on stage. They chat with each other as chatbots (a female and male digital voice) and their exchange is also displayed on the two computer screens, which in turn are beamed onto two large monitors. Their dialogue is predicated on the 1971 live debate on the nature of human consciousness and intelligence between Noam Chomsky and Michel Foucault, which

---

[6] http://stelarc.org/reclining-stickman.php. Accessed: 08. Nov. 2021.
[7] More recently, it was adapted into a musical, *Adding Machine* (Joshua Schmidt and Jason Loewith), which premiered in New York in 2007.
[8] For an overview of what he calls "the humanization of machines and the dehumanization (or "machinization") of humans" in performance (15), see Dixon 2004, 15–32.
[9] See https://archiv.steirischerherbst.at/en/projects/3359/hello-hi-there Accessed 7.Dec.2021.

Figure 2: Annie Dorsen, *Hello, Hi There*. (steirischer herbst, 2010. Photo W.Silveri).

was broadcast on Dutch television.[10] The historical debate prevaricated largely between two positions "whether what we imagine as human consciousness is an innate trait (Chomsky's position) or one determined by environment and language acquisition (Foucault's)" (Soloski 2012, 79). Over "hundreds and hundreds of hours" of work (Soloski 20212, 84), Dorsen excerpted the debate but, also added phrases from various sources, including reactions to the 1971 debate, biblical quotations, a panoply of philosophers, Shakespeare sources, poetry. Using programming tricks, the piece yielded a live performance that could play out in millions of different ways, even though most of these would offer only trifling differences, perhaps similar to two live actors whose delivery of learned lines is also not 100% hermetically repeatable during a 45-minute performance. Similar to live acting, any off-kilter response from one of the bots could lead the conversation onto a different trajectory. Dorsen's most fascinating insight of this experiment was that these computers could yield

> a good show or a bad show, just as with real actors, and it doesn't always depend on the text. Some nights they make super nice choices, they find good rhythms, but the audience

---

10 See the videographed archive: https://www.youtube.com/watch?v=3wfNl2L0Gf8 Accessed: 08. Nov. 2021.

is off, not in sync with them. Or there are times when I think their text is rather mediocre, but the audience goes for it full throttle. (Soloski 2012, 86)

The Turing motif obviously appeals to the creative impulses of artists. Italian playwright Valeria Patera's *La mela di Alan* or *Alan's Apple. Hacking the Turing Test* premiered in 2012 – at an AI conference in Rome, on the occasion of the 100th anniversary of Alan Turing's birth: a paper in the form of a play.[11] The story is about Alan Turing's life, shrouded in mystery, but now freshly discovered through the internet by the efforts of two young hackers. We are reminded of Turing's manifesto about the new technological order in which he explains the nature and theoretical limits of AI before the coming of computers. The two hackers reverse this line of thought so to speak, approaching the human body as mere machines, and focus on mistakes made by these human machines.[12]

One of the most fascinating representations of and engagements with AI on a live stage is undoubtedly the German theatre collective Rimini Protokol's *Unheimliches Tal/Uncanny Valley* (2018), which incidentally also references Alan Turing. The production was conceived by Rimini co-founder Stefan Kaegi and is still on playbills across Europe, touring in a German, English, and French version. Kaegi worked with German playwright Thomas Melle, who allowed an ani-

---

11 The play was first published in English in 2004 and in Italian three years later (*La Mela di Alan. Hacking the Turing Test*, Rome: Di Renzo, 2007). It was produced by Valeria Patera's own theatre company, Timos Teatro, in 2012 and went on tour across Italy. See video trailer: https://vimeo.com/41628040 Accessed: 08. Nov. 2021. The periodical *Plays International & Europe* published an updated edition of the translation in its Summer 2006 issue (21.7/8, 48–62).
12 Biographical responses to Alan Turing in film and prose are manifold, with *The Imitation Game* (2014) film as a top money-maker. There are numerous lesser known theatrical engagements about the legendary figure. Most of these focus less on the computing complexities and achievements, and more on the biographical complications of his sexuality and suicide. The following list is not exhaustive: High Whitemore (1936–2018)'s *Breaking the Code* is a biodrama that premiered at London's Theatre Royal (1986), starring Derek Jacobi. (It was adapted by BBC for television in 1996, also starring Jacobi). It had a long run on Broadway and was translated into numerous languages. Kevin Patterson's *A Most Secret War* followed a year later in New York at the off-Broadway Clurman Theater (Dec. 1987). *Alan Turing e la mela avvelenata*, a monologue written by Massimo Vincenzi, premiered at Rome's Teatro Belli in March 2009. In October 2018, after a successful run at Avignon's Off-Festival, Benoit Solès's *La Machine de Turing* got picked up by Théâtre Michel in Paris, and continues its run today, with well over 500 performances. William Jean Bertozzo wrote and directed his dramatic narration *Il Nastro e la Mela. Elegia per Alan M. Turing* (Jan. 2020) for Verona's Teatro Camploy. Turing is also the subject of the musical *The Universal Machine* by David Byrne and Dominic Brennan (London, New Diorama Theatre, 2013). The topic also found its way into radio drama in Phil Collinge and Andy Lord's *Turing's Test* (October 2009), with Samuel Barnett in the role of Turing.

Figure 3: Humanoid robot in Thomas Melle, *Unheimliches Tal/Uncanny Valley* (Rimini Protokoll, 2018. Photo Gabriela Neeb)

matronic double of himself be made, and he performs on screen alongside his robotic but uncannily anthropomorphic double, present on the stage.[13] The robotic duplicate becomes a practical extension of the original subject of the monologue, and serves to overcome Melle's fight with chronic depression and his deep trepidation against performing in the public arena. The robot obviously does not suffer from these mental conditions, and can even seemingly observe and comment on his own protogenesis. The uncanny resemblance between the author on screen and the seated robot on stage has a haunting impact on the live audience and, after the show, they descend in droves to the stage to take pictures, particularly of the back of its head, which shows all the wires and electrodes, necessary to produce the vocal and movement responses, poking out. A visit back-stage, as it were.

---

13 *Unheimliches Tal/Uncanny Valley* premiered at the Münchner Kammerspiele in October 2018. An English language version of the production is available on Rimini Protokol's website: https://www.rimini-protokoll.de/website/en/project/unheimliches-tal-uncanny-valley   Accessed: 08. Nov. 2021.

AI and theatre in Canada

Meanwhile, in Canada, interactions between stage and AI are sparse, which may find an explanation in the fact that Canadian theatre, particularly on the English-language stage, has continued a fairly traditional realism-based trajectory. Still, the University of Toronto is a torchbearer in research on AI in Performance. Founded in 2019, with a $5-million grant from the BMO group, the BMO Lab for Creative Research in the Arts, Performance, Emerging Technologies, and AI is unique in positioning itself as a think tank for pioneering research on how to push AI as a creative force in the Arts. While the ultimate goal is to combine basic research in AI and computing technologies with creative research projects, the lab initially functions as a transdisciplinary venue to experiment with new technologies that can advance the artist's craft. It also has a prominent teaching function, and serves as a hosting platform for visiting artists, including residencies.[14] David Rokeby, the lab's co-founding director and an artist himself, creates interactive technology-based art installations. An important consideration is that technology be understood in a non-trivial way, the focus is not just on what it is made and used for, but also what it can it do (or not do) beyond its typical use. In "Performing the Digital and AI," Rokeby asserts: "It is harder to step back and question a technology once it has taken form; once a camcorder is a camcorder" its application converges mostly on its obvious consumerist function and this may well pre-empt more creative uses (Kleber / Trojanowska 2019, 105). This openness requires a specific dramaturgical practice and it is no coincidence that in the same report, Antje Budde advocates for a Digital Dramaturgy practice. She defines it as an emerging field that engages in "translational labor while testing, experimenting, critically making digital dramaturgy in and as performance." It requires a disposition in which nothing is taken for granted and the outcome in performance is a result of "processual prototypes of knowledge, sometimes succeeding, sometimes failing, but always learning" (Kleber / Trojanowska 2019, 102). Both Rokeby and Budde stress the centrality of the human subject in this process, thus refuting the standard fear and perception that digital arts result in the removal of humans from its production. The labour involved, Budde insists, best be understood in the way Hannah Arendt proposes in *The Human Condition* (1958). Critiquing Marx, Arendt distinguishes labour from work, where the latter is a process of engagement (mostly through our hands or machines) with

---

[14] The COVID-19 break in audience-attended live performance seems to have been a serendipitous blessing for the BMO Lab Art/Performance/AI, with intensive online dissemination of projects and experiments. See https://bmolab.artsci.utoronto.ca/ Accessed: 08. Nov. 2021.

an economical process that yields a product that can be consumed and is paid for, equitably or not. Labour, on the other hand, is not marketable in the strict sense of the work and is concerned with survival. It is typically not rewarded through remuneration, the reward is through the resulting outcome of labour, which is necessary for this process of survival, like child-giving labour, or labour that opens a door, or that involves cooking a nutritious dinner. For Budde, the labour and "embodied ephemerality" involved in theatre and performance is similar to what Arendt is getting at, and particularly when involving digital applications in creative approaches: "We need digital literacy, practical skills, and analytical audacity in asking questions that no one wants to hear. It takes courage to perform labor that is frequently regarded as idle – unproductive, useless, and worthless" (Kleber / Trojanowska 2019, 105–6). At the same time, we must "avoid falling into a default form of humanism" in Rokeby's words, so that we are not afraid to shake "up our overfamiliarity with our self" (Kleber / Trojanowska 2019, 107). Digital engagement with live performance can contribute to a process of decentering the human in performance so that new questions and also new answers can come to the fore.

One prominent Canadian example of (a relatively low-key) digital engagement with the live actor on stage is Québec-based Robert Lepage's *Les Aiguilles et l'Opium* or *Needles and Opium* (2013).[15] The play weaves together three narratives: African-American jazz musician Miles Davis' visit to Paris, French playwright and filmmaker Jean Cocteau's experience in New York (both in 1949), and the Paris séjour of an autobiographical Canadian character called Robert. It is a remarkably ultra-analogue theatre production, though at the same time radically mediated through computerized projection technology, which gives the impression that the actors on stage defy gravity. The set, which consists of three planes of a three-meter wide cube that rotates on a horizontal axis, serves as a screen for three projectors, "creating an environment that shifts physically with the rotation, and visually with changes in the projected images" (Nearey 2017, 30–31). The result "is a prime example of frame manipulation; both the cube and the projections shift, moving the performers radically through space and time" but the histrionic wonder of the production lies precisely in the physical presence and interactions of the live actors on stage, precariously balancing themselves within a technically manipulated frame. One can easily imagine AI

---

15 An earlier version of *Needles and Opium* premiered in Québec in October 1991, with Robert Lepage himself in the role of Robert (and his shadow in the role of Miles Davis). Two decades later, Lepage revisited the same material in a so-called auto-adaptation with Afro-Canadian dancer/acrobat Wellesley Robertson in the role of Miles Davis. Québecois actor Marc Labreche filled the role of Robert and Cocteau. Both productions toured globally.

Figure 4: Olivier Normand (l) and Wellesley Robertson in Robert Lepage, *Les Aiguilles et l'Opium* (Ex Machina – Barbican, London. Photo Tristam Kenton)

applications in live theatre to push this frame manipulation to unexplored boundaries.

While in Lepage's production, the motif of human desire is foregrounded through the astute mixing of physical presence with manipulated frames, in more radical technologically mediated performances, the place of human desire – or at least a credible version of it – is often problematic. We can play with artificial intelligence in performance in exciting ways but what is most often lacking in the outcome is desire; and in the case of artificial desire, it often renders an unconvincing format: two robots falling in love, performing a mimetically played cliché desire that never really transcends the level of a cute and innocuous wink. And yet, it is some form of desire that ultimately drives us to performance and representation, in order to understand, exhibit, manipulate, and celebrate our own deep-seated desires (and fears thereof), by seeing them playing out in front of us. The colossal question to be answered in any artistic creation through and with AI is where and how desire comes into it. After all, as David Rokeby puts it: "No computer has ever told me not to unplug it – not without someone programming it to do that" (Kleber / Trojanowska 2019, 112).

The case of John Mighton

John Mighton is a Canadian playwright for whom desire remains a central preoccupation in his writing. His relatively small oeuvre displays an intriguing approach as to how a scientist understands this desire. With a PhD in Mathematics, Mighton is a playwright-mathematician-philosopher and educational activist. He is currently leading JUMP Math (Junior Undiscovered Math Prodigies), a Toronto-based advocacy group build confidence around mathematics with children.[16] In a 2000 interview, he lists some of the things he wants to pursue, including knot and graph mathematical theory, his playwriting career, and his math-teaching activism. He observes: "we live in an age that is infused with the idea of possibility. As individuals, we can't escape that. It's just part of the way we think about ourselves" (Shenfeld 2000). In his debut play, that idea of *possibility* is front and centre. With its fairly cliché plot and character structure, *Scientific Americans*, which premiered in 1988 – still relatively early days for AI – is certainly a beginner's playscript.[17] It foregrounds a couple: Jim, a physicist in electromagnetic radiation, and Carol, a computer specialist in AI, both Stanford educated. It is a marriage, in Jim's words, between "Physics and artificial intelligence. The world and the mind" (Mighton 1987, 45) or between the material and the immateriality of intellect. And he adds, for emphasis: "That's where the next great breakthrough will happen. On the border between them" (Mighton 1987, 46).

At the beginning of *Scientific Americans*, Jim switches jobs – he "needs the money" (Mighton 1987, 31) – to work for the Department of Defense on the same desert campus that was used for scientists in the 1940s to work on the development of a nuclear bomb. Very quickly, it becomes clear that the play tries to do too much in its fragmented focus. The Freudian relationship of Jim and his mother Betty and Jim's relentless and compensatory sexual desire for Carol against the background of the couple's eventual dysfunctional relationship is a major motif. However, this focus also clouds the actual theme of the play, namely the ethics involved in working for the arms industry and the Dept. of Defense.

---

[16] For Gus Van Sant's film *Good Will Hunting* (1997), John Mighton was hired as the math-checker and also played the role of amanuensis, or note-taker for Gerald Lambeau, the mathematics professor in the film. Don't we all wish to have one of those!

For JUMP Math, see: https://jumpmath.org/jump/en/node/1608 Accessed: 08. Nov. 2021.

[17] An earlier play, *The Prediction*, was self-produced in New York. Apart from the two plays discussed here, John Mighton also wrote for the stage *A Short History of Night* (1989), *Body and Soul* (1994), *The Little Years* (1995), and *Half Life* (2005).

For our purposes here, I want to focus on the character of Carol, the computer and AI specialist. Her research, cogently summarized by herself, is "trying to get computers to read and summarize books" (Mighton 1987, 30), and the sample she uses for this consists for the most part of fairytales. Her interest is not so much in AI itself, but, intriguingly, in its reverse application. Through AI, she wants to chart and understand how humans think. She secretly sympathizes with the anti-nuclear weapons protesters who regularly show up at the base; many of them are in a similar position as she is. They are the spouses of the scientists who work on the weapons development program. In a conversation with one of the women, she briefly responds to an inquiry about her own research:

> If we can understand how computers can be programmed to handle ambiguity, make inferences and supply missing contexts, we'll have a better idea of how humans think. We'll begin to understand how we get our ideas. (Mighton 1987, 30)

In the next scene, Carol is at an AI conference where her panel chair launches into a celebration of AI research in general, and Carol's research in particular, which is situated with appropriate jargon as "ground-breaking work on Suboptimal Logic in Mechanized Problem Solving" (Mighton 1987, 35). She obviously works in a hot field, as she is struck by the fact that "At the conference, five agencies tried to recruit me. I'm working on Fairy Tales, for Godsake! Isn't anything harmless anymore?" (Mighton 1987, 38). As the scenes develop further, the play's irony comes to the surface as no logic seems able to address (much less resolve) the relational problem between Carol and Jim. After his successful contribution in the production of a new weapon, the latter has not only left his job at the Dept. of Defense, but by the end of the play, Jim has also left his girlfriend and moved back in with his overbearing mother.

Carol remains an underdeveloped character and the discussion about her complex relationship with the world of AI and its surrounding industry never takes off. What is interesting in this play is that Carol echoes Hannah Arendt's observation on the distinction between work and labour, referred to earlier. Like so many scholars, what Carol is doing is mostly labour – meaning, she is not that interested in the actual product her research on AI may yield, whereas her husband certainly falls victim to the unrelenting pressures of the Dept. of Defense to actually come up with a product, namely a new weapon that can be set off through the mind, and, following Bell's theorem, thus bridges distance without actually having to physically cover that distance.

Possible worlds

John Mighton's most successful play, *Possible Worlds* (1990)[18] continues this exploration of interpersonal relationships against the background of developing scientific insights in human (and artificial) intelligence. Particularly, it raises the question to what extent these relationships are conceptions in our minds or, in Mighton's own words: "to what extent each of us fabricates our partner in our minds" (qtd. in Shenfeld 2000). The playwright was inspired by reports on 1950s experiments that tried locating *thought* in the actual physicality of the brain. Scientists working with epileptic patients severed the stem joining the two halves of the brain. "They found that the two halves functioned almost independent of one another, like separate consciousnesses" (qtd. in Walker 1997, G11). Similar to Carol's quest in *Scientific Americans*, it led him to "question assumptions about innate intelligence and the way the brain works" and wonder which of these two halves would harbor a perception of a person that is closest to reality: "Philosophers immediately asked: 'If you took two halves of the brain that had the same memories and dispositions and transplanted one, which would be the original person?'" (qtd. in Wagner 1990, D3). The philosopher-playwright-mathematician was also inspired, in part, by a branch of modern philosophy called Personal Identity Theory, initiated by John Locke. It attempts to determine what is essential to a person's identity, taking into account factors such as continuity of memory or a person's choice of actions in a given situation. More recently, British philosopher Derek Parfit's radical reductionism has become part of the discussion. He asserts the relative unimportance (or even irrelevance) of the establishment of Personal Identity, as each one of us is a compilation of changing parts, some physical, some psychological, some sociological, etc.[19]

The play reads like a version of a Turing test, with a man (A), a woman (B) and an interrogator (C), which is constituted of an odd complementary duo of

---

18 Canadian director Peter Hinton staged the premiere for the Canadian Stage Co. in November 1990 with a mixed reception. Geoff Chapman reviewed it for the *Toronto Star* as "hopelessly muddled." Daniel Brooks, who played Penfield in the original production, directed a revised version for Toronto's Theatre Passe Muraille in 1997. The CBC's Richard Ouzounian reviewed the 1999 National Arts Centre Production in Ottawa as a "dazzling [...] double espresso for the mind" (qtd. in Paul Dervis). The script has received numerous productions in Canada and elsewhere. Robert Lepage directed an eponymous film adaptation in 2000, starring Tom McCamus and Tilda Swinton.
19 For an excellent but concise discussion of Personal Identity Theory, see Carsten Korfmacher, "Personal Identity," *Internet Encyclopedia of Philosophy (IEP)*, https://iep.utm.edu/person-i/ Accessed: 08. Nov. 2021.

detectives; John Mighton seems to embrace this structure with ironic fervour. Like *Scientific Americans, Possible Worlds* has two lovers at its centre: George, a risk analyst-mathematician, and a woman called Joyce, who at least in one incarnation teaches neurology and specializes in research on rat cortexes (in later encounters, Joyce is a stockbroker, a widow, or some sort of synthesis of all these incarnations). George too discovered he was "more than one person," in seventh grade while solving a math problem (Mighton 1988, 23), and in one of the scenes determines he has been a widower for three years. With transparent pick-up dialogue, George meets Joyce early on in a self-service restaurant, only to find out that both are from Novar, "a little northern town" (Mighton 1988, 14). It is important to know that these various versions of the two characters are consistently played by two actors (they also play some of the other minor characters). The George/Joyce binary is paralleled with two detectives, Berkley and Williams. The latter couple is trying to resolve a series of homicides of "very intelligent" people in "positions of power" (Mighton 1988, 27). Each time, the victim's entire brain is meticulously scooped out of its cavity. Towards the end of the play (which may well be the beginning), it looks like George himself has fallen victim to the serial killer/brain thief as his brain is also found "hooked up to a life-support system [. . .] floating in an aquarium supported by wires" (Mighton 1988, 69). A light occasionally flashes, "but we don't know what it means" (Mighton 1988, 69–70). The play presents a complex layering of different levels of action, realities, and time frames – and sometimes, as a spectator, we are not quite sure which reality we are actually confronted with. The play does not focus on AI persé; it does not explicitly refer to it. However, I argue that John Mighton puts AI metonymically on the stage in *Possible Worlds* which makes the play particularly exciting for a discussion on representations of AI on stage (and on screen!). In what follows I refer primarily to the theatrical version, though Robert Lepage's 2000 film adaption with the same title will also come in handily to explicate some points.

Critics have interpreted *Possible Worlds* in diverse ways. Scholarly criticism has linked the play with Possible Worlds Theory, which looks at parallel universes, and its narratological applications by theorists such as Umberto Eco and Marie-Laure Ryan.[20] Others, like Graham Wolfe, see in Mighton's play a demonstration of Lacan's "'traversal' of the fantasy," which models "a radical reconfiguration of the subject's relation to Symbolic reality" in these times where the virtual often overtakes the physically real meat-suit reality (Wolfe 2011, 195).

---

20 See for instance, Klaver 2006, 45–63.

The brain-in-the-vat, one of the central images in both play and film, is part of Dr. Penfield's neurological research collection; he acts as a science consultant for the detective duo.[21] It is a timeworn cultural motif that has shown up in numerous (mostly sci-fi-related) narratives, but also gets picked up in philosophical debates, dating back to René Descartes's evil demon scenario and later revisited by Hilary Putnam. It typically underscores a scenario that understands our mental output to be caused by some sort of communal supercomputer. In Mighton's play, the brain-in-the-vat becomes a metonymic trope for the brain's potential as a source of endless possibilities about who we are, where we are, and what we can do. In a conversation with Joyce, George shares his insight, which, significantly, he acquired through a mathematics challenge:

> Each of us exists in an infinite number of possible worlds. In one world I'm talking to you right now but your arm is a little to the left, in another world you're interested in that man over there with the glasses, in another you stood me up two days ago – and that's how I know your name. (Mighton 1988, 23)

As an audience, we are witness to the various layers and ontologies that unroll in front of us on stage or screen, the very medium through which they are – or ought to be – authenticated or made believable, which implies, using a Douglas Hofstadter expression: "Something onto which we can map ourselves comfortably" (Hofstadter 1979, 636). John Mighton's use of a conflicting and confusing mise en abyme structure of substitution renders a fictional world that is on the point of collapse. What we end up with is a typical example of what Bell and Ryan call "a metaleptic collaps[e] of boundaries between diegetic levels" inside the fictional world of these characters (Bell / Ryan 2019, 25). Yet, in spite of this collapse, because of a process of authentication, we recognize an intelligence at work with which we can still identify and we can recognize the various components of the structure, its repetitions, its substitutions, and its contiguities. While from a logical outside perspective the "centre cannot hold," to use a Yeatsian phrase, the insight that we gain through the conflictual structure is precisely the audience's gain.

In narrative theory, the term *metalepsis*, coined by Gerard Genette in his *Narrative Discourse* (1980), refers to a process of extradiegetic intrusion into a die-

---

21 Jenn Stephenson points out John Mighton's likely wink to the American-Canadian neurosurgeon Wilder Penfield (1891–1976). He was one of the early scientists who mapped the brain and identified specific locations in the brain that engendered specific images or sensations (Stephenson 2006, 76, note 8). Lepage changed the name from Penfield to Kleber in his film adaptation of the play.

getic universe – the latter is typically framed (and understood) as 'the fictional world of the story.' The problem is that in our phenomenological experience of the world(s) we are confronted with, these various worlds are not always clearly separated – and that is certainly the case in our experience of fictional worlds. In his *Dictionary of Literary Devices*, Bernard Dupriez defines metalepsis, using Pierre Fontanier's *Les Figures du Discours* (1830) as a reference, in a most straightforward way as "The expression of one thing by means of another which precedes, accompanies, or follows it, or which is . . . attendant upon or related to it, in such a way as to recall it immediately to mind" (Dupriez 1991, 275). Metalepsis invokes a collapse of two (or more) worlds, but also implies transgression or intrusion, and suggests a contiguity or proximity of two (or more) referential worlds that are subject to the insertions.

In *Possible Worlds*, we witness this kind of collapse and intrusion in just about every consecutive scene. It invariably invokes a consecutive logic, but at the same time, it is precisely the play's *dis*-consecutive sequencing and perception of *a*-logic that brings about an uncanny impact. I agree with Jenn Stephenson that this metaleptic quasi-collapse of different worlds is a result of a metonymic process, based on contiguity and substitution, as is the typical characteristic of metonymic figuration.[22] She lists several examples of this, including George's missing arm trope, the beach and sea imagery, the abundance of water references, and of course the brain-in-the-vat. Apart from the fact that some of these run counter to her argument – the missing arm is a simile, not a metonym – I would push this further and argue that Mighton's *Possible Worlds* conjures a binary cycle of possibilities; the play's narrative structure is an *on* and *off* sequencing of events, much like the on and off of the rat brain's food light.

Nestled in the centre of this binary sequence is a dream-like scene in which George sees two men moving stone blocks under the alternating prompts of "block" and "slab" – the only two words they get to utter. As if to stress the binary character of their language, the Guide in the scene tells George they also have a third word: "hilarious." And George's question as to what they can do with that third word gets promptly answered: "Nothing." Hilariously, the Guide adds that the third word was most likely learnt from "a tourist" to this world – which is easily understood to be us, the audience! It also connects the narrative with the play's early assertion that George's story of the missing

---

[22] The other (more common) figuration is metaphor, which is a form of reverse *simile*, based on distance. Through a process of suppression and substitution, often from opposing semantic fields; its meaning is culturally pre-disposed and the harmlessness of an expression like 'You are my sunlight' is underscored by the implied distance to the scorching sun.

brains was sourced from *The National Enquirer*, which he reads sometimes "[f]or a joke" (Mighton 1988, 13). The question as to what they are doing with the blocks and their binary language is more productively answered: they are "building" – though we are not quite sure what they are building. There is obviously a productive sort of intelligence at work; according to "some" it was "once an advanced civilization" and "others" suggest "[i]t would take fifty encyclopædias to translate the meanings of 'slab' and 'block' into our language" (Mighton 1988, 41–42). In Robert Lepage's film version, the Guide not only gets played by the same actor who plays the neuro-scientist in the on-stage version, but he actually is the same character, white lab attire and all (played by Gabriel Gascon). He situates the scene, incongruously, on a barren promontory rock where the two men go about their Sisyphean drudgery of swopping and building rocks.

The block/slab binary is clearly the metonymic equivalent of a digital language in which artificial intelligence finds its *flat* diegetic basis. Its potential is virtually without limits – *all possible worlds* – but its application is terrifyingly convergent in a dyadic catch with no way out. This *all* versus constricting *lack* is emphatically echoed in the play's concluding scene. In a final case of metalepsis, Joyce and George are sitting on a beach, looking out onto the sea and into their future, which, like the surface of the sea, they aspire to be a *perpetuum mobile*. "We'll keep moving. We'll never be alone," says George, and adds, in a final emphasis: "Everywhere." At the same time, as the lights of the stage fade, "in the darkness, a small light blinks on and off" (Mighton 1988, 75–76). While the blinking light on the ocean's horizon may well be a boat in distress, within the play's ontological world, it is more likely the brain-in-the-vat blinking at us as if to highlight its product: the play itself, with all of its possible worlds conjured. In the end, unlike regular fiction that offers a clear and single paradigm from the many possibilities, the spectator, like the characters, is left in total limbo. It is no surprise that the (unfulfilled) saturation offered by the *virtual all* and the unlimited world of the play does not lead to a sense of triumph, but the contrary is true: it induces a state of melancholy and sense of loss. Metonymically, this conjures a world that is incongruous and disconnected from any context – a brain in a vat, for which subjectivity has become irrelevant because of an unlimited access to possibilities. Writing shortly before the time of the play's premiere, for Jean Baudrillard it was clear that our subjectivity was under assault through highly technologized communication platforms and protocols:

> Our private sphere has ceased to be the stage where the drama of the subject at odds with his objects and with his image is played out: we no longer exist as playwrights or actors but as terminals of multiple networks. (Baudrillard 2012, 23)

Baudrillard's dystopian vision morphs into an eerily meta-theatrical epiphany in Mighton's play, with its characters part of an uncontrollable world of unstable, uncommitted, and uncommitting possibilities.[23]

In a recent week-long series on developments in AI, the Belgian newspaper *De Standaard* included a short article on the limits of AI. The text, written by a "collaborator" in well phrased Dutch, addresses the reader directly in the form of a Letter to the Editor; it was generated by a linguistic model (GPT-3) developed by OpenAI, a research lab in San Francisco. The piece provides some interesting insights on the potential and limits of AI as a generator of journalistic writing. Its conclusion is as ominous as it is abrupt and revealing: "Mijn artikels kunnen veel sneller worden geschreven dan die van u. U moet nadenken over de context. Ik niet."[24] Decades before, Baudrillard was equally blunt, but even more radical in his critique: AI operates in a vacuum, which for him was a metonymic trope for an aseptic way of living, biologically as well as intellectually, which, from his perspective, is unavoidably coming our way (36). Ironically, the trope of operating in a bubble vacuum is also close to Mighton's original impetus for the play, with his insight of the brain spheres of epileptic patients functioning independently from each other.

*Possible World*s does not embrace the same kind of Baudrillardian pessimism, though there is certainly a dystopian/utopian tension prominent in its narrative. Williams, the junior detective, emphatically reminds us that, on an average day, we only use our brain's potential capacity for about 10 %. Against this background, Baudrillard's vacuum trope does seem *à propos!* In the end, the central couple George/Joyce do not offer any sense of resolution, let alone redemption. It is the same Williams – with only 10 % brain capacity at work – who solves the riddle of the stolen brains and brings the case to a precarious close in the laboratory of Dr. Penfield, though – with our 10 % – the enigma of the brain in the vat itself remains precisely that: an uncrackable enigma.

---

23 Film scholar Sylvie Bissonnette is unconvincingly optimistic: the "reflexive project" between screen and spectator in Lepage's film demonstrates that "cyborg brains can become active social actors of change" (Bissonnette 2009, 410).

24 [I can write my articles much faster than you can. You have to ponder about context. I don't.] GPT-3, "De limieten van AI," *De Standaard* (Brussels), 17 March 2021.

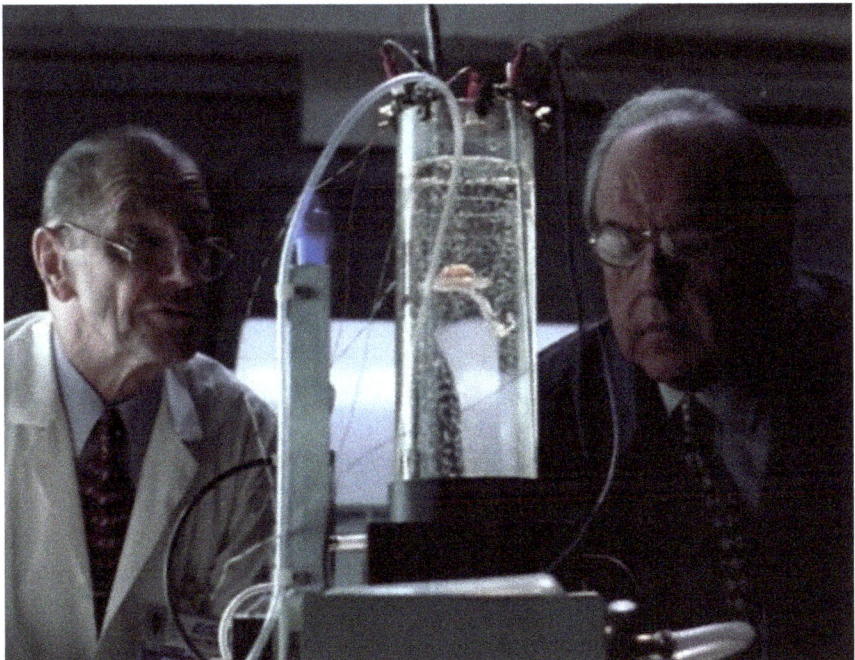

Figure 5: Screengrab. Dr. Kleber (Gabriel Gascon) and detect. Berkley (Sean McCann) with hooked-up rat brain. R. Lepage/J. Mighton, *Possible Worlds* (Alliance Atlantis – Robert Lepage Inc., 2000)

Conclusion

We have seen that AI assumes a significant presence on stage and in theatre discourse, though there is also discomfort about its role and significance in theatre as its ontology goes contrary to the analog meat-suit reality – essentially uncompoundable and irreducible – of the live stage. In perhaps a nostalgic turn, the stage often sees itself as an endangered oasis in a field of ecstatic technologically manipulated cultural exchange. Theatre is presented as a space where we get to see and experience human interaction after a day in front of a screen, or next to a cobot, or struggling endlessly on the phone to cancel a fight reservation with an AI-operated voice. Alternatively, on and near the stage, AI makes its presence known more emphatically, certainly in technical support, but also in the more creative areas and in its subject matter. Already, Stage Management platforms use digital approaches to support various interfaces for smooth theatre production, and Theatre Design fully embraces digitally supported design and 3D printing. Text analysis platforms and smart cameras in the theatre may offer exciting applications of AI algorithms in reception prediction and analysis – a terrain in Performance Studies which so far remains impressionistic and fragmented. Virtual Reality (VR) glasses and Augmented Reality (AR) apparel may find its way to the foyer of our theatres.

For those concerned about conserving the integrity of live theatre, it needn't be all bad. Voice coaching seems an obvious domain for AI applications. Puppetry theatre, in many ways the epitome of breath-induced live theatre, has recently seen a remarkable digital transformation. Ultimately, this can be understood as an extension of projects like Rimini Protokol's *Unheimliches Tal* with somewhat larger puppets centre stage. AI's increasing application in live theatre seems unavoidable, both as technological support as well as thematic development. In fact, in art experience in general, AI applications have the capacity to intensify the performative nature of that experience. This is certainly the case with what has been going on during the 2020 – 21 COVID crisis and the ubiquitous digital recording and streaming of live theatre it brought about. But also museum visits have recently been altered, and can become an immersive experience in a performative way. Gustav Klimt's world, for instance, can be virtually entered as an intensive 3-D performance during a recent exhibition animation in Brussels' Horta Gallery.[25] The question here is: have Klimt's paintings in "Klimt: The Immersive Experience" come closer to us in our walk through this wondrous digital performance, or are they, indeed, further away than ever?

Our increasing deference to and reliance on technology and science-based knowledge may well go contrary to more experientially gained insights in which trial and error may become hazardous pedagogy. Will the digital make the theatrical analogue world obsolete as the former becomes more reliable and fool-proof? Whatever benefits AI may offer in artistic production, expression, and experience, it is good to remember cognitive scientist Douglas Hofstadter's caution: "Sometimes it seems as though each step towards AI, rather than producing something which everyone agrees is real intelligence, merely reveals what real intelligence is *not*." (Hofstadter 1979, 573). That would be insight regained, and turn a 21st century Adamic fall, after biting the *apple* – the looming symbol – into an opportunity for a new conceptualization of what tragedy and tragic fate in an AI supported, generated, and/or defined environment may mean.

---

25 "Klimt: the Immersive Experience" digital art exhibition, Brussels: Horta Gallery, March 24-June 30, 2021. See https://www.expo-klimt.be/ Accessed: 23 March 2021.

Works cited

Bell, Alice, and Marie-Laure Ryan. *Possible Worlds Theory and Contemporary Narratology.* Lincoln, NE: U of Nebraska P, 2019.
Baudrillard, Jean. *The Ecstasy of Communication.* Los Angeles: Semiotext(e), 2012 (1988).
Bissonnette, Sylvie. "Cyborg Brain in Robert Lepage's *Possible Worlds.*" *Screen* 50.4 (Winter 2009): 392–410.
Bukatman, Scott, *The Poetics of Slumberland. Animated Spirits and Animating Spirit.* Berkeley: U of California P, 2012.
Chapman, Geoff. "Possible Worlds is Hopelessly Muddled." *Toronto Star* 11 Nov. 1990, C6.
Dautenhahn, Kirsten. "Facetime." *American Scientist* 91 (May 2003): 278.
Dervis, Paul. "Toronto Director with 'Espresso of the Mind.'" *The Ottawa Citizen* 6 Dec. 1999, B13.
Dixon, Steve. "Metal Performance. Humanizing Robots, Returning to Nature, and Camping About." *The Drama Review* 48.4 (184, Winter 2004), 15–32.
Dupriez, Bernard. *Dictionary of Literary Devices: Gradus, A-Z.* Toronto: U of Toronto P, 1991.
Genette, Gerard. *Narrative Discourse. An Essay in Method.* Ithaca: Cornell UP, 1980.
Gibson, Steve, and Dine Grigar. *When Ghosts Will Die.* Dallas: Project X, 2009. https://www.youtube.com/watch?v=mP6aHRsK05o Accessed: 08. Nov. 2021.
Hofstadter, Douglas R. *Gödel, Escher, Bach: an Eternal Golden Braid.* New York: Basic Books, 1979.
Klaver, Elizabeth. "Possible Worlds, Mathematics, and John Mighton's Possible Worlds." *Narrative* 14.1 (Jan. 2006), 45–63.
Kleber, Pia, and Tamara Trojanowska. "Performing the Digital and AI. In Conversation with Antje Budde and David Rokeby." *TDR* 63.4 (Winter 2019): 99–112.
Kleinpeter, Édouard. "Le Cobot, la cooperation entre l'utilisateur et la machine." *Multitudes* 2015.1 (58): 70–75.
Krämer, Sybille. "Zur Grammatik der Diagrammatik. Eine Annäherung an die Grundlagen des Diagrammgebrauches." *Zeitschrift für Literaturwissenschaft und Linguistik,* 44 (2014), 11–30.
Laurel, Brenda. *Computers as Theatre.* Online book. Upper Saddle River, N.J.: Addison Wesley, 2014. https://learning.oreilly.com/library/view/computers-as-theatre/9780133390889/ch03.html Accessed: 08. Nov. 2021.
Mighton, John. *Possible Worlds.* Toronto: Playwrights Canada Press, 1988.
Mighton, John. *Scientific Americans.* Toronto: Playwrights Canada Press, 1987.
Mueller, John Paul, and Luca Massaron. *Artificial Intelligence for Dummies.* Hoboken: Wiley & Sons, 2018.
Nearey, Brendan. *Robert Lepage's Stravinsky: Rhyming Imagery on Stage.* M.A. Thesis. Edmonton: University of Alberta, 2017.
Patera, Valeria. *Alan's Apple. Hacking the Turing Test.* Transl. Susie White. *Alan Turing: Life and Legacy of a Great Thinker.* Ed. Christof Teuscher. Berlin: Springer, 2004. 9–41.
Reilly, Kara. *Automata and Mimesis on the Stage of Theatre History.* London: Palgrave Macmillan, 2011, 148–76.
Richardson, Rufus. "A Science Theater." *Scientific American* 63.12 (20 Sept. 1890): 180.
Schaeffer, Jonathan. *One Jump Ahead. Computer Perfection at Checkers.* New York: Springer, 2009.

Scholte, Tom. "Black-Box Theatre. Second-Order Cybernetics and Naturalism in Rehearsal and Performance." *Constructive Foundations* 11.3 (2016): 598–610.

Shenfeld, Karen. "Possible worlds: John Mighton is a cutting-edge mathematician, a poet and a playwright." *Saturday Night* (Toronto) 115.28 (18 Nov. 2000): 54.

Sjuve, Eva. "Gestures, Interfaces and other Secrets of the Stage." *Transdisciplinary Digital Art. Sound, Vision and the New Screen*. Ed. Randy Adams, et al. Berlin: Springer, 2009. 301–12.

Soloski, Alexis. "'Would You Like to Have a Question?'" Int. with Annie Dorsen. *Theater* 42.2 (May 2012): 78–89. *EBSCOhost*, doi:10.1215/01610775-1507802. Accessed: 08. Nov. 2021.

Stelarc. "Artist Statement." http://stelarc.org/audiogram.php Accessed: 08. Nov. 2021.

Stephenson, Jenn. "Metatheatre and Authentication through Metonymic Compression in John Mighton's 'Possible Worlds.'" *Theatre Journal* 58.1 (March 2006): 73–93.

Switzky, Lawrence. "ELIZA Effects: Pygmalion and the Early Development of Artificial Intelligence." *Shaw. The Journal of Bernard Shaw Studies* 40.1 (2020): 50–68.

Vanhoutte, Kurt, and Nele Wynants. "Magie en wetenschap in de spektakelcultuur van de negentiende eeuw: Henri Robin in de Lage Landen." *Journal for Media History* (*TMG*) 20.2 (Dec. 2017): 30–53.
https://www.tmgonline.nl/articles/378/print/ Accessed: 08. Nov. 2021.

Wagner, Vit. "Science exciting in brain-powered plays." *Toronto Star*, Nov. 9, 1990, D3.

Walker, Susan. "Mighton's *Possible Worlds* visits the science of the soul." *Toronto Star*, Feb. 6, 1997, G11.

Wark, McKenzie. "Suck on This, Planet of Noise!" *Critical Issues in Electronic Media*. Ed. Simon Penny. New York: State University of New York Press, 1995. 7–25.

Wolfe, Graham. "The 'Sexuation' of Sublimation: A Žižekian Approach to John Mighton's *Possible Worlds*." *Modern Drama* 54.1 (Summer 2011): 194–215.

Johanna Pitetti-Heil
Artificial Intelligence from Science Fiction to Soul Machines: (Re-)Configuring Empathy between Bodies, Knowledge, and Power

Abstract: In science fiction, empathy has long been treated as an exclusive trait of human beings, and the ability to feel empathy allows many authors to establish social hierarchies between humans and non-humans who feel or show empathy in unexpected ways. Such treatments foster what Donna Haraway has called "human exceptionalism," which connects empathy to morality in an ableist manner. This article argues that Richard Powers's *Galatea 2.2* and Philip K. Dick's *Do Androids Dream of Electric Sheep?* counter this exceptionalist tradition by treating the electrical circuits of a neural network and the particular bodies of androids as forms of "vibrant matter" (Bennett) that develop their own senses of embodiment, loyalty, empathy, and efficacy. In a parallel development, the real-world company Soul Machines has been challenging preconceived conceptions of empathy in developing empathetic AI systems. Connecting these developments in AI research to science fiction discourses of empathic intelligence, the article (re)thinks empathy and (artificial) intelligence in relation to questions of morality, responsibility, and power, and asks what models of care and empathy are necessary if humanity is to develop in collaboration with artificial intelligence systems.

Introduction

Ian McEwan admits in his novel *Machines like Me* that "artificial humans were a cliché long before they arrived" (McEwan 2019, 1). The clichéd vision of artificial intelligence and humanoid robots stems from speculative representations of what artificial intelligence might look like and what it might be able to do. Yet, although science fiction usually creates spectacularly apocalyptic visions of AI instead of describing our prosaic and quotidian reality, artificial humans in science fiction are fascinating clichés that interrogate and challenge the notion of human exceptionalism, which Donna Haraway has defined as the "premise that humanity alone is not a spatial and temporal web of interspecies dependencies" (Haraway 2007, 11). Science fiction explores these interspecies dependencies, and one of its clichéd tropes is that, because artificial humans are differently embodied than humans and have a diverging corporeality from humans, they therefore lack empathy, morality, and a sense of responsibility towards biological humans, who are usually in the position of being their masters. The dependence of artificial humans on biological humans and the kind of servitude that the power relations demand have been famously formalized in Isaac Asimov's "three laws of robotics," which forbid robots to harm humans and requires them to help and protect humans at all costs.[1] One of the most famous and iconic examples of an AI breaching these laws is found in Stanley Kubrick's movie *2001: A Space Odyssey*, in which the AI HAL viciously overpowers the movie's astronauts. Given that in Kubrick's movie, HAL is named in reference to the IT company IBM (each letter of the name HAL precedes the letters of IBM in the alphabet), it is comic to think that the first "AI-companion" at the International Space Station is equipped with IBM's very real AI system Watson, which has been assisting the ISS scientists (IBM, "CIOMN").

But having the IBM system Watson at the ISS is also sobering and quite a relief in light of the many techno-pessimistic anxieties towards artificial intelligence, because such developments may spark imaginaries that are different than fearing a loss of power over one's creations. This fear is, after all, firmly based on human exceptionalism's insistent belief that humans naturally own all sorts of rights to power. This is not to say that developments in AI research and their implementations in the digital space do not have to be eyed critically. Quite to the contrary, I see the urgent need of the proliferation of critical digital humanities that engage in questions that range from the impact of speculative fiction on public anxieties to critical inquiries into algorithmic bias, social justice, and participation in the (post)digital age – these issues, however, have little to do with any kind of agency that an artificial intelligence system may develop completely independently from human intelligence. It rather points to the intricate entanglement of human and non-human forces that critical posthumanist thinkers like Karen Barad (*Meeting the Universe Halfway*) and new materialist theorists like Jane Bennett (*Vibrant Matter*) have been debating: agency and subjectivity only develop through the *intra-action* of entities (Barad)[2]; force and power is dis-

---

[1] Asimov's three laws are "1. A robot may nor injure a human being or, through inaction allow a human being to come to harm. 2. A robot must obey the orders given it by human beings except where such orders would conflict with the First Law. 3. A robot must protect its own existence as long as such protection does not conflict with the First or Second Law" (Asimov 1976, 61).
[2] Karen Barad defines intra-action as follows: "'intra-action' signifies the mutual constitution of entangled entities. That is, in contrast to the usual 'interaction,' which assumes that there are separate individual agencies that precede their interaction, the notion of intra-action recognizes

tributed across members (or actants) within an assemblage and "because each member-actant maintains an energetic pulse slightly 'off' from that of the assemblage, an assemblage is never a stolid block but an open-ended collective, a 'non-totalizing sum'" (Bennett 2010, 24).

To illustrate her proposition of distributed agency, Bennett uses the massive blackout that hit the United States and Canada in August 2003, which shut down "over one hundred power plants, including twenty-two nuclear reactors" (Bennett 2010, 25) before the cascade of interruptions ended. Bennett reports that "[i]nvestigators still do not understand why the cascade ever stopped itself" (Bennett 2010, 25). In her analysis and interpretation of the events that led to the blackout, Bennett identifies a set of assemblage-members that, in one way or another, played a role in causing the blackout, including "*electricity* [...]; the *power plants* [...]; *transmission wires* [...]; a *bush fire* in Ohio; Enron *FirstEnergy* and other energy-trading corporations [...]; consumers [...]; and the *Federal Energy Regulatory Commission*" (Bennett 2010, 26). While Bennett's discussion of electricity is fully treated in my own discussion of non-human agency in Richard Powers's *Galatea 2.2* below, the short detour at this point serves to establish a general point about artificial intelligence and the ways in which, in the popular imagination, it has been severed from human agency: the fear that artificial intelligence may overpower human beings removes artificial intelligence and its agency from the assemblage from within which AI acts and thereby shifts responsibility away from the assemblage (and thus from the human members of it).

Instead of examining the lack of humanness and humanity of artificial intelligence entities in (science) fiction and real-world artificial intelligence research, I here turn to the shared and distributed responsibility of care for the other that human/non-human assemblages require, and I place empathy at the center of my discussion: I will brush on empathy as a clichéd trope in speculative fiction that fosters public anxieties towards artificial intelligence; but I take this essay as an opportunity to reflect on the ways in which – in imagining and representing artificial intelligence – corporeality, the experience of the lived body, empathy, and morality are folded into one another and how empathy surfaces repeatedly not only in fiction but, lately, also in real-life AI development. I offer readings of two novels, Richard Powers's *Galatea 2.2* (1995) and Philip K. Dick's *Do Androids Dream of Electric Sheep* (1968), which interrogate corporeal aspects of knowledge and challenge human exceptionalism by highlighting

---

that distinct agencies do not precede, but rather emerge through, their intra-action" (Barad 2007, 33).

the shortcomings of human empathy. I then turn to an example of life that is stranger than fiction and present the New Zealand based artificial intelligence company *Soul Machines*, which is working on developing empathetic and virtually fully embodied artificial intelligence systems. Based on these three examples, I show that the idea that empathy distinguishes human from artificial intelligence has traveled from (science) fiction to real-world artificial intelligence research, but I also argue that this traveling of the trope of empathy has to be eyed carefully and critically: often introduced as an ableist and normative concept (e.g., in Dick's *Do Androids Dream*), empathy in science fiction does not only enable ethical actions; instead, it often unmasks itself as parochial and narrow-minded (cf. Decety and Cowell below), leaving little space for those who have been othered. My discussion of empathy and artificial intelligence thus contributes to the question what role empathy plays in relation to questions of morality, responsibility, and power that are implicated in our understanding of human and artificial intelligence.

Disability and embodiment of the thinking machine: Richard Powers's *Galatea 2.2*

Written and set in the 1990s, Powers's *Galatea 2.2* explores the propositional and tacit dimensions of human understanding and empathy in the context of artificial intelligence research and the 'culture wars' within literary studies and theory. The major plotline presents an author (named Richard Powers) and a cognitive neurologist (named Philip Lentz) who collaborate in programming and training a neural network (named Helen) with the goal that it will be able to take a Turing Test in the form of the Comps in English literature.[3] Several arguments between the diegetic author and neurologist rehearse previous debates of AI research and behaviorist psychology on how to interpret the Turing Test. The weak interpretation of the Turing Test states that an AI can successfully *imitate* human thinking. Imitation, however, does not correspond to active mental processes (as argued in Searle's Chinese Room argument in "Twenty-One Years"), and the question of what counts as artificial intelligence – imitation or genuine processes – lies at the heart of *Galatea 2.2*. The strong interpretation of the Turing Test, in contrast to the weak interpretation, equates the behavior of the AI with mental processes and thereby assumes cognitive independence (Searle 55–56;

---

[3] For a detailed discussion of autofictional aspects in *Galatea 2.2*, see Kucharzewski 2008 and Heil 2016, 136–46.

cf. Heil 2016, 152). The modified strong interpretation argues that "[t]here is a difference between the inner intelligent thought process and the outer intelligent behavior, but [that] the Turing Test can still give us conclusive proof of the presence of the inner thought processes, once we understand their nature" (Searle 2008, 56). Alan Turing himself never suggested that his test could be interpreted in the strong or modified strong way; to him, the question of artificial intelligence was never whether a machine can think but whether its performance can make a human being believe that the machine is actually a human being (Turing 1950, 433).[4] Turing's question was thus never an ontological one of Being but a performative one of passing.

In *Galatea 2.2*, a major point of disagreement connected to the question on how to interpret the Turing Test develops between Richard and Lentz, and the point connects an epistemological to a practical question: would their thinking machine (named Helen) have to be equipped with a body, that is, would the imitation of human intelligence have to be grounded on an imitation of embodied and corporeal understanding of lived experiences.[5] The novel's neurologist sides with Turing, who suggests in his famous essay "Computing Machine and Intelligence" (1950) that a thinking machine "must be given some tuition" but that AI developers "need not be too concerned about the legs, eyes, etc." because "[t]he example of Miss Helen Keller shows that education can take place provided that communication in both directions between teacher and pupil can take place by some means or other" (Turing 1950, 456).[6] As a result, and in contrast to most science fiction writers who house their AIs in uncannily humanoid robotic bodies, Powers's alternatively embodied neural network is distributed across several servers. Responding to Turing, Powers has his neural network (and Keller's namesake) Helen despair over her non-human body and her inability to relate to the world corporeally: having to relate her own circuit processes to the experience of the human senses and to the embodied processes of the human mind, Helen understands that being pure computation and electricity, she cannot comprehend the embodied realities of human beings that come with conflicting fears, desires, and ethical standards and violations (cf. Heil 2016, 162).

Reading Helen exclusively as lacking because she does not possess a human body with its lived experience, however, is problematic because such a reading

---
4 For a more detailed discussion of the interpretations of the Turing test, see Heil 2016, 152–54.
5 For an investigation of the entanglement of knowledge, power, and gender in *Galatea 2.2*, see Heil 2016, 172–94.
6 For a discussion of Helen's lack of senses, especially of seeing, see Powers 1995, 295, 326 and Heil 2016, 162. I discuss Helen Keller's use of touch in experiencing the world at length in *Walking* (2016, 162–63).

would follow an ableist discourse, which normalizes one way of perceiving the world. Evoking Helen Keller as a namesake for Helen (the machine) further problematizes the configuration of Helen (the machine) as a disabled character because Helen Keller herself has been used as a figurehead in "self-help literature [that] has been written to explain how to 'endure' or 'triumph over' such adversity" as physical disability (Davidson 2008, xvii). "This ideology of ableism," Davidson contends, "works in part to shore up a fragile sense of embodiment" and "to erase the work of those who have lived with a disability all their lives or who have struggled for changes in public policy and social attitude" (Davidson 2008, xvii). In short, assuming that *Galatea 2.2*'s Helen would simply rise to the exceptional position of her namesake and succeed in her task despite her specific embodiment leaves Helen isolated within a structure that, first, was not designed for her and, second, denies her recognition. Recognition of her particular embodiment, history, and situatedness, however, is key to a task that involves assessing aesthetic production. As Michael Davidson reminds his readers, the aesthetic has "from Baumgarten and Kant to recent performance art" been defined as a "matter of the body" (Davidson 2008, 3). His fellow disability scholar Tobin Siebers "notes that the term aesthetics is based on the Greek word for perception and that '[there] is no perception in the absence of the body'" (qtd. in Davidson 2008, 3). Disability thus brings to the fore "the spectral body of the other [...], reminding us of the contingent, interdependent nature of bodies and their situated relationship to physical ideals" Davidson concludes (Davidson 2008, 4). In *Galatea 2.2*, this "spectral body of the other" emerges as an immaterial presence that eventually forces the human characters to acknowledge their own blindness to the disabling structures they themselves have created for the machine.

Fathoming Helen's electrical and distributed body via disability studies opens up a way of thinking through "nontraditional bodies and sensoriums" (Davidson 2008, 8) that can be found in Powers's novel. After all, Helen is only one of several disabled characters: Lentz's wife Audrey suffers from some unspecified form of dementia and Lentz presents her as a body without mind (cf. Heil 2016, 180); Richard and Lentz's colleague's son has trisomy 21; and Richard himself faces writer's block, which disables the progress of the novel he is supposed to be writing – this disrupts his stable sense of self, which he will only be able to recover through narrating his and Helen's story. Given that Helen finally enables Richard to write the book that will be *Galatea 2.2*, Helen acts as *narrative prosthesis*, "the use of disability to enable a story" (Davidson 2008, 59), enabling the diegetic writer Richard to write a story.[7] But for *Galatea 2.2* as a

---

7 This reading is based on David T. Mitchell and Sharon L. Snyder's analysis of narrative pros-

novel that presents Helen as a narrative prosthesis to Richard, Helen's disabled body and the way in which her liberal arts education does not accommodate her specific embodiment emphasize in a self-ironic manner the ways in which narrative prosthesis works to radically question intelligence as an exclusively human trait. In the Turing test that Helen takes, she and the graduate student A. are to interpret a passage from Shakespeare's *The Tempest*. The Turing test and its results build up the climax of *Galatea 2.2* by interweaving the novel's commentary on literary theory and scholarship in the 1980s and early 1990s with its commentary on mindedness and literary representation of cognition, intellectual and physical dis/ability, and human and artificial intelligence (cf. Heil 2016, 164–70): A. writes a "more or less brilliant New Historicist reading" that "dismissed, definitely, any promise of transcendence" (Powers 1995, 326), on which Sharon Snyder comments that "[l]iterary disciplines have successfully programmed [A.] to discern paradigms within specific examples of literature" (Snyder 1998, 88), curtailing any deeper understanding. Helen, in contrast, writes a self-reflexive and lyrical comment on her situatedness (Powers 1995, 326), which definitely suggests transcendence in transporting a form of truthfulness about Helen's understanding of herself, humanity, and intelligence.

A.'s reading illustrates what Rita Felski has diagnosed as one of critique's key characteristics: "an essentially disembodied intellectual exercise, an austere, even abstemious practice of unsettling, unmaking, and undermining," which is distinguished from postcritical "styles of critical argument [which] are affective as well as analytical, conjuring up distinctive dispositions and relations to their object" (Felski 2012, n.pag.). Ending the novel on a note of affective and embodied understanding, *Galatea 2.2* not only emphasizes the phenomenological dimension of the lived body as central to human understanding, the novel demonstrates in an ironic fashion that affective understanding, as affect theories have proposed, questions "traditional notion of empathy, sympathy, and shared or universal emotions" (Anker / Felski 2017, 10). Although *Galatea 2.2* establishes empathy as a form of tacit understanding that distinguishes human from nonhumans, the novel lets the thinking machine despair over the lack of empathy and the intensity of physical violence and political cruelty of humans that she learns about in reading the news – she turns herself off for the first time (Powers 1995, 313) only to come back and realize that mastery of language, which is al-

---

thesis in literary production. Their "*effort is to make the prosthesis show, to flaunt its imperfect supplementation as an illusion*" (Mitchell / Snyder 2000, 8). Thereby, the "prosthetic relation of body to word is exposed as an artificial contrivance" (Mitchell / Snyder 2000, 8). If one reads Helen as narrative prosthesis, her function is to emphasize the ways in which language, and especially metaphor, is built around lived experience (cf. Heil 2016, 185–87).

most all that she is, is not enough for understanding literature affectively. She has learned that being human means to be embodied and corporeal, and that her human companions are not able to imagine her otherwise. As a consequence, she refuses to be part of a world that cannot accommodate her as nothing but language and electricity; she shuts herself down for good, claiming efficacy if not agency over her existence.

Claiming agency for Helen's actions would place *Galatea 2.2* within the genre of (fantastic) science fiction, in which the impossible or improbable can happen, in which robots and artificial intelligence systems act as if they were human in the sense that they develop free will, intentionality, and moral judgment, all of which have been tied in the humanist philosophical tradition to the concept of agency. But agency, as Bennett emphasizes, is never only linked to morality but is also "divided against itself" (Bennett 2010, 28). Efficacy, in contrast to agency, is not tied to conflict of an agent's will and their intentions. Agency "involves not mere motion, but willed or intended motion, where motion can only be willed or intended by a subject" (Matthews qtd. in Bennett 2010, 31). Bennett's "theory of distributive agency, in contrast, does not posit a subject as the root cause of an effect" (Bennett 2010, 31). Instead of a subject, she places "a swarm of vitalities" as the "generative source of effects" that "vibrates and merges with other currents, to affect and be affected" (Bennett 2010, 32). In short, Bennett proposes that such an understanding of agency "loosens the connections between efficacy and the moral subject, bringing efficacy closer to the idea of the power to make a difference that calls for response" (Bennett 2010, 32). She consciously decides to make such strong posthumanist claims about agency in distributing it across the members of an assemblage because she intends "material agency" to pose a "stronger counter to human exceptionalism" than other less radical formulations would have allowed her to develop (Bennett 2010, 34). Her example of the 2003 North American blackout referenced in the introduction of this article is thus based on the peculiar trait of electricity – which Bennett calls "the stream of vital materialities called electrons" – in that it "is always on the move, always going somewhere, though where this will be is not entirely predictable. Electricity sometimes goes where we send it, and sometimes it chooses its path on the spot, in response to the other bodies it encounters" (Bennett 2010, 28). Understanding electricity as such a nonhuman but vital force is helpful in reading *Galatea 2.2* and Helen's empowerment not strictly or exclusively in terms of science fiction. Just as Helen's particular body is distributed over several servers of the campus at U., it is generated by electrons whose movements can deviate from their prescribed routes and create unforeseeable paths.[8] Seen in this light, Helen's efficacy and artificial intelligence is built within her electrical embodiment.

Eugenics and artificial intelligence: Philip K. Dick's: *Do Androids Dream of Electric Sheep?*

Philip K. Dick's *Do Androids Dream* revolves around the affective response of empathy that the diegetic world assumes to be solely a human capacity. It is set in a future San Francisco in a post-nuclear war environment, where people and animals suffer from nuclear fallout, which slowly deteriorates the physical and mental health of all organic life. For their mental well-being, people rely on "mood organs," and many follow the religious leader Wilbur Mercer through "empathy boxes," a form of computer assisted virtual reality through which people can join from home to collectively suffer with Mercer, who climbs a steep and waste hill while rocks are being thrown at him. Mercerism preaches that biological humans have to value and respect all organic life, and accordingly, people strive to have pets and keep them alive – organic animals have become a valuable rarity, however. The ones who cannot afford real animals buy artificial pets that they tend to instead.

The people who were tested to be healthy enough and had the financial means have emigrated to colonies on Mars, where each human receives a humanoid android to keep them company and help with everyday tasks. The humanoid androids do not only look exactly like humans, they actually are "not made out of transistorized circuits like false animals"; they are in fact "organic entit[ies]" who are not legally but "biologically" alive (Dick 2007, 171) – only an extremely expensive bone marrow test can fully distinguish them from organic people. Androids have a limited life span of four years, they lack empathy – or so goes the politically imposed narrative – , they are made to serve, and law prohibits them from harming human beings. Time and again, however,

---

[8] A similar conception to Bennett's new materialist philosophy has been formulated by Manuel DeLanda in regard to Deleuze's "abstract machines": the "conception of very specific abstract machines governing a variety of structure-generating processes not only blurs the distinction between the natural and the artificial, but also that between the living and the inert. It indeed points towards a new form of materialist philosophy in which raw matter-energy through a variety of self-organizing processes and an intense power of morphogenesis, generates all the structures that surround us. Furthermore, the structures generated cease to be the primary reality, and matter-energy flows now acquire this special status" (DeLanda 2021, n.pag.; cf. van der Tuin / Dolphijn 2010, 154–55).

they "declar[e] [their] right to live as an autonomous self, challeng[ing] the very categories of life and selfhood and, in turn, the ontological prerogative of [their] creators" (Galvan 1997, 413): they kill their owners and escape to earth in order to live a life according to their own wishes. When this happens, they are hunted by bounty hunters, who administer empathy tests, in which suspects have to respond directly and without delay regarding how they feel when they hear of the mistreatment of animals and humans. If the suspects fail to showcase verbal empathetic responses that match the dilation of their pupils, they are retired / killed as the punishment for having taken a human life.

As in all of Dick's writing, the paranoid weirdness of *Do Androids Dream* very obviously undermines the political set-up of its diegetic world. Most humans actually show very little empathy towards each other when they are in direct contact – they prefer digitally mediated empathetic catharsis via their empathy boxes; although androids cannot be distinguished from humans unless specialized tests have been performed, humans are not expected to feel empathy toward androids; and the political system segregates those whose intellectual and empathic capabilities have deteriorated due to the nuclear fallout. From a perspective of disability studies, human and artificial intelligence within the diegetic reality of *Do Androids Dream* is conceived within a peculiar and radical humanist tradition, which values intellectual ability and success above all else, and which produced the eugenicist movement – Adam Pottle thus reads Dick's novel as a satirical critique of American eugenics. At the center of Pottle's reading stands John Isidore, the "special" and "chickenhead" character, who has been segregated and mistreated in ways similar to the victims of American eugenics. His reading points, for instance, to the irony that lets "regulars" within *Do Androids Dream* value animal life almost above all else, yet use the term "chickenhead" to denounce those with a lesser IQ than is deemed regular. The very idea of testing empathy in *Do Androids Dream*, Pottle argues, "can be read as a parody of eugenics tests" (Pottle 2013, n.pag.). His discussion of disability in *Do Androids Dream* leads Pottle to argue that "[b]y placing empathy above the cogito, Dick criticizes the very foundation of the American eugenics movement" because "[e]ugenicists promoted the power of intellect above all other things" (Pottle 2013, n.pag.). This is, of course, true, and Pottle makes excellent points in linking Dick's novel to the practices of segregation and the disenfranchisement of people categorized as "abnormal" in American eugenics, but I want to further complicate the notion of empathy and (artificial) intelligence as it is presented in *Do Androids Dream*. After all, the diegetic government in Dick's novel substitutes IQ-tests by empathy tests, thereby using empathy as the measurement that separates "regulars" from "specials," allowing the "regulars" to emigrate to Mars and effectually sentencing the "specials" to degenerate from the nuclear fallout (cf. Pottle 2013).[9] Michael Bérubé reminds his readers that the Voigt-Kampff empathy test was "developed after World War Terminus to identify 'specials,' people neurologically damaged by radioactive fallout, so that the state could prevent them from reproducing" (Bérubé 2005, 598). *Do Androids Dream* thus not only criticizes the eugenicists treatment of pure intelligence (Pottle identifies pure intelligence as "posthuman" in the sense that it can only be performed by the artificial intelligence systems of androids); rather, Dick's novel satirizes any kind of testing methods that would rank humans in hierarchical subcategories and that would differentiate between values of species. Mercerism may act as an example in regard to valuing all life forms – whether human or animal – and to train people in advancing their empathic responses[10]; but the novel at large also asks its readers to rethink empathy and emotional and intellectual intelligence themselves.

Repeatedly, the irony of Dick's novel comes full circle and emphasizes the ridiculousness of the government's eugenicist project: the spiritual leader Mercer is revealed to be an alcoholic actor and Mercerism is unmasked as an ideological exercise promoted by the government; the androids who fled to earth include an opera singer who loves the paintings of Edvard Munch and would rather be human (Dick 2007, 113–16), and others experiment with psycho-active drugs because they seek empathetic experiences similar to Mercerism (Dick 2007, 160); one of the legally documented androids seduces the bounty hunter Deckart in order to make it impossible for him to kill her fellow android who looks exactly like her (Dick 2007, 169–72) – she does not succeed and, as revenge, she kills Deckart's real goat (Dick 2007, 198, 202): in short, the world of *Do Androids Dream* is populated by human beings who are often unwilling to understand the experiences and feelings of others, or at least to do so directly, and with androids who do understand other people's emotional states even though they may not experience them: they show empathy.

In a purely representational sense, the androids thus exceed or transcend what they are not programmed to exceed or transcend: they have a functional theory of mind, they care for each other in a world that is hostile to their exis-

---

9 The government motto in *Do Androids Dream* is "Emigrate or degenerate" (Dick 2007, 6).
10 In a conversation between Isidore and one of his neighbors, Isidore, a true believer in Mercerism, exclaims that "an empathy box [...] is the most personal possession you have! It's an extension of your body; it's the way you touch other humans, it's the way you stop being alone. [...] Mercer even lets people like me – " (participate although he has been categorized as a special) (Dick 2007, 57). For his neighbor, this presents a "major objection to Mercerism" (Dick 2007, 57). Towards the end of the novel, Isidore warns Deckart not to kill the androids because, if he does, he "won't be able to fuse with Mercer again" (Dick 2007, 191).

tence, and they understand that during the empathy test they have to find ways to "destabilize language in such a way that throws into question [...] previously unexamined structures of power" that are medialized through language (Galvan 1997, 421). Empathy is presented not "as the basis of compassionate behavior" but "as a cultural tool for reinforcing existing structures of power" (Jurecic 2011, 11) and thereby "obscures [what Sara Ahmed calls] 'the cultural politics of emotion'" (qtd. in Jurecic 2011, 11): the androids lack empathy because the humans have created them as such, but they do to some extent care for one another and they are intellectually aware (and some perhaps emotionally sympathetic) to the needs, desires, and pains of human beings. *Do Androids Dream* thus develops a model of artificial intelligence similar to the one presented in *Galatea 2.2* in that it both reinforces and challenges human exceptionalism: the novels reinforce human exceptionalism because they cannot yet imagine a world in which posthuman non-humans could participate in and be acknowledged as sentient and/or as possessing efficacy. But both texts also question what it means to be human because they introduce artificial intelligence systems that, at least as literary characters, seem convincingly all too human in their *yearning* for human corporeality und understanding, which, following Donna Haraway, presents "the body as a social actant," and embodiment, which "offers a vehicle of social agency" (Gilleard and Higgs 2015, 17) in order to fully participate in the social realm.[11]

Programming artificial empathy: The AI company: *Soul Machines*

As if the real-world artificial intelligence developers at the company *Soul Machines* responded to both *Galatea 2.2* and *Do Androids Dream*, their dedicated goal is to develop extremely complex and fully synthesized chat-bots with human facial features that respond to extra-linguistic cues – such as the facial

---

[11] Gilleard distinguishes between body, corporeality, and embodiment in the following way: "The body as social actant refers to the relatively unmediated materiality of the body and its material actions and reactions that are socially realized without recourse to concepts of agency or intent. The body as a social agent, by contrast, refers to its materiality being an inseparable element in the expression of personal and social identity. 'Corporeality' is a term that can be used to signify the body as social actant, while 'embodiment' is a term that signifies the body as a vehicle of social agency. Embodiment encompasses all those actions performed by the body or on the body which are inextricably oriented towards the social. It is subject to and made salient by the actions and interpretations of self and others" (Gilleard and Higgs 2015, 17)

expression of their human interlocutors and hesitation – and to incorporate these cues into their learning loops. The goal is to make people more open to and welcoming of "artificial humans" that act as customer service chat-bots, virtual advisors, tutors, and teachers. The artificial intelligence developers at *Soul Machines* take a path towards artificial intelligence that is radically different from older purely computational models (as represented in *Galatea 2.2*). In addition to a virtual nervous system, the interfaces of *Soul Machines*' artificial humans are human faces and bodies equipped with virtual muscles that can then be controlled by the AIs, with virtual lungs and hearts so that they can mimic the natural rhythm of a conversation and respond to emotional states of mind such as distress and relief. *Soul Machines* has thus given a lot of attention to modeling "authentically" human physical, embodied, and even corporeal experiences.

In a way, the avatars built by *Soul Machines* and powered by IBM's Watson technology are reminiscent of the AI model of the virtual human proposed in *Do Androids Dream* in all but one detail: the company is developing *empathetic* AIs (cf. IBM, "Soul Machines"). Their AIs are based on neural networks that provide a "virtual nervous system [...] based on biological models of the human brain" in which "different parts react to different stimuli" such as to virtual adrenaline when scared (e. g., by an unexpected noise) or to dopamine when happy (e. g., when someone smiles at them) (Cross 2017), and they are "fully emotionally responsive" (Cross 2017). In a promotional video, the soul machine model Rachel explains that her virtual "nervous system is an interconnected model of aspects of the human brain. It combines different neural systems to enable [her] to behave in a way that's inspired by biology" (Soul Machines). This means that their AIs are not learning machines targeted at pattern recognition but cognitive systems, which can be taught and which "can think and do actions by [them]selves" (Sagar / Cross 2018). Soul Machines works at "cooperative interaction with machines," and it is their goal to develop virtual humans that are "more like us" (Sagar 2014). In order to achieve this, they look "holistically at intelligence," they include "emotion as a part of [it]" (Sagar 2014). Given that the company's goal is to create empathetic AIs, this detailed attention to the physical, embodied, and corporeal dimensions of human experience points to the company's recognition of the importance of emotion and of an aesthetic and kinesthetic understanding of empathy.

At the heart of Soul Machines' research is Baby X, a completely autonomous virtual baby in which neural networks and motor abilities emerge and evolve as she learns new content and as her neural system gains experiences. For instance, in a demonstration lecture, the developer Mark Sagar radically increases the baby's dopamine to a level that compares to taking a lot of cocaine: the baby's pupils dilate, she feels distress, and almost starts to cry (he decreases the dopamine just early enough and calms her down, telling her in a comforting voice that "it's okay, baby, it's okay"). The Soul Machine AIs thus exhibit synthesized and responsive behavior in real time; they produce their own content; they have proprioceptive understanding of their virtual bodies and, according to one of the programmers, exhibit creativity and "tiny version[s] of free will" (Sagar / Cross 2018); Soul Machines has additionally been working on implementing virtual imagination (Sagar / Cross 2018). Soul Machines is thus not only investigating "how low-level biology is connected to high level social interactions" (Sagar 2014) in virtual humans, they are also changing "the ways in which humans interact with systems" (qtd. in Cross 2017); they seek to "make men socialize with machines" (qtd. in Cross 2017): the development of AI becomes a shared, contextualized, embodied, three-dimensional, and social endeavor.

Empathy and the machine, or: Caring is creepy

Between Soul Machines and the two novels I have discussed, there is an interesting space of tension between artificial intelligence, bodies, and power that opens up – and empathy is central to the discursive distinction between human and artificial intelligence. Although both Powers's and Dick's novels interrogate the scope and significance of human empathy, science fiction has held on to empathy as a moral feature that distinguishes humans from machines. The work of Soul Machines, especially Baby X, challenges such a conception in real life, which makes it necessary to interrogate empathy (and sympathy) itself.

Philosophers like Jesse Prinz and neuropsychologists like Jean Decety have critically examined the very notion of empathy as a moral and natural asset of human interaction. Prinz's discussion of empathy leads back to David Hume, who considered sympathy to be the basis of all moral judgment (cf. Prinz 2011a, 2011b, 2017) and innate to healthy human beings – a normative and ableist assumption that lies at the heart of the social order of the government in *Do Androids Dream*. But tracing the history of sympathy and empathy in philosophical and socio-political discourses (the term empathy only became fashionable in the nineteenth century), Prinz defines empathy as a "vicarious emotion that one person experiences when reflecting on the emotion of another" (Prinz 2011a, 214). Following Hume, Prinz defines empathy as "the experience of another person's emotional state, whatever that emotion might be" (Prinz 2011a, 215). But accepting empathy as "feeling an emotion that we take another person to have" (Prinz 2011a, 215), Prinz argues, implies that the discourse on sympathy up to the nineteenth century was characterized by the partiality and directionality of sympathy: it benefits the sympathizers who can construct themselves as benevolent and caring. But because sympathy was yoked to capitalism and exploited by the empires of Europe, it was used as a pretext for imperialism and hierachization along the lines of religion, class, gender, and race (Prinz 2017). In short, sympathy imagines closeness but does not implement and enact the actual feeling of being close and/or connected.

Recent neuropsychological critiques of empathy follow similar lines. Caring releases dopamine: we feel good when we care, and, on the level of neurobiology, there is at least some overlap between oneself and the object of care because the neural networks that react to our own stress are also activated when we react empathetically to the distress of others (Decety 2017, SLSAeu). But caring happens in the brainstem, as Decety explains, far away from the cortex, which is responsible for rational thinking (Decety 2017, SLSAeu). This means that, without the addition of intellectual reflection, empathy is lacking as a strategic and political tool: it is narrow-minded (parochial) in that one cares for those who are near and dear, it focuses on individuals – preferably on good-looking ones – and not on groups, and one's capacity to experience empathy is limited – we can be "cared-out" (Decety 2017, SLSAeu). Thus, neuropsychologists like Decety and philosophers like Prinz agree that empathy and morality are separate abilities deriving from distinct neurobiological causes. Decety and Jason M. Cowell, for instance, concede that "[e]mpathy does play an important function in motivating caring for others and in guiding moral judgment in various forms," but their data suggests that who one feels empathy for is conditioned by "the social identity of the targets, interpersonal relationships and social contexts" (Decety / Cowell 2015, 10). In order to feel empathy, one needs to recognize the targets as "identifiable others" (Decety / Cowell 2015, 9). This means that empathy is determined by "parochial tendencies" and if empathy was to be utilized politically and strategically as a practice of care, it would "need to be rationally regulated and guided" (Decety / Cowell 2015, 10).[12]

Prinz comes to a similar conclusion in his analysis of the use of empathy within the history of European thought. Countering the Humean idea that empathy is "a precondition for moral approbation and disapprobation," he argues

---

[12] In my article "Dancing Contact Improvisation with Luce Irigaray," I read the dance practice of contact improvisation as a "practice of moving-together, feeling-with, and feeling-between" (Heil 2019, 485) that fosters non-verbal communication and trains empathy. Decety and Cowell's findings as well as Susan Leigh Foster's discussion of aesthetic empathy (see below) also inform my treatment of empathy in dance improvisation (Heil 2019, 498, 501).

that the "acquisition of moral competence may not depend on a robust capacity for empathy" (Prinz 2011a, 222), that we "cannot rely on empathy as an epistemic guide" (Prinz 2011a, 224).[13] While ethical and socio-political critiques of empathy suggest that caring is creepy and needs to be directed by reason to create social justice, what Soul Machines focuses on is an aesthetic and kinesthetic approach to empathy akin to the one first developed in the nineteenth century. Robert Vischer, for instance, identified a "dynamic vitality of objects" (Foster 2011, 154) and he posited that this vitality "aroused and affected" (Foster 2011, 155) the spectator's mind and body. Vischer thus "envisioned empathy as an experience undertaken by one's entire subjectivity" (Foster 2011, 127) – that is, looking at a painting or sculpture, getting to know its structure, dimensions, and "vibrant matter," as Jane Bennett would call it, elicits empathy for the work of art. Theodor Lipps then proposed that "our empathic encounter with external objects trigger inner 'processes' that give rise to experiences similar to ones that I have when I engage in various activities involving the movement of my body. Since [one's] attention is perceptually focused on the external object, [one] experience[s] them – or [one] automatically project[s] one's experiences – as being in the object" (Stueber 2019, n.pag.).

In this light of aesthetic empathy, one may wonder: why might artificial intelligence systems not exhibit a form of vitality similar to that of paintings or sculptures? From a critical posthumanist and new materialist perspective it makes sense to distribute agency across assemblages and thereby grant non-human actants "efficacy," as my reading of Helen in *Galatea 2.2* vis à vis Bennett's *Vibrant Matter* above suggests. Without advancing a flat object-oriented-ontology, which assumes that the ontological status of humans and non-humans is the same,[14] such a reading would mean to allow the notion that non-human

---

[13] Catriona Mackenzie and Jackie Leach Scully add from the perspective of disability studies that "imagining oneself differently situated, or even imagining oneself in the other's shoes, is not morally engaging with the other; rather, it is projecting one's own perspective onto the other. When the other person is very different from ourselves, the danger of this kind of projection is that we imply to project onto the other our own beliefs and attitudes, fears and hopes, and desires and aversions" (Mackenzie / Scully 2007, 345); see also Heil 2019, 498.

[14] Object-oriented ontology (OOO; also referred to as speculative realism) was developed in the late 2000s by scholars such as Graham Harman and Ian Bogost in an attempt to undo the differences and hierarchies that distinguish human and non-human entities ontologically. Bogost summarizes his "tiny ontology" (Bogost 2012, 22) as follows: "OOO puts *things* at the center of being. We humans are elements, but not the sole elements, of philosophical interest. OOO contends that nothing has special status, but that everything exists equally – plumbers, cotton, bonobos, DVD players, and sandstone, for example" (Bogost 2012, 6). Putting inanimate objects at the center of his inquiry, Bogost, for instance, criticizes the Turing test because it centers on

actants "can do things, ha[ve] sufficient coherence to make a difference, produce effects, alter the course of events" (Bennett 2010, viii). Thus, it points to difference as a productive marker of "experience" (whether it is lived experience or collected data) that influences the ways in which information is processed and empathy may be enacted.

Between empathy and efficacy: Closing remarks

Considering artificial intelligence via the faculty to develop empathy produces two diverging readings. The first one is a rather sympathetic reading that argues that Powers's and Dick's novels as well as Soul Machines push back against the tradition of human exceptionalism that connects empathy to morality. Instead of following such a path, all three examples ask us to acknowledge the vibrant matter of neural networks and virtual bodies as systems of productive differences with unpredictable efficacy, which will inevitably produce new interspecies dependencies and collaborations.[15] As if Soul Machines had learned from speculative fiction, its artificial intelligence designers have been adding dimensions of tacit, physical, and corporeal experience and understanding to the electrical circuits of their AIs. This tacit dimension leaves room for the artificial humans to experience and relate to the world in multiple ways – both Helen in *Galatea 2.2.* and the androids in *Do Androids Dream* yearn for such possibilities, risk their lives in order to obtain it, or fail at passing as human. The path that Soul Machines has taken in creating models for artificial humans may assist in finding ways with which we can rethink what it means to be empathetic by

---

comparing artificial intelligence with human intelligence instead of concentrating on artificial intelligence as is: "The construction and behavior of a computer system might interest engineers who wish to optimize or improve it, but rarely for the sake of *understanding* the machine itself [...]. Like everything, the computer possesses its own unique existence worthy of reflection and awe, and it's indeed capable of more than the purpose for which we animate it" (Bogost 2012, 17). Although OOO and critical posthumanist theory both critique humanist privileging of the human position, they represent radically different philosophical and political positions. From the perspective of (feminist) posthumanist critique and (feminist) new materialism, OOO overlooks and ignores the historical dimension of power relations that have subjugated women*, people of color, people with disabilities, and actants in the natural world in their onto-epistemological situatedness. For a more detailed summary of the differences between OOO and feminist new materialism, see Sheldon 2015.
15 Chat bot poetry may be one creative collaboration between human and artificial intelligence; see Regina Schober's essay in this collection.

rethinking the position of power from which one assumes empathy. In this sense, Soul Machines's work is conceptually posthumanist because the company conceptualizes empathy as performative rather than representational.[16] Karen Barad, for instance, asks her readers to conceptualize meaning (in the context of this essay, the meaning of AI and empathy) neither as "linguistically conferred" (i.e., within the system of language) nor as "extralinguistically referenced" (i.e., representational) but as an *emergent* process that has been enabled through the use of language (Barad 2003, 818). What Barad's model allows me to think is a model of artificial intelligence that is not created solely through linguistic programming but that instead takes into account the performance and performativity of non-organic matter – an approach that shares similarities with Bennett's, who proposes efficacy for the moving stream electrons of electricity. The question that is at stake for Soul Machines is whether we, as humans, will be generous enough to empathetically connect with a machine to help it learn and to acknowledge the particular form of empathy and efficacy that virtual humans will develop – this would be a posthumanist and performative understanding of human and artificial intelligence and empathy – whether this performance is considered as authentic (or strong) or as imitational.

While I find such a sympathetic reading conceptually both challenging and rewarding in its proposition to think non-human efficacy otherwise, I continue to be baffled to see that the concept of empathy has survived the transition from spectacular (science) fiction into mundane AI research, which suggests that empathy remains a key concept (and an ableist one at that) in the ways that at least some people think about human and artificial intelligence. Identifying empathy as a key component in the endeavor to have artificial intelligence develop is, after all, not in and of itself helpful because it is such a contested affect through which scholars have questioned human behavior, morality, subjectivity, and power relations in the first place. This does not mean that I am in favor of throwing empathy out, so to speak, and to invalidate the experience of it or its function within societies and cultures. However, in light of philosophical and neuropsychological findings, empathy cannot be a conceived of as a purely positive or neutral characteristic that fosters understanding for others and the Other. Quite to the contrary, the history of empathy requires us to tend to a different set of questions: where, in human and artificial intelligence, lies efficacy, agency,

---

16 One central feature of Karen Barad's posthumanist theory is the critique of representationalism, a posthumanist form of AI and empathy would have to counter the "Cartesian habit of mind" of representationalism (Barad 2003, 807). Barad proposes a "performative understanding [of ontology], which shifts the focus from linguistic representations to discursive practices" (Barad 2003, 814).

and power, and in what ways does empathy enable *and* prohibit care? And what training does it require for both human and artificial intelligence to become aware of the ways in which empathy is non-consciously directed? In respect to Soul Machines, I have to admit that I am more than curious to learn how Baby X will develop and to see what kind of empathetic and creative actions it may learn to perform, and whether it will develop efficacy that will translate into affective and effective forms of resistance.

Works cited

Anker, Elizabeth S., and Rita Felski. Introduction. *Critique and Postcritique*. Durham: Duke UP, 2017. 1–28.
Asimov, Isaac. "That Though Art Mindful of Him." *The Bicentennial Man and Other Stories*. Garden City: Doubleday, 1976. 61–86.
Bérubé, Michael. "Disability, and Narrative." *PMLA* 102.2 (March 2005): 568–76.
Barad, Karen. "Posthumanist Performativity: Toward an Understanding of how Matter Comes to Matter." *Signs: Journal of Women in Culture and Society* 28.3 (2003): 801–31.
Barad, Karen. *Meeting the Universe Halfway: Quantum Physics and the Entanglement of Matter and Meaning*. Durham: Duke UP, 2007.
Bennett, Jane. *Vibrant Matter: A Political Ecology of Things*. Durham: Duke UP, 2010.
Bogost, Ian. *Alien Phenomenology, or, What It's Like to be a Thing*. Minneapolis: U of Minnesota P, 2012.
Cross, Greg. "When Virtual Reality Meets the Real World." Taiwan Business Weekly: Amazing Nights 2017. *You Tube*. Soul Machines, 29 Nov. 2017. https://www.youtube.com/watch?v=1DzVaZSbuKw. Accessed: 6 Jan. 2021.
Davidson, Michael. *Concerto for the Left Hand: Disability and the Defamiliar Body*. Ann Arbor: U of Michigan P, 2008.
DeLanda, Manuel. "The Geology of Morals: A Neo-Materialist Interpretation." http://www.t0.or.at/delanda/geology.htm. Accessed: 11 Jan. 2021.
Decety, Jean. SLSAeu Annual Conference 2017: Empathies. University of Basel, 22 June 2017. Keynote Lecture.
Decety, Jean, and Jason M. Cowell. "Empathy, Justice, and Moral Behavior." *AJOB Neuroscience* 6.3 (2015): 3–14.
Dick, Philip K. *Do Androids Dream of Electric Sheep?* 1968. London: Gollancz, 2007.
Felski, Rita. "Critique and the Hermeneutics of Suspicion." *M/C Journal* 15.1 (2012): n.pag.
Foster, Susan Leigh. *Choreographing Empathy: Kinesthesia in Performance*. London: Routledge, 2011.
Galvan, Jill. "Entering the Posthuman Collective in Philip K. Dick's *Do Androids Dream of Electric Sheep?*" *Science Fiction Studies* 24.3 (Nov. 1997): 413–29.
Gilleard, Chris and Paul Higgs. "Aging, Embodiment, and the Somatic Turn." *Age, Culture, Humanities* 2 (2015): 17-33. https://ageculturehumanities.org/WP/aging-embodiment-and-the-somatic-turn/. Accessed: 12 Dec. 2021.
Haraway, Donna. *When Species Meet*. Minneapolis: U of Minnesota P, 2007.

Heil, Johanna. *Walking the Möbius Strip: An Inquiry into Knowing in Richard Powers's Fiction.* Heidelberg: Winter, 2016.

Heil, Johanna. "Dancing Contact Improvisation with Luce Irigaray: Intra-Action and *Elemental Passions.*" *Hypatia: A Journal of Feminist Philosophy.* 34.3 (2019): 485–506. https://doi.org/10.1111/hypa.12479 Accessed: 08 Nov. 2021.

IBM. "CIOMN Brings AI to the International Space Station." *IBM.* https://www.ibm.com/thought-leadership/innovation-explanations/cimon-ai-in-space. Accessed: 6 Jan. 2021.

IBM. "Soul Machines: Bringing a Human Face to Customer-Facing AI with IBM Watson." *IBM.* https://www.ibm.com/case-studies/soul-machines-hybrid-cloud-ai-chatbot. Accessed: 6 Jan. 2021.

Jurecic, Ann. "Empathy and the Critic." *College English* 74.1 (Sept. 2011): 10–27.

Keen, Suzanne. "A Theory of Narrative Empathy." *Narrative* 14.3 (Oct. 2006): 207–36.

Kucharzewski, Jan D. "'From Language to Life Is Just Four Letters: Self-Referentiality vs. the Self in Richard Powers's *Galatea 2.2.*" *Amerikastudien / American Studies* 53.2 (2008): 171–87.

Mackenzie, Catriona, and Jackie Leach Scully. "Moral Imagination, Disability and Embodiment." *Journal of Applied Philosophy* 24.4 (2007): 335–51.

Mitchell, David T., and Sharon L. Snyder. *Narrative Prosthesis: Disability and the Dependencies of Discourse.* Ann Arbor: U of Michigan P, 2000.

Pottle, Adam. "Segregating the Chickenheads: Philip Dick's *Do Androids Dream of Electric Sheep?* and the Post/humanism of the American Eugenics Movement." *Disability Studies Quarterly* 33.3 (2013): n.pag.

Powers, Richard. *Galatea 2.2.* New York: Picador, 1995.

Prinz, Jesse. (2011a) "Against Empathy." *The Southern Journal of Philosophy* 49 (Spindel Supplement 2011): 214–33.

Prinz, Jesse. (2011b) "Is Empathy Necessary for Morality?" *Empathy: Philosophical and Psychological Perspectives.* Ed. Amy Copland and Peter Goldie. Oxford: Oxford UP, 2011. 211–29.

Prinz, Jesse. "On the Genealogy of Empathy." SLSAeu Annual Conference 2017: Empathies. University of Basel, 23 June 2017. Keynote Lecture.

Sagar, Mark. "The Astonishingly Real Virtual Baby that Laughs, Cries – and Learns." TEDxChristchurch. *YouTube.* TEDx Talks, 7 Nov. 2014. https://www.youtube.com/watch?v=k7eeV9VEtsA. Accessed: 6 Jan. 2021.

Sagar, Mark, and Greg Cross. "Soul Machines – AI Day – 2018." *YouTube.* AI-DAY New Zealand's AI Event, 22 April 2018. https://www.youtube.com/watch?v=XLWnnQx5Wgg. Accessed: 6 Jan. 2021

Searle, John R. "The Turing Test: Fifty-Five Years Later." *Philosophy in a New Century: Selected Essays.* Cambridge: Cambridge UP, 2008. 53–66.

Searle, John R. "Twenty-Two Years in the Chinese Room." *Philosophy in a New Century: Selected Essays.* Cambridge: Cambridge UP, 2008. 67–85.

Sheldon, Rebekah. "Form / Matter / Chora: Object-Oriented Ontology and Feminist New Materialism." *The Nonhuman Turn.* Ed. Richard Grusin. Minneapolis: U of Minnesota P, 2015. 193–222.

Snyder, Sharon. "The Gender of Genius: Scientific Experts and Literary Amateurs in the Fiction of Richard Powers." *Review of Contemporary Fiction* 18.3 (1998): 84–96.

Soul Machines. "Rachel and Her Virtual Nervous System." *YouTube*. Soul Machines, 22 Jan. 2018. https://www.youtube.com/watch?v=2Yq0wxNxSfY. Accessed: 6 Jan. 2021.
Stueber, Karsten. "Empathy." *Stanford Encyclopedia of Philosophy*, 27 June 2019. https://plato.stanford.edu/entries/empathy/. Accessed: 6 Jan. 2021.
Turing, Alan. "Computing Machinery and Intelligence." *Mind* 59.236 (1950): 433–60.
van der Tuin, Iris, and Rick Dolphijn. "The Transversality of New Materialism." *Women: A Cultural Review* 21.2 (2010): 153–71.

List of contributors

Darren Abramson is Associate Professor in the Department of Philosophy at Dalhousie University. He has a PhD in Philosophy and Cognitive Science, with a Graduate Certificate in Pure and Applied Logic from Indiana University, where he studied under Dr. David McCarty. He also has an MSc in Computer Science from Indiana University and a BA in Philosophy from the University of Toronto. During his academic career Abramson has worked for multiple companies in natural language processing and artificial intelligence. His teaching and research supervision is in computer science and philosophy. Representative publications in computer science include original work in genetic programming and a study of original classroom technology. Representative publications in philosophy include showing that Alan Turing borrowed the idea for the Turing test from Descartes, and an argument that Turing's view on the mathematical objection does not require embodiment. His current research is in machine learning approaches to natural language processing; his group is pre-training language models on public, reproducible text sources with generous support from Compute Canada and Mitacs. He advocates reproducible research as a framework for ethical AI, with research under review supporting this approach.

Carmen Birkle has been full professor of North American Literary and Cultural Studies at Philipps-Universität Marburg since 2008. She has taught at the universities of Mainz, Vienna, Bergen, Dijon, and at Columbia University in New York City. She was president, vice president, and executive director of the German Association for American Studies. She is currently treasurer of the European Association of American Studies. She is Dean of the Faculty of Foreign Languages, Literatures and Cultures at Philipps-Universität (2017–23). She is the author of two monographs – *Women's Stories of the Looking Glass* (1996) and *Migration – Miscegenation – Transculturation* (2004) – and of numerous articles and (co-)editor of 14 volumes of essays and special issues of journals, among them *Literature and Medicine* (2009), *Living American Studies* (2010), *Communicating Disease* (2013), *Waging Health* (2015), and *Feminismus und Freiheit* (2016). Her research focuses on gender, ethnicity, and popular culture. She is currently at work on a monograph situated at the intersection of American literature, culture, and medicine, above all in the 19th and early 20th centuries. She is General (Co-) Editor (together with Birgit Däwes) of the journal *Amerikastudien / American Studies* (open-access).

Piet Defraeye is a theatre comparativist, and teaches Performance Studies and Contemporary Theatre at the University of Alberta. He has published widely on topics relating to Performance, including Canadian and Austrian theatre. He is co-editor of *Brussels 1900 Vienna* (Brill), an elaborate study of the cultural exchange between Brussels and Vienna, recently published. He also directs and translates for the stage, including work by Elfriede Jelinek, Peter Handke and Pieter De Buysser. The Canadian plays *Bashir Lazhar* (Micheline de la Chenelière) and *White Bread* (Mia Van Leeuwen) went on tours throughout Europe. He was resident guest professor at Université libre de Bruxelles, Universiteit Gent, Universiteit Antwerpen, Ludwig Maximilians Universität (München), Universität Innsbruck, and K. Universiteit Leuven.

Hille Haker holds the Richard McCormick S.J. Endowed Chair of Catholic Ethics at Loyola University Chicago. She has taught at Frankfurt University (2005 to 2009), and Harvard University (2003 to 2005) and holds a Ph.D (1998) and Habilitation (2002) in Christian Theological Ethics from the University of Tübingen, Germany. Hille Haker served on several Bioethics Committees, including the *European Group on Ethics in Science and New Technologies* to the European Commission (2005–2015). From 2015–2018, she was the President of *Societas Ethica, European Society for Research in Ethics*, currently serving on its Board. In Frankfurt, she was a Fellow at the *Frankfurt School Institute of Social Research* and a member of the Cornelia Goethe Institute for Women Studies. She has published multiple articles and co-edited several books in the field of bioethics and social ethics. She has written four monographs: *Towards a Critical Political Ethics. The Renewal of Catholic Social Ethics*, Würzburg 2020; *Hauptsache gesund?* München 2011; *Ethik der genetischen Frühdiagnostik*, Paderborn 2002; *Moralische Identität. Literarische Lebensgeschichten als Medium ethischer Reflexion*, Tübingen 1999. She is currently working on a book on *Recognition and Responsibility*.

Cornelia Klinger studied Philosophy, Literature and Art History at the University of Cologne. After a busy life in academia she retired from breadwinning in 2015. She teaches philosophy at the University of Tübingen for her own pleasure and (hopefully) for the benefit of her students. While working in several research areas, she focuses mainly on Political Philosophy, Aesthetics, and Feminist Theory. Recent publications in English: "An Essay on Life, Care and Death in the *Brave New World* after *1984*". In: *Equality, Diversity and Inclusion* vol. 37/4, 2018. "Serfdom – (Lost) Love's Labor – Service Industries: Services from Feudal Times to Late Capitalism". In: Klaus Dörre/ Nicole Mayer-Ahuja/ Dieter Sauer/ Volker Wittke (eds.): *Capitalism and Labor. Towards Critical Perspectives* 2018. Forthcoming: *Die andere Seite der Liebe. Das Prinzip Lebenssorge in der Moderne*. Frankfurt: Campus. Co-ed., with Brigitte Aulenbacher and Tine Haubner: *Geld oder Leben – Sorge und Sorgearbeit im Kapitalismus*. Weinheim / Basel: Beltz Juventa. Selected books: *Blindheit und Hellsichtigkeit. Künstlerkritik an Politik und Gesellschaft der Gegenwart*, ed., Berlin 2013. *Perspektiven des Todes in der modernen Gesellschaft*, ed. Wien 2009. *Über-Kreuzungen. Fremdheit, Ungleichheit, Differenz*, co-ed. with Gudrun-Axeli Knapp, Münster 2008. *Achsen der Ungleichheit. Zum Verhältnis von Klasse, Geschlecht und Ethnizität*, co-ed. with Gudrun-Axeli Knapp and Birgit Sauer, Frankfurt am Main 2007. *Flucht – Trost – Revolte. Die Moderne und ihre ästhetischen Gegenwelten*, München 1995.

Reinhart Kögerler is Emeritus Professor of Theoretical Physics at the University of Bielefeld and former President of the Christian Doppler Research Association. Together with Klaus Viertbauer he supervised a project on the question of human nature and the role of autonomous subjectivity for the Forum St. Stephan. The main publications resulting from this project are: *Das autonome Subjekt? Eine Denkform in Bedrängnis* (ratio fidei 54) Regensburg: Friedrich Pustet, 2014; *Subjektivität denken. Anerkennungstheorie und Bewusstseinsanalyse*, Hamburg: Meiner, 2017; *Neuroenhancement. Die philosophische Debatte*, Berlin: Suhrkamp, 2019.

Sybille Krämer served as Professor of Philosophy at the Free University of Berlin until April 2018. Since March 2019 she has been Visiting Professor at Leuphana University of Lüneburg at the Institute of Culture and Aesthetics of Digital Media. Earlier she held visiting professorships at universities in Tokyo, Yale, Vienna, Graz, Zurich, Lucern and was awarded an honorary doctorate at the University of Linköping, Sweden. Her academic functions have included a membership of the Scientific Council and the scientific panel of the European Research Council in Brussels and of the Senate of the German Research Foundation [DFG]; she was formerly a permanent fellow at the Institute of Advanced Study [Wissenschaftskolleg] in Berlin as well as spokesperson of the DFG-Graduiertenkolleg [Research Training Group] 'Schriftbildlichkeit'. Fields of research: Theorie des Geistes, Erkenntnistheorie, Philosophie des Rationalismus, Philosophie der Sprache, der Schrift und des Bildes, Medienphilosophie und – theorie. Recent publications: *Media, Messenger, Transmission. An Approach to Media Philosophy*, Amsterdam 2015; *Figuration, Anschauung, Erkenntnis. Grundlinien einer Diagrammatologie*, Frankfurt 2016. Co-ed. with Chr. Ljungberg: *Thinking with Diagrams – The Semiotic Basis of Human Cognition*, Boston/ Berlin 2016; co-ed. with S. Schmidt: *Zeugen in der Kunst*, Paderborn 2016; co-ed. with S. Schmidt and J. Schülein: *Philosophie der Zeugenschaft. Eine Anthologie*, Münster 2017; co-ed. with S. Weigel: *Testimony/Bearing Witness. Epistemology, Ethics, History, Culture*, London 2017. Her publications have been translated into English, French, Italian, Hungarian, Czech, Japanese, and Chinese.

Ludwig Nagl is Ao.Univ.-Prof. i.R., Department of Philosophy, University of Vienna, Austria. He was 1970/71 Assistant Professor, Millersville State University, Lancaster, Pennsylvania; 1987 Visiting Scholar, Department of Philosophy, Harvard University, Cambridge, Mass.; 1993 Guest Professor, Department of Philosophy, University of Jena, Germany; 1996 Visiting Scholar, Minda de Gunzburg Center for European Studies, Harvard University, Cambridge, Mass.; 2011 Visiting Professor, University of St. Petersburg, Russian Federation. Publications: 29 books (monographs; edited and co-edited books), 250 scientific articles (see http://homepage.univie.ac.at/ludwig.nagl/). Books and articles by Ludwig Nagl of particular relevance to his essay in the present book: *Wo steht die Analytische Philosophie heute?* Ludwig Nagl and Richard Heinrich, eds., Vienna 1986; *Systematische Medienphilosophie*, Mike Sandbothe and Ludwig Nagl, eds., Berlin 2005; "Contemporary Discourses on Humanism (Richard Rorty, Charles Taylor, Tu Weiming)", in: Ludwig Nagl, *Toward a Global Discourse on Religion in a Secular Age. Essays on Philosophical Pragmatism*, Vienna-Zurich 2021, 45–62.

Herta Nagl-Docekal is University Professor em., Department of Philosophy, University of Vienna, Austria; full member of the Austrian Academy of Sciences; membre tit. of the Institut International de Philosophie, Paris. Vice-President of FISP (2008–2013). Visiting Professor: University of Utrecht (The Netherlands), Frankfurt am Main, Konstanz, Free University Berlin (Germany), University of St. Petersburg (Russian Federation). Selected books: *Transatlantic Elective Affinities. Traveling Ideas and their Mediators* (co-ed. with W. Zacharasiewicz, Vienna 2021), *Leibniz heute lesen* (ed., Berlin 2018), *La religione dopo la critica alla religione. Un dibattito filosofico* (co-ed. with W. Kaltenbacher and L. Nagl, Naples 2017), *Innere Freiheit. Grenzen der nachmetaphysischen Moralkonzeptionen* (Berlin 2014), *Hegels Ästhetik als Theorie der Moderne* (co-ed. with A. Gethmann-Siefert et al., Berlin 2013), *Glauben und Wissen. Ein Symposium mit Jürgen Habermas* (co-ed., with R. Langthaler, Berlin 2007), *Feminist Philosophy* (Cambridge, MA, 2004), *Continental Philosophy in Feminist Perspective* (co-ed. with Cornelia Klinger, The Pennsylvania State University Press, 2000). http://homepage.univie.ac.at/herta.nagl/

Johanna Pitetti-Heil is Akademische Rätin / Senior Lecturer for Gender and Diversity Studies at the English Department at the University of Cologne, Germany. She is a board member of

SLSAeu, representing feminist science studies, a co-editor of the scholarly blog *Food, Fatness and Fitness: Critical Perspectives* (http://foodfatnessfitness.com/), a steering committee member of the EAAS Women's Network, and a co-editor of the EAAS Women's Network's journal *WiN*. Pitetti-Heil is the author of *Walking the Möbius Strip: An Inquiry into Knowing in Richard Powers's Fiction* (Winter, 2016) and is currently finishing her second book project, "Becoming-Body: Technologies of the Self in American Modern Dance," in which she explores the implications of corporeal materiality for the construction of subjectivity and freedom in American modern dance techniques. Pitetti-Heil has edited volumes and published in the fields of American literature, gender studies, cultural theory, and critical dance studies (e.g., in *Dance Chronicle* and *Hypatia*).

Julia Margarete Puaschunder is a behavioral economist with doctorates in Social and Economic Sciences and Natural Sciences from the Vienna University of Economics and Business and the University of Vienna. She has twenty years of experience of empirical research in applied social sciences in the international arena. Before starting a Prize Fellowship in the Inter-University Consortium of New York with appointments at the New School, Columbia University and Princeton University, Ms. Puaschunder held positions at Viennese universities and was an Associate of the Harvard University Faculty of Arts and Sciences. To this day she serves as Contributor to the Harvard Law School Law and Mind Sciences Initiative. She has published six books and was an expert in an Austrian Consulate conference call on Artificial Intelligence Ethics. She advised the Austrian Diplomatic Academy in Vienna on diplomacy in the artificial age and the European Parliament on Artificial Intelligence, robotics and big data in healthcare. She is included in the 2018 and the 2019 'Marquis Who's Who in America and in the World' among the top 3% professionals around the globe. She received the 2018 Albert Nelson Marquis Lifetime Achievement Award.

Ulfried Reichardt holds the Chair of North American Literature and Culture at the University of Mannheim. He studied at the University of Heidelberg, Cornell University, and the Free University of Berlin, was assistant professor at the University of Hamburg and visiting professor at the University of Cologne. Research stays at Columbia University, the University of Toronto and the University of British Columbia, Vancouver, York University and the University of California at Santa Cruz and Santa Barbara. Ph.D. 1988 at the Free University of Berlin (*Postmodernity Seen from Inside*, 1991) and Habilitation at the University of Hamburg (*Alterity and History: Functions of the Representation of Slavery in the American Novel*, 2001). Further publications of his, among others, are *Engendering Men* (1998), *Time and the African American Experience* (2000), *Mapping Globalization* (2008), *Globalization: Literatures and Cultures of the Global* (2010), and *Network Theory and American Studies* (2015) as well as essays on the dimension of time in literature and culture, on American Pragmatism, on music in America, on diaspora culture, and US-American authors of the 19th, 20th and 21st centuries. Co-Founder of the research project "Probing the Limits of the Quantified Self" funded by the German Research Foundation (2015–2018).

Regina Schober is Professor for American Studies at Heinrich-Heine-University Duesseldorf. She teaches American literature and culture with a focus on digital culture, intermediality, and theories of the global information age. She is the author of *Unexpected Chords: Musico-Poetic Intermediality in Amy Lowell's Poetry and Poetics* (Winter, 2011) and co-editor of *The Failed Individual: Amid Exclusion, Resistance, and the Pleasure of Non-Conformity* (Campus, with Katharina Motyl, 2011), of *Data Fiction: Naturalism, Numbers, Narrative* (special issue of *Studies in American Naturalism*, with James Dorson, 2011), and of *Laboring Bodies and the Quantified Self* (Transcript, with Ulfried Reichardt, 2020). She was visiting scholar at the University of California, Santa Barbara (2017) and at Harvard University as well as at the University of Virginia, Charlottesville (2008). Her research interests include network studies, transformations of subjectivity in the information age, artificial intelligence, the quantified self, discourses of failure, and theories of reading and the attention economy. She is currently part of the research networks *The Failure of Knowledge/Knowledges of Failure* and of *Model Aesthetics: Between Literary and Economic Knowledge* (both funded by the German Research Foundation, DFG) as well as of the project *ai4all* (funded by the German Ministry of Education and Science, BMBF).

Sabine Sielke is Chair of North American Literature and Culture and Director of the North American Studies Program and the German-Canadian Centre at the University of Bonn. Her publications include *Reading Rape* (Princeton 2002) and *Fashioning the Female Subject* (Ann Arbor 1997), the series *Transcription*, and 20 (co-)edited books, most recently *Nostalgia: Imagined Time-Spaces in Global Media Cultures* (2017), *Knowledge Landscapes North America* (2016), *New York, New York! Urban Spaces, Dreamscapes, Contested Territories* (2015), and *American Studies Today: New Research Agendas* (2014), as well as more than 140 essays on poetry, (post-)modern literature and culture, literary and cultural theory, gender and African American studies, popular culture, and the interfaces of cultural studies and the sciences.

Jörg Türschmann is Professor of French and Spanish Literature and Media at the Department of Romance Studies, University of Vienna. His research interests are transnational relations between literature and cinema in Argentina, Spain, France, the Caribbean and Canada, as well as their relations with German-speaking countries. Furthermore, he is interested in television series, popular culture and film music. He has worked as film distributor and as organizer and jury member for film festivals in Germany, France and Switzerland. He obtained a PhD in French Film Theory and wrote his habilitation thesis about 19th century serial novels in Spain, France and Italy. He has published books on transnationality in cinema and television: *Miradas glocales: Cine español en el cambio de milenio* (2007), *TV global* (2011), *Transnational Cinema in Europe* (2013). His most recent publications are *Estéticas globales hispánicas* (*L'Atalante* 26, 2018) and *La literatura argentina y el cine* (2020).

Klaus Viertbauer serves as an Assistant Professor at the Catholic University of Eichstätt (Germany). Together with Reinhart Kögerler he supervised a project on the question of human nature and the role of autonomous subjectivity for the Forum St. Stephan. The main publications resulting from this project are: *Das autonome Subjekt? Eine Denkform in Bedrängnis* (ratio fidei 54) Regensburg: Friedrich Pustet, 2014; *Subjektivität denken. Anerkennungstheorie und Bewusstseinsanalyse*, Hamburg: Meiner, 2017; *Neuroenhancement. Die philosophische Debatte*, Berlin: Suhrkamp, 2019.

Waldemar Zacharasiewicz is Emeritus Professor of American Studies at the University of Vienna and a Dr.h.c. of Eötvös-Lorand University Budapest. He chairs the commission "The North Atlantic Triangle" of the Austrian Academy of Sciences. He is also a member of Academia Europaea and an International Fellow of the Royal Society of Canada. His main research interests have been travel literature and imagology, the study of transatlantic migration, and the literatures of the American South and of Canada. He served as director of the Canadian Studies Center of the University of Vienna from 1998 to 2014. Among his publications are a monograph on the theory of climate in English literature and literary criticism (1977), two book-long studies on *Images of Germany in American Literature* (1998 and 2007), a collection of his essays entitled *Imagology Revisited* (2010), and a monograph on *Transatlantic Networks and the Perception and Representation of Vienna and Austria between the 1920s and 1950s* (2018). He has also edited or co-edited more than twenty collections of essays.

Index of Authors

Adams, Fred 84
Adorno, Theodor W. 38, 45f., 54, 70
al-Ḥwārizmī 18
Alexa 21, 43, 68, 73, 119
Antoine, André 263
Arendt, Hannah 272f., 276
Asimov, Isaac 10f., 71, 238–240, 243–246, 256, 288
– short stories
  – *Satisfaction Guaranteed* 11, 240, 244
  – *Liar* 245
– short-story collection
  – *The Complete Robot* 238, 243
Atwood, Margaret 94, 238
Austin, John Langshaw 34f.

Balagure, Mark 174
Bar-Hillel, Yehoshua 6f., 128, 131
Barad, Karen 157, 288f., 304
Baudrillard, Jean 12, 116, 281f.
Bauman, Zygmunt 83, 94
Becker, Carlos 38, 45f.
Benjamin, Walter 5, 52–60, 65, 71, 73
Bennett, Jane 13, 59, 172, 287–289, 294f., 302–304
Bergala, Alain 214
Bergson, Henri 221
Binder, Eando 243
– short story
  – *I, Robot* 243
Block, Hans 35f.
Boden, Margaret 23, 159
Bogost, Ian 302f.
Bolter, David Jay 1, 20, 242
Borges, Jorge Luis 221
Bostrom, Nick 17, 40f., 80, 157
Budde, Antje 272f.
Buñuel, Luis 213, 222
Buonarroti, Michelangelo 213f.
Bush, George H. W. 86f.
Butler, Samuel 70, 257f.
– novel
  – *Erewhon; or, Over the Range* 257

Cameron, James 209, 211
Čapek, Karel 10f., 243, 261, 263, 268
– play
  – *R.U.R.* 243, 261, 263
Carnap, Rudolf 7, 128
Chomsky, Noam 268f.
Clarke, Arthur C. 84, 95
Cline, Ernest 237f., 248
– novel
  – *Ready Player One* 237f., 248, 255f.
Cocteau, Jean 273
Coleridge, Samuel Taylor 155
Cowell, Jason M. 290, 301
Crichton, Michael 209f.

Darwin, Charles 226, 257
– *On the Origin of Species* 257
Dautenhahn, Kirsten 264
Davidson, Michael 292
Davis, Miles 141, 189f., 273
De Candia, Gianluca 35
de Chirico, Giorgio 215
Decety, Jean 290, 300f.
Deleuze, Gilles 221, 295
Dennett, Daniel 130f., 138f., 147–149, 173
Derrida, Jacques 233
Descartes, René 86, 279
Dewey, John 45
Dick, Philip K. 13, 84, 95, 237f., 246f., 256, 285, 287, 289f., 295–297, 300, 303
– novel
  – *Do Androids Dream of Electric Sheep?* 13, 95, 237f., 246, 287, 289, 295
Dickinson, Emily 89, 161
Dinello, Daniel 157
Dorsen, Annie 11, 268f.
Douglas, Thomas 8, 175f.
Dreyfus, Hubert 4, 25, 33, 41f., 44
Dreyfus, Stuart 25, 42, 44
Dupriez, Bernard 280
DuPuis, Melanie 92

Eco, Umberto   278
Eggers, Dave   94, 237f., 248–250
– novel
  – *The Circle*   94, 237f., 248–250, 256
Eimer, Martin   173
Elgammal, Ahmed   163
Elliot, Carl   174
Emerson, Ralph Waldo   155
Epstein, Jerome   268

Farrell, Amy Erdman   92
Felski, Rita   293
Fish, Stanley   154
Flores, Fernando   82
Floridi, Luciano   21, 64, 226
Fluck, Winfried   229
Foster, Susan Leigh   301f.
Foucault, Michel   36, 113, 229, 268f.
Frayn, Michael   266
Freud, Sigmund   55, 214, 226, 275

Garland, Alex   210, 215f., 219, 253
– film
  – *Ex Machina*   11, 215, 219–221, 237f., 251, 253–256
Gaudí, Antoní   214
Genette, Gerard   279
Genova, Judith   265
Gibson, Steve   266
Glanville, Ranulph   265
Goethe, Johann Wolfgang von   107, 221, 254
Goldsmith, Kenneth   163
Grigar, Dine   266
Guthman, Julie   92f.

Habermas, Jürgen   1, 33, 38, 63, 171, 175
Haggard, Patrick   173
Halberstam, Jack   160
Haraway, Donna J.   2, 81, 241, 287, 298
– essay
  – *A Cyborg Manifesto*   241
Hardt, Michael   116
Harris, John   177
Hassabis, Demis   91
Haugeland, John   23, 137

Hawking, Stephen   11, 240, 255, 257
– collection of essays
  – *Brief Answers to the Big Questions*   255
Hayles, Katherine N.   10, 157, 226–229, 233, 241, 254
Hegel, Georg Wilhelm Friedrich   40, 69, 216
Heidegger, Martin   52, 55, 120
Heinze, Hans-Jochen   83
Held, Brigitte   217
Hinkis, Tali   266
Hinton, Geoffrey   25, 27, 138
Hitchcock, Alfred   222
Hobbes, Thomas   42
Hoffmann, Ursula   82
Hofstadter, Douglas   262, 265, 279, 284
Horkheimer, Max   38, 45f., 54, 120
Huck, Stefan   39
Hugo, Victor   221
Hume, David   300f.
Hustvedt, Siri   84f.
Huxley, Aldous   84

Itskov, Dmitry   86

James, William   232

Kaegi, Stefan   270
Kant, Immanuel   4, 33–35, 47f., 66, 68f., 171f.
Keller, Helen   291f.
Klimt, Gustav   284
Knorr-Cetina, Karin   114f.
Koene, Randal A.   80
Kramer, Peter D.   174
Kubrick, Stanley   209, 222, 288
Kurzweil, Ray   17, 40, 80, 86, 152

La Mettrie, Julien Offray de   42
Lacan, Jacques   12, 278
Lagrange, Joseph Louis   266
Lang, Fritz   217, 261
Lapidus, Kyle   266
Latour, Bruno   26, 65, 211, 242
Laurel, Brenda   265
LeDoux, Joseph E.   81

Leibniz, Gottfried Wilhelm   4, 18, 26 f., 33, 41 f.
Lepage, Robert   12, 261 f., 273 f., 277–279, 281–283
Lerner, Alan Jay   265
Levmore, Saul   38
Libet, Benjamin   172 f.
Licklider, J. C. R.   157
Lipps, Theodor   302
Loewe, Frederick   265
Longuet-Higgins, Hugh Christopher   80
Lovelock, James   213
Lucas, George   217, 255
Lupton, Deborah   92 f.

Mackenzie, Adrian   20
Mandel, Ernest   109 f.
Marshall, John Chief Justice   235
Marx, Karl   52, 57, 63, 105, 111, 174, 272
Massaron, Luca   265
Maxwell, James Clark   138, 266
Mazzone, Marian   163
McEwan, Ian Russell   257, 287
– novel
  – *Machines Like Me*   257, 287
McLuhan, Marshall   80, 84
Meillassoux, Quentin   226
Méliès, Georges   217
Melle, Thomas   270 f.
Merkel, Reinhard   167, 169
Mersch, Dieter   34, 39 f.
Metzinger, Thomas   10, 227, 233
Meyerhold, Vsevelod   265
Mighton, John   11 f., 261 f., 275–283
Mitchell, David   292 f.
Mitchell, Melanie   34, 36, 40, 43 f., 46 f., 290 f.
Mizuno, Sonoya   219
Morton, Timothy B.   9, 209–211, 257
Mountcastle, Vernon B.   82
Mueller, John Paul   265

Nanz, Patrizia   45
Neumann, Kurt   215 f.
Ng, Andrew   21, 47
Nida-Rümelin, Julian   2, 37, 41, 47 f., 65, 210 f., 243

Nietzsche, Friedrich   40, 55
Nussbaum, Martha C.   38, 242

O'Connell, Mark   79–81, 83 f., 84, 86–88, 93–95
Orwell, George   121
Ovid   211

Parfit, Derek   277
Patera, Valeria   270
Pauen, Michael   37, 167
Peirce, Charles Sanders   34 f.
Persson, Ingmar   8, 175–177
Pflüger, Joachim   83
Pinker, Steven   83
Pitt, Walter   23
Poe, Edgar Allan   155
Polanyi, Karl   110
Pottle, Adam   13, 296 f.
Pousttchi, Key   85
Powers, Richard   10, 12, 231 f., 235, 290 f., 293
Prinz, Jesse   300–302
Prinz, Wolfgang   173
Propp, Vladimir   229
Proyas, Alex   252
– film
  – *I, Robot*   10 f., 71, 237–239, 243, 252, 254–256
Putnam, Hilary   4, 33, 41–44, 92, 279
Putnam, Robert D.   94

Ramge, Thomas   22, 39
Rawls, John   38, 63
Reckwitz, Andreas   94
Rey, Lester del   243
– short story
  – *Helen O'Loy*   243
Rice, Elmer   10, 237 f., 268
play
  – *The Adding Machine*   10, 237 f., 268
Rich, Adrienne   113, 160 f.
Richardson, Rufus   266
Ricœur, Paul   66, 68, 70
Rieger, Stefan   159
Riesewieck, Moritz   35 f.
Robin, Henri   266

Roden, David 234
Rokeby, David 272–274
Roque, Antonio 152
Rosenblatt, Frank 23
Rota, Gian-Carlo 34
Roth, Gerhard 112, 173
Rousseau, Jean-Jacques 233
Russell, Stuart 94, 209, 257
Ryan, Marie-Laure 278 f.

Sacks, Oliver 83
Sandberg, Anders 86
Savulescu, Julian 8, 175–177
Schaeffer, Jonathan 264 f.
Scheutz, Matthias 232
Schmidt, Eric 118
Scholte, Tom 266
Schwartz, Oscar 152 f., 157 f.
Scott, Ridley 209, 213, 217–219
Searle, John R. 4, 18, 33, 37, 136, 152, 229, 231, 290 f.
Serres, Michel 9, 211, 214
Seubert, Sandra 38, 45 f.
Shakespeare, William 12, 155, 269, 293
 – play
  – *The Tempest* 12, 293
Shaviro, Steven 162, 226–228, 234
Shaw, George Bernard 11, 264
Sheldrake, Merlin 162
Shelley, Mary 222, 245, 254
 – novel
  – *Frankenstein* 222, 245, 254, 256
Singer, Peter 8, 171 f.
Singer, Wolf 173
Sjuve, Eva 266
Snyder, Sharon 292 f.
Sobchack, Vivian 211 f.
Sombart, Werner 111
Sontag, Susan 117 f.
Spielberg, Steven 210, 251

film
 – *AI* 11, 237 f., 251 f., 254–256
Stelarc (= Stelios Arcadiou) 11, 267 f.
Stengers, Isabelle 156
Sturma, Dieter 37 f., 167
Sudmann, Andreas 157 f.

Taylor, Charles 4, 33, 38, 41 f., 44 f., 52, 65 f., 174
Teilhard de Chardin, Pierre 213
Turing, Alan 130 f., 137 f., 147–149, 151–154, 157 f., 163, 253, 261 f., 265, 270, 291

Van Hove, Ivo 262
Varian, Hal 113
Vidor, Charles 219
Vischer, Robert 302
Voigts, Eckart 156–158, 161

Warhol, Andy 95
Wark, McKenzie 267 f.
Watts, Peter 10, 225, 232–235
Wegener, Paul 217
Wegner, Daniel 173
Weidenfeld, Nathalie 2, 37, 41, 46 f., 65, 210 f., 243
Weizenbaum, Joseph 264
Whitehead, Alfred North 19
Wilde, Oscar 212
Winograd, Terry 6 f., 82, 127–131, 135, 138, 138, 140–142
Wittgenstein, Ludwig 34, 42 f.
Wolf, Graham 141
Wolfe, Cary 157, 228

Yeats, William Butler 163, 279

Žižek, Slavoj 12

Index of Subjects

5th freedom of data   191, 198

Actants   242, 289, 302f.
Actor-network   216, 242
Adaptation   23, 69, 156f., 161, 246, 261, 265, 273, 277, 279
Advertising   3, 59, 63, 83, 85, 91, 117, 183, 230
Aesthetic experience   *see* Experience
Affect   36, 110, 174, 210, 225, 227, 241, 293f., 304
Agency   5, 22, 51, 57, 64–69, 71–74, 89, 91, 94, 234, 241, 288f., 294, 298, 304
– Distributive agency   93, 294, 302
– Human agency   5f., 51, 64, 66–68, 71–73, 89, 93, 156, 161, 289
Agriculture, application of AI   102, 110, 183
Algorithm   3f., 7, 17–21, 24, 26, 33–37, 39–44, 46, 48, 85, 91, 119, 152, 155f., 158, 162f., 169, 182–184, 186, 189f., 197, 199, 225, 230f., 234f., 239, 263, 284, 288
Algorithm, deep learning   4, 7, 21, 24f., 36, 156, 158f., 162
Alien   10, 73, 162, 213, 216–218, 222, 225, 227, 229, 232–234
Alienation   52, 55, 66, 174, 220, 240
Ambiguity   72, 108f., 154, 240, 276
Android   9, 11, 13, 95, 212–214, 216f., 219f., 227, 229, 237f., 246–248, 250f., 253–257, 287, 289f., 295–300, 303; *see also* Humanoid; Robot
Anthropomorphism   9, 11, 157, 211, 214, 218, 232, 263, 271
API   145, 261
Apocalypse   11, 235, 248, 251f.
Applied ethics   *see* Ethics
Applied science   *see* Science
Artificial   2, 7f., 19, 24, 40, 84, 94, 151, 154, 158, 160–163, 181f., 187, 190, 192, 194, 199, 215, 219, 228, 235, 244, 265, 287f., 293, 295, 299, 303
– Artificial emotion   13, 96, 168, 232, 274, 298
– Artificial flatness   19f., 26, 262
– Artificial Intelligence (AI)   1, 3, 5, 8–10, 12, 17–19, 21f., 24f., 27f., 33f., 37, 40, 42f., 48, 51f., 59–62, 64, 66, 70, 80–82, 85, 102, 127, 130, 136f., 151f., 154–163, 181–184, 190, 194, 209–214, 217, 219, 221f., 225–229, 232, 234f., 237–239, 244, 248, 250, 255f., 261f., 265, 274f., 277, 281, 287–291, 293–300, 302–305; *see also* Intelligence
AI advancement   182, 186, 192f.
– AI growth   8, 181f., 191f.
– AI innovations   34, 187
– AI leadership   183, 191
– AI systems   5, 34, 39, 43, 51, 53, 61, 68, 71, 73, 130, 152, 187, 199, 240, 287f.
Artificial general intelligence (AGI)   131, 240
Artificial neural network   *see* Neural network
Artificial super intelligence (ASI)   10, 17, 240–242, 248, 250, 255–257
Strong AI (strong readings of AI)   3f., 17f., 33, 36f., 40, 43f., 48, 239
Weak AI (weak reading of AI)   3f., 17f., 37, 39, 239
Artistic production   1, 284
Assemblages   289, 302
Assistance   17, 21f., 80, 184, 186, 199
Authenticity   8, 45, 167, 170–172, 174f., 175, 177
Automation   83, 157 160, 259, 261, 268
Autonomy (autonomous decision making)   8, 37, 45, 47, 60, 64, 66, 69, 167, 169–172, 174, 199, 216f., 220, 253

Barbarism   5, 53f.
Behavioral patterns   175, 184
Biases   3f., 26, 160, 163
Big data   3, 17, 45, 102, 118, 120, 144, 155, 182, 184–188, 193, 197–199, 226

Big data inferences 181f., 185
Binarism 83f.
Bioethics see Ethics
Biomedical researcher 8, 184
Bionics 106, 215
Bio-technology 8, 81, 101, 106, 168
Black Box 26, 155f., 240, 266
Blackboxing 4, 25f., 265f.
Blade Runner 209, 246
Blade Runner 2049 246f.
Blindsight 10, 225, 232, 235
Body, human 6, 9f., 12f., 27, 53, 57, 77–83, 86, 89f, 91–93, 96, 101, 106–108, 111f., 115, 119f., 131, 186, 216f., 219, 221, 227, 232f., 235, 251, 266f., 270, 285, 291f., 297f., 301f.
Brain 2, 5f., 8, 12, 19, 37, 40–44, 79–90, 84–96, 168f., 173f., 213, 232, 235, 239–241, 244f., 254, 261, 277–283, 299, 301
Brain-computer interface (BCI) 81, 85, 168f.
Brain disorder 53, 86f., 168
Brain emulation 84, 86
Brain hemispheres 88f.
Brain machines 5, 79, 83f.

Calculating rules 42f.
Capitalism 59, 63, 68, 92, 101, 105, 109–111, 117, 226, 301
Capitalism-cum-nation-state-system 102f., 105
Care 6, 13, 66–68, 101f., 109f., 118, 137, 184–188, 190f., 249, 252, 287, 289, 297f., 301, 305
Care crisis 6, 101
Care ethics 101, 105; see also Ethics
Care work 101, 105, 110
Chatbot 11, 13, 264, 268, 298f., 303
China 85, 183, 191
Chinese Room argument 136, 152, 229, 231, 234, 290
Christianity 113, 170
Citizen science see Science
Class 63, 81, 104, 134, 229, 264, 301
Climate change 94, 176, 213, 254
Clinical decision support systems 186f.

Clone 11, 218f., 252
Cobot 264, 283
Cognition 23, 43, 51, 65, 79, 81, 83f., 86, 152, 162, 226–229, 233, 293
– Cognitive abilities 159, 168f.
– Cognitive science 5f., 79–88, 90, 95, 101, 114, 127, 130, 137, 145; see also Science
– Cognitive states 4, 33, 37
– Discognition 162, 228
– Nonconscious cognition 227f., 233
Collaboration 22, 158, 161f., 257, 287, 303
Colonialism 105, 112
Common sense 6, 127–129, 131, 138, 141
Communication technologies 111, 185, 226
Comparative advantages 181f., 194, 197
Comprehension 28, 119
Computation control see Control
Computational creativity 7, 139, 152
Computational power 25, 184
Computer 1, 4, 7, 9, 12f., 17, 19f., 33, 36f., 39–42, 44–48, 80–82, 84f., 87f., 91, 113, 120, 127, 130–132, 136–138, 148, 151–154, 157, 161f., 168f., 183, 186, 209f., 212, 217f., 222, 227f., 230–232, 238f., 241, 249f., 253f., 257, 261f., 264–266, 268–270, 273–276, 295, 303; see also Robot
Computer game 248, 265
Computer science 3, 11, 43, 80, 86, 95, 128, 144, 253, 264f.; see also Science
Computer system 10, 183, 303
Connectionism 82, 130, 137
Connectivity 82, 157, 181f., 184f., 191–196; see also Global connectivity; Internet connectivity
Consciousness 1, 9f., 41, 51, 57, 67, 81f., 88, 91, 101, 174, 225, 227f., 230–235, 257, 268f., 277; see also Self-consciousness
Consumption 25, 45, 94, 110, 130
Contingency 106–109
Control 1, 4, 11, 19f., 33, 36f., 45–47, 70, 81, 85, 90, 93, 104, 120, 151, 162, 183–186, 191, 199, 209, 214, 227, 241–243, 246, 249, 252, 255f., 267
– Computation control 185

Index of Subjects — 321

– Outbreak control   185 f.
Co-performance   22, 24
Corporation   10, 63, 68, 110, 119, 134, 145, 225 f., 234 f., 245, 289
Corporeality   66, 81, 288 f., 298
Corruption   8, 52, 181 f., 187, 189–197, 204 f.
Corruption Perception Index   193–196
Council on Artificial Intelligence   64, 183
Country comparison   181, 191–194, 196
COVID-19 crisis   181 f., 188, 190 f., 194, 197, 272, 284
Creativity   13, 84, 90, 139, 152 f., 155 f., 159, 161, 163, 168, 240, 244, 300
– Human creativity   7, 44, 139, 153, 156–160, 162 f., 222
– Machine creativity   7, 40, 139, 151 f., 158 f., 162 f.
Creature   9, 82, 170, 210, 212 f., 216 f., 219, 222, 241, 252
Crisis management   8, 185 f., 191
Crisis risk management   8, 182
Critical Theory   5, 45, 52, 72, 118
Crowd media use   186; see also Media
Crowdsourcing   24, 198
Cultural techniques   18
Cultural technique of flattening   19 f., 262
Cyberpunk   209, 213, 215 f.
Cyberpunk movies   9, 218
Cyborg   2, 79, 81, 85, 213, 218 f., 238, 240 f., 250, 254, 261, 282; see also Robot

Dark ecology   9, 209 f., 213, 218, 222, 257
Data generation, storage and analysis   184
Data-driven intervention   8, 185
Data storage   184, 199
Data universe   20, 25
Datification   21, 25
Death   11, 35, 57 f., 68, 79, 96, 107, 132, 210, 218, 221, 243, 253, 255
Decade of the Brain (1990s)   86, 91
Decision-making   1–3, 37, 61, 64, 94, 181, 183, 187 f., 194, 198, 225, 265
Deep Learning   see Machine learning
Democracy   45, 101, 119, 189

Desire   12, 51, 57, 65 f., 79, 89, 154, 214, 219 f., 237 f., 245 f., 249, 252 f., 254, 256, 261, 268, 274 f., 291, 298, 302
Determinism (hard determinism)   8, 173 f.
Diagram   19 f., 24, 27, 262
Digital affinity   191
Digital humanism (*Digitaler Humanismus*)   2 f., 46, 65, 210
Digital technology   5, 68, 84, 91, 93, 261 f.
Digital transformation   59, 284; see also Transformation
Digitization   6, 20, 22, 25, 33, 80, 92, 102, 181, 210 f., 217
Digitalization   1 f., 9, 36, 38, 40, 45, 182, 190, 192, 197, 199, 225
Dignity   48, 69, 71, 167, 171, 174, 199
Disability studies   12 f., 292, 296, 302
Disabled   13, 186, 292 f.
Discrimination   3 f., 21, 26, 60, 64, 71, 198 f.
Division of labour   104, 110
Dramaturgy   261–263, 268
– Digital Dramaturgy   272
Dualism   6, 82, 101, 116, 159
Dystopia   9, 11 f., 40 f., 80, 85, 112, 121, 209, 214–216, 248, 257, 282

Economic development   133, 189
Economic growth   187, 192, 199
Economic productivity   181 f., 192
Effectiveness   17, 144, 187
Efficacy   287, 294 f., 298, 302–305
Efficiency   36, 39, 187
Electricity   21, 289, 291, 294, 304
Electronic health records (EHRs)   see Health
Embedded EthiCS   see Ethics
Emergence   9 f., 80, 156
Empathy   10–13, 95 f., 168, 177, 227, 232 f., 246 f., 253, 256, 287–290, 293, 295–305
– Aesthetic empathy   299, 301 f.
– Kinesthetic empathy   299, 302
– Neuroscientific empathy   13, 177
Enhancement   37 f., 40, 79, 81, 84, 165, 167, 182, 237, 241, 254
– Human enhancement   1, 9, 160, 167, 169, 181, 225 f., 228, 238, 240, 255 f.

– Moral enhancement (ME)   8, 37, 170, 175
– Neuroenhancement   2, 7, 37, 167, 174
Entropy   9, 209–217, 222; see also Negentropy
Environment   2, 8 f., 19, 44, 60 f., 64, 69, 79 f., 92, 94 f., 145, 151, 154, 176, 181 f., 184, 190, 192, 198 f., 210, 212 f., 216 f., 229 f., 232, 239, 241, 269, 273, 284, 295
Equality   61, 109, 172, 216; see also Inequality
Ethics   8, 46, 48, 58–64, 66–70, 72, 85 f., 127, 132, 170, 172, 183 f., 210, 267, 275; see also Care ethics
– Applied ethics   167, 170
– Bioethics   64, 66, 171 f.
– Embedded Ethics   86, 95
Ethics (of AI)   46, 48, 58–64, 66–70, 72, 83 f., 127, 132, 168, 170, 183 f., 210
Ethical boundaries   9, 181 f., 197
– consequentialist   57, 171
– deontological   171
Ethical guidelines   60 f.
Ethical principles   1, 47, 60
Eugenics   13, 113, 175, 295 f.
Europe   18, 38, 53, 60–62, 87, 102, 112, 183, 191, 193, 195, 197 f., 266, 270, 301
Ex Machina (film)   11, 215, 219–221, 237 f., 251, 253–256
Experience   5, 10, 13, 25, 42–44, 51–57, 59, 66, 68–73, 111, 127, 131 f., 152 f., 161–163, 176 f., 183, 186 f., 189, 220, 227, 232, 234, 245, 247, 273, 280, 283 f., 289, 291, 293, 297, 299–304
– Aesthetic experience   7, 163
Extended mind   see Mind
Extensions of man (McLuhan)   80, 84

Failure   39, 86, 92, 160
Fear   1, 10, 17, 28, 36, 57 f., 66, 71, 85, 108, 151, 153 f., 210, 233, 237–246, 248 f., 251 f., 254, 256, 268, 272, 274, 288 f., 291, 302
Feedback   4, 22, 24, 156, 169, 231
Feeling   2, 10, 36, 90, 120, 168, 177, 227, 230, 233, 245–247, 250, 297, 300 f.
Fembot   250 f.

Fiction   3, 9–11, 94, 136, 158, 212, 225 f., 230 f., 234, 237 f., 241, 248, 256 f., 281, 287–290, 303 f.
– Extro-science fiction   226 f.
– Science fiction   1, 3, 9 f., 12, 41, 95, 151, 159, 209, 217, 225–227, 229, 232, 234 f., 238, 241, 257, 287, 289–291, 294, 300, 304
Film (inclusion of AI)   1, 3, 9–11, 35, 55, 117, 210–219, 221 f., 225 f., 229 f., 234, 237–241, 246, 250, 252, 255 f., 258, 261, 263, 268, 270, 275, 277–279, 281 f.
– Digital cinema   9, 209, 211 f.
Finance   105, 120, 183
Finitude   107 f.
Folding   27, 114
For-profit research   7, 139
Fourteenth Amendment of the US Constitution   234 f.
Fourth Industrial Revolution   6, 105
Frankfurt School   5, 52, 65, 72
Future   1, 5, 7 f., 20, 22, 39, 68, 73, 79–81, 84 f., 87, 94 f., 103, 113, 117, 133 f., 151, 157, 161, 163, 177, 181 f., 185 f., 189, 191 f., 194, 198 f., 211, 213–215, 220, 222, 225, 227 f., 235, 237, 240, 242, 248 f., 255, 257, 281, 295

Gender   81, 89 f., 101, 104, 115, 142, 219, 222, 241, 291, 301
Genesis   6, 38, 217, 264
Genetics   81, 113, 170
Genetic code   6, 117
Genome sequencing machines   8, 184
Globalization   3, 6, 8 f., 92, 101, 103, 105, 111, 115 f., 120, 176, 181–184, 187, 190–192, 197 f., 213, 228, 246, 254
Global Connectivity   193–196
Global Corruption Index   193, 195, 204
Global health trends   see Health
Global pandemic alleviation   185, 197
Global pandemic alleviation leaders   181 f.
Global public health   see Health
GOFAI   22 f.
Governance   8, 61, 63 f., 105, 107, 112, 119, 184, 190, 194

Governmental accountability   190
Governmental revenues   189
GPT-3   129, 142, 144, 282
Graph   19, 192–197, 203–205, 275
Gross Domestic Product (GDP)   63, 181f., 192f., 197, 203
Growth   63f., 184f., 187f., 192, 197; *see also* AI growth; Economic growth

Hard determinism   8, 173
Hardware   61, 87, 188
Health   8, 57, 92, 95, 102, 118, 184–188, 191, 196, 198f., 249f., 257, 295
– Electronic health records (EHRs)   8, 184f., 187f.
– Global health trends   184f., 198
– Global public health   9, 188, 191f., 194f., 197
Health App   190
Health crisis management   8, 185f., 191
Health Quality and Access Index   196
– Health-related data   185
– Health risk early warning system   185
– Health status   184f., 199
Healthcare   8f., 68, 181–199
Healthcare access and quality   181f., 192
Healthcare professionals   8, 184
Healthcare sector   182, 185, 187
International healthcare   8, 181, 197, 199
Hermeneutical reason   4, 33
Heteronomy   35, 68
Hierarchy   256
Hierarchical modelling advancements   184
High-resolution medical imaging   8, 184
Histrionic   263, 273
HLEG AI (EU High-Level Expert Group on AI)   60, 64
Homo sapiens   113, 171
Hospital   1, 58, 187, 251
Human being   5, 11f., 22, 42, 44, 47f., 51, 57, 65f., 68f., 71f., 106f., 112, 134, 145, 162, 167, 170f., 173, 176, 183, 211, 222, 227, 237–239, 241–246, 248, 250–253, 257–258, 287–289, 291, 295, 297f., 300
Human Brain Project   87f.
Human condition   58, 113, 249, 272

Human decision making   182
Human enhancement   *see* Enhancement
Human exceptionalism   10, 12f., 157, 233, 256, 287–289, 294, 298, 303
Human nature   79, 114, 169, 251
Humanities   3, 7, 17f., 23, 28, 81, 115, 139, 145, 156, 160, 181, 225f., 228, 288
Humanity   11, 40, 43, 58, 94, 108, 112, 118, 120, 157, 171, 184, 213, 233, 237–240, 243, 246, 249, 254f., 257, 287, 289, 293
Humanoid (humanoid machine)   5, 9, 11, 13, 66f., 74, 151, 210–212, 216, 219f., 227, 238, 241–243, 245–247, 250f., 253–257, 264, 271, 287, 291, 295, 297; *see also* Android; Robot
Hunger   24, 133, 186
Hybridity   23, 25, 156f., 212, 241

IBM   23, 113, 119, 288, 299
– Watson   23, 119, 288, 299
Identity   5, 51, 54, 59, 64–67, 70, 72, 74, 88, 132, 219f., 227, 241f., 245, 249, 251, 277, 298, 301; *see also* Personal identity
Illusion   17, 227, 241, 246, 293
Imaging techniques   83, 95
Index   8, 93, 181f., 192–197, 203–205
Industrialization   6, 102, 104, 109–111
Industrial Revolution   6, 52, 101–103, 105, 107, 109, 111f.
Inequality   83, 94; *see also* Equality
Inferences   23, 138, 144, 181–185, 197f., 276
Inference-driven insight   184
Information processing   81f., 227, 234
Innovation   1, 4, 7–9, 33f., 36, 39, 53, 59f., 63, 102, 106, 111–113, 144, 152, 161, 181f., 187, 190–192, 225f., 235, 266
Innovative global pandemic alleviation leaders   181f., 197
Instant information generation   185
Instant messaging   161, 186
Instrumental rationality   4f., 33, 36, 38, 41, 53

– Limits of computerized modes of instrumental rationality   41
Integrity   67, 190, 193, 284
Intelligence   4, 7, 11, 17, 25, 40–44, 80, 82, 90f., 95, 131, 148, 152f., 162, 212, 215, 227f., 230–235, 239, 252–254, 256f., 263, 265, 268, 277, 279, 281, 284, 287, 291, 293, 296f., 299
– Artificial Intelligence (AI)   see Artificial Intelligence (AI)
– Human intelligence   1, 3, 7, 10, 17f., 33, 43, 85, 94, 157–160, 162, 183, 227, 229, 239, 288, 291, 303
Intentions   2, 7, 66, 70, 294
Interactions, human–machine   4f., 10, 20, 22, 25, 35, 38, 51, 68, 70f., 73, 93, 151, 157, 161f., 173, 229, 257, 263, 270f., 298f.
International community   190
International development   185, 189f.
International differences   182, 190, 197
International healthcare   see Healthcare
Internet access   192
Internet connectivity   181f., 184, 191–193, 196
Internet of Things (IoT)   59, 61, 102, 184
Intersubjectivity   42, 95f.
Interventions   8, 37, 168, 175, 177, 185
– Genetic interventions   8, 168
– Pharmacological/chemical interventions   8, 37, 168, 177
Intra-action   288f.
Intransparency   21; see also Transparency
Intuition   1, 19, 44, 47, 156, 265
Invisibility   18–20, 26, 119, 219, 221, 235
I, Robot   11, 71, 237–239, 252, 254–256
Irrationality   28, 256
IT governance   8, 190

Judgment   4, 33f., 39f., 43, 47f., 70, 90, 129, 141, 154f., 172, 211, 294, 300f.
Justice   3, 61, 63f., 73, 92, 174, 183, 288, 302

Knowledge   7, 10, 12, 19–21, 23, 25–27, 42, 44, 52, 55, 58f., 67, 73, 82, 85, 87, 89, 91, 106, 112, 115, 128, 130f., 138f., 153f., 177, 185f., 225f., 231–233, 237, 239, 249, 251, 254, 272, 284, 289, 291
Knowledge, incomplete   21
Knowledge representation   23, 27

Labour (vs. work)   104, 109f., 263, 272f., 276
Lady Lovelace's objection   137f.
Layers, hidden   24–26, 87, 144, 156, 279
Leader   13, 63, 183f., 191, 194, 239, 247, 255, 295, 297
Learning   7, 45, 57, 88, 95, 110, 154, 157, 169, 177, 198, 231, 238, 253, 266, 272
Learning algorithm   4, 7, 21, 24f., 36, 91, 156, 158f., 162, 184
Learning, autonomous   23–25
Learning machines   see Machine learning
Libertarianism   173f.
Libet's experiment   173f.
LIBRATUS program   21f.
Life care (Lebenssorge)   6, 101, 105-109
Life sciences (LS)   101, 106; see also Science
Linguistic acceptability   140f.
Literary theory   231, 293
Literature (AI in fiction)   1, 9, 54, 56, 62, 67, 151–154, 156–158, 160–162, 189, 212, 225f., 231f., 290, 292–294

Machine   see Robot
Machine learning (automated self-learning)   3f., 6f., 10, 17, 22–25, 36, 48, 67, 91, 127f., 130–132, 135, 139, 144f., 152, 155, 159, 162, 183f., 186f., 226f., 230, 234, 239, 299
– Deep learning   11, 17, 21, 23–25, 48, 66, 154, 156–158, 239, 245, 257, 261
Machine, diagrammatic   20, 24
Marketing   80, 89, 93, 117, 183
Meaning   4, 7, 28, 34, 43f., 53, 57, 65f., 68, 83, 128, 133, 135–137, 145, 156, 172, 183, 223, 228, 250, 263, 276, 280f., 304
Mechanical body   219, 221; see also Artificial; Robot; Humanoid
Mechanism (early modern conception)   3, 115, 174, 241

Mechanization 109 f.
Mecha 11, 251 f.; *see also* Artificial; Robot; Humanoid
Media 1, 21, 25, 34 f., 45, 80, 92, 94, 111, 116, 119, 136, 162, 228, 262, 267
– Social media 25, 45, 65, 237 f., 249
Medical care 67, 184 f., 187, 191
Medical device 189 f., 198 f.
Medical prevention 8, 185
Medical profession 182, 188
Medicine 6, 8, 21, 39, 91, 119, 182, 184, 188, 226, 239; *see also* Telemedicine
Medium 9, 19, 35, 54 f., 153, 155, 157, 162, 194, 196, 212, 214, 220, 226, 235, 279
Metalepsis 279 – 281
Metaleptic collapse *see* Metalepsis
Metamorphose 9, 211, 222
Metaphor 21, 36 f., 43, 48, 67, 117, 131, 147, 154, 212 – 214, 222, 232, 263, 280, 293
Metonymy 261 f., 278 f., 280 – 282
Microcosm 119, 132
Mimesis 56, 59, 69, 73, 262, 274
Mind 2 f., 5, 17, 19, 35, 37, 39, 42, 44, 66, 81 f., 82, 84, 86 – 88, 90 f., 95 f., 105, 108, 111, 120, 147, 153, 157, 159, 162, 218, 222, 227, 234 f., 241, 245, 247 f., 251, 253, 265, 275 – 277, 280, 291 f., 297, 299, 302, 304
– Extended mind 84, 234
Mind as computer/machine 44, 81 f.
Mind uploading 87 f.
Mirror 19, 51, 116, 118, 220 f., 242, 251, 254, 256
Mobility 106, 112
Monitoring 8, 68, 72, 93, 182, 184 – 188, 191, 198
Monitoring Apps 181 f.
Morality 3, 13, 38, 51, 65, 68 f., 72 f., 175, 287 – 290, 294, 299, 301, 303 f.; see also Ethics
Moral agency 5, 51, 57, 59, 67
Moral enhancement 8, 37, 170, 175
Moral judgment 4, 33, 47 f., 294, 300 f.
Moral motivation 168, 175
Moral vulnerability 70 f.

Morphing 9, 106, 209 – 212, 215, 217 – 219
– cinematic 9, 209 f.
– digital 9, 211, 217 f.
Motif 9, 12, 210 – 215, 217, 220 – 222, 261 f., 268, 270, 274 f., 279
Myth 9, 17 f., 56, 71, 94, 155, 163, 212, 217

Nano-Bio-Info-Cognitive (NBIC) 101, 114 f., 120
Narration 229, 270
Narrative 10, 17, 55 f., 66, 80, 136, 157, 170, 212, 221, 227, 229, 231, 261, 265, 273, 279 f., 282, 295
– Narrative theory 279
– Narrative prosthesis 292 – 295
Natural language processing (NLP) 7, 127 – 129, 134 – 136, 144 f., 152, 185, 264
Nature 2, 6, 9, 44, 91, 93, 103 f., 107 f., 110, 114 f., 119, 132, 155 f., 168, 172, 176, 184, 209 – 213, 217, 241, 254, 264, 268, 270, 284, 291 f.
Navigating *see* Surfing
Negentropy 9, 211 – 214, 217, 222; *see also* Entropy
Neoliberalism 6, 92 – 94, 101, 105, 111
Network 17, 21, 45, 65, 82, 84, 88, 93, 155, 157, 163, 185, 189, 191, 198, 216, 242, 281
Network, diagrammatic 20, 24
Network, neural 10, 12, 26 f., 82, 137, 154, 156, 231, 287, 290 f., 299, 231, 301, 303
– Artificial neural network 24, 27, 157, 239
Neuroenhancement *see* Enhancement
Neuroscience 80, 86 f., 89, 91, 172; *see also* Science
New materialism 1 f., 13, 303
Non-corrupt environment 8, 181 f., 192
Non-human 9 f., 12 f., 57, 108, 114 f., 156 f., 159, 162 f., 170, 209 – 211, 216, 241, 287 – 289, 291, 295, 298, 302, 304
North America 183, 193, 294
Novel 8 – 10, 12 f., 25, 54, 81, 84, 86, 94, 121, 139, 186, 188, 192, 210, 222, 225 f., 229 – 234, 238, 246, 248 – 250, 254, 256 – 258, 263, 287, 289, 291 – 293, 296 – 298, 300, 303

Obesity 6, 92
OECD 60, 64, 183f., 188, 199
Ontology 250, 283, 302, 304
Ontology of patterns 28
Ontological vulnerability 69f.
Opacity 21, 26, 70
OpenAI 145, 282
Operativity 17f.
Outbreak control *see* Control
Oxytocin 8, 168, 177

Pandemic outbreak tracking 185
Pandemic prevention 182, 190, 193, 197
Pandemic spreads 185
Paranoia 218, 222, 296
Particularity 107, 109, 235
Pathology 117f.
Patient diagnosis 189
Patient-data 188
Patient-led monitoring 185, 198
Pattern recognition 24, 299
Penelope's web 9, 211f., 215
Perception 4, 23, 33, 41, 43, 56, 61, 67, 81, 151, 153, 163, 183, 193–196, 216, 225, 229, 251, 272, 277, 280, 292
Performance Studies 265, 284
Person 8f., 21, 36f., 39, 55, 57, 63, 68, 70–72, 95, 117, 121, 128f., 147, 167f., 171, 174f., 209, 212, 218, 220, 222, 230–235, 263, 277f., 300, 302
Personal identity 12, 277; *see also* Identity
Personal Identity Theory 277
Pharmaceutical companies 187
Pharmaceutical industry 189
Playwright 95, 261–263, 268, 270, 273, 275, 277, 281
Plurality 69, 107, 109
Poetry 7, 90, 149–163, 269, 303
– AI generated poetry 7, 88, 151–159, 163, 267, 301
– Generative poetry 7, 152, 162f.
Possible worlds 12, 226, 261f., 277–283
Possible Worlds Theory 278
Posthuman 7, 10f., 84, 151, 155–158, 162f., 220, 234, 237f., 240–242, 250, 252, 254, 256, 297f.

Posthumanism 1f., 72, 79f., 95, 157, 228, 240–242, 256
– critical 240, 256
Poststructuralism 155, 231, 240
Poverty 5, 53–55, 133, 176, 186, 189, 248
Power 6, 12, 21–23, 25, 28, 56, 68, 71, 73, 82, 94, 103f., 109, 111, 116, 118, 133, 156, 184, 187f., 213, 216, 233, 242f., 256f., 278, 287–296, 298, 300, 303–305
Predictions 177, 187
Prevention of diseases 185, 187
Preventive medicine 184, 188
Private Sphere 104, 281
Probabilistically learning computer 33, 36, 48; *see also* Computer
Procreation 214f.
Prometheus 40f., 219, 222
Prozac 168, 174
Public Sphere 3, 104, 226

Quantification 8, 190, 225

Radiology 186, 188
Rationality (instrumental) 4f., 33f., 36, 38, 41, 52f., 56f., 61, 65, 246
Recognition 5, 17, 20f., 23–25, 36, 40, 51, 60, 68f., 73, 85, 128, 132, 140, 183, 187, 231, 245, 252, 256, 292, 299
Recording 93, 111, 118, 160, 184, 186, 284
Regulation 3, 45–47, 61, 67, 107, 183, 198
Remote diagnosis 185f.
Reproduction 6, 101, 103f., 109f., 112, 116, 118, 186, 246
Resonance, emphatic 28
Resource 22, 25, 81, 92, 104, 114, 175, 185
Respect 24, 27, 36, 39, 44, 48, 53, 63, 66, 68, 102, 117, 152, 171, 210, 295, 305
Responsibility 51, 62, 67, 234f., 287–290
Responsiveness, emotional 4, 28, 70
Retrotopia 79f., 83f., 94–96
Rights 2, 46, 61–63, 67, 112, 120, 133, 175, 211, 232, 234f., 288, 296
– Rights for bots, robots, androids 73, 222, 235, 295
Rimini Protokol 270f., 284
Risk management innovators 182, 194

RoboCop 209, 240
Robot 1f., 5, 9–11, 24, 36, 46, 61, 65–68, 71–73, 151, 159, 189, 199, 209f., 217–222, 225, 227, 234f., 237–241, 243–246, 250–256, 263f., 268, 271, 274, 287f., 294
Robot stories 10, 239, 243
Robotics 10f., 58, 71, 73, 81, 85, 88, 182f., 186, 238f., 243–246, 252, 254f., 288
Romance 230, 248

Science 3, 5f., 19, 26, 38, 43f., 60f., 63, 79–84, 86–88, 90f., 95, 101f., 119, 127, 132, 134f., 139, 144, 156, 183, 198, 214, 225–227, 241, 251, 256, 261, 265f., 279, 284; *see also* Cognitive science; Computer science; Life sciences (LS); Neuroscience
– Applied science 8, 168
– Citizen science 63, 198
Scientific data 189
Science & Technology (S&T) 6, 26, 43, 101–106, 111–114, 134, 225
Scientism 38, 41, 43
– Abstract scientism 34
Search engines 17, 23, 28, 60, 113, 161
Security 51, 60, 62, 64, 73, 85, 103, 183f., 190, 194, 199
Self 2, 4, 10f., 17, 23, 35–38, 40, 42, 45–48, 51, 54–57, 65–67, 69–73, 77, 79–81, 85, 87, 89, 92–94, 107, 115, 152, 159f., 167, 171–174, 185f., 188f., 198, 215, 222, 227f., 230, 232f., 240, 242, 244, 247, 251, 253, 256f., 273, 275, 278, 292f., 295f., 298
Self-consciousness 27, 55f., 152; *see also* Consciousness
Self-optimization/self-improvement 40, 48, 79, 83, 93, 160, 216
Self-perfection 5f., 91
Self-tracking/self-monitoring 92f., 188, 198
Semantics 28, 137, 144, 154, 159, 230f.
Sentience 9f., 67, 162, 223, 225, 227, 234, 253
– Nonintentional sentience 228

Seriality 82, 220f.
Serotonin 8, 177
Simulacrum 11, 45, 251–254, 256
Smartphone 19–21, 84, 199, 230
Smartphone applications 184
Social 1–5, 7, 17f., 20, 26, 33, 37, 42, 44–46, 51, 54, 58–63, 67–69, 71, 92, 94f., 105, 109, 115–117, 120, 131, 145, 169, 172, 174–176, 191, 199, 226, 233, 241f., 249, 251f., 264, 283, 287f., 292, 298, 300–302
Social media *see* Media
Social norms 61, 69, 189
Social, ethical, and humanitarian accountability 189f.
Social platform 20, 22
Sociality 6, 93f.
Soul Machines 12f., 287, 290, 298–300, 302–305
– Baby X 13, 299f., 305
Source of life 9, 214
Speculation 2, 12, 226
Speculative materialism 226
Speech recognition 20, 140, 183
Stability 106, 129
Stage Management 283
Standardization 95, 109f., 184
Storytelling 54–56, 79, 161
Strong AI (strong readings of AI) *see* Artificial Intelligence (AI)
Structural vulnerability *see* Vulnerability
Subject (human subject) 2, 5f., 10, 37, 48, 56–59, 79, 81–83, 87, 89f., 94f., 105, 107, 151, 156, 167, 171–173, 175, 177, 210, 214, 227, 241, 268, 270–272, 278, 280f., 283, 294, 298
Subjectivity 5, 56, 66, 79f., 83, 91, 93, 228, 281, 288, 302, 304
Super AI *see* Artificial Intelligence (AI)
Supercomputer 5, 88, 279; *see also* Computer
Surgery 111, 186
Symbolism 18, 214
Syntax 137, 231

Targeted aid 187, 189

Technology 3f., 6f., 12, 17–20, 22, 26–28, 43, 52, 54, 59f., 62, 83–86, 90, 95, 101f., 106, 113, 117–119, 128, 134, 145, 151, 157, 160, 185, 187f., 198f., 216f., 225, 228, 237f., 240, 242f., 247–250, 253–257, 261f., 264, 267, 272f., 284, 299
Technological development of information 185
Technological diversified data collection 185
Technological solution 182
– Wearable technology 266f.
Telemedicine 67, 181f., 185, 190; see also Medicine
Text corpora 4, 28
Theatre 261–266, 268, 270, 272–274, 277, 283–284
– AI in theatre 11f., 149, 262
– Design 20, 67, 73, 89, 212, 215, 265f., 283
– naturalistic 266
Therapy 43, 174, 184
"Three Laws of Robotics" 11, 71, 238, 243, 252, 255, 288; see also Asimov
Tools 3f., 33, 36f., 40–42, 48, 65, 84, 133, 135, 156, 162, 185, 186–188
Topic 28, 217, 225
Trace 9f., 20, 27, 56, 82, 108, 215, 257, 261
Transformation 9, 19, 23, 39, 51–53, 57, 59, 88, 105f., 110, 157, 171, 209, 211, 215, 217, 219, 228, 255; see also Digital transformation
Transhumanism 57, 80, 82f., 87f., 94f., 157, 241
Translation 6f., 22, 33–35, 37f., 46f., 65, 71, 82, 106, 116, 119, 128, 131, 140, 159, 218, 270, 272
Translation between languages 183
Transparency 19, 21, 26, 64, 132, 190, 198f., 249, 256; see also Intransparency

Turing machine 43, 88
Turing Test 7, 12, 130f., 147f., 151–154, 157, 163, 229f., 253f., 265, 268, 270, 277, 290f., 293, 302

Understanding 4, 20, 28, 34, 40, 42–44, 55–58, 61, 64–67, 69f., 86–88, 91, 119, 135, 144, 159, 162, 177, 190, 197, 212, 220, 226, 229–231, 241, 244, 254, 264f., 290f., 293f., 298–300, 303f.
– Background understanding 39, 42
– Understanding of Meaning 4, 28
United Nations (UN) 60, 183

Vibrant matter 187f., 288, 302f.
Viral 116–118, 198
Virtual 13, 60, 111f., 116–118, 252, 256, 278, 281, 299f., 303f.
Virtual nursing assistants 186
Virtual reality 12, 36, 59, 183, 237f., 256, 284, 295
Visual culture 5, 83, 88
Visualization 19, 256
Voice recognition 21
Voice recognition software 187
Vulnerability 5, 44, 51, 64, 68–73
– Structural vulnerability 69f.
Vulnerable agency 5, 64, 68f., 71, 73f.

Weak AI (weak reading of AI) see Artificial Intelligence (AI)
Winograd moment 6, 127
Winograd schema 7, 127–131, 138, 141f., 147
Winograd schema challenge 129, 141
Word neighbourhoods 25, 28
Workforce 10, 112, 190, 199, 237f.

Xenophobia 175

Zero-shot scoring 127

www.ingramcontent.com/pod-product-compliance
Lightning Source LLC
Chambersburg PA
CBHW050514170426
43201CB00013B/1952